Lecture Notes in Computer Science   14791

Founding Editors

Gerhard Goos
Juris Hartmanis

The series Lecture Notes in Computer Science (LNCS), including its subseries Lecture Notes in Artificial Intelligence (LNAI) and Lecture Notes in Bioinformatics (LNBI), has established itself as a medium for the publication of new developments in computer science and information technology research, teaching, and education.

LNCS enjoys close cooperation with the computer science R & D community, the series counts many renowned academics among its volume editors and paper authors, and collaborates with prestigious societies. Its mission is to serve this international community by providing an invaluable service, mainly focused on the publication of conference and workshop proceedings and postproceedings. LNCS commenced publication in 1973.

Joel D. Day · Florin Manea
Editors

Developments in Language Theory

28th International Conference, DLT 2024
Göttingen, Germany, August 12–16, 2024
Proceedings

Springer

Editors
Joel D. Day (iD)
Loughborough University
Loughborough, UK

Florin Manea (iD)
University of Göttingen
Göttingen, Germany

ISSN 0302-9743   ISSN 1611-3349 (electronic)
Lecture Notes in Computer Science
ISBN 978-3-031-66158-7   ISBN 978-3-031-66159-4 (eBook)
https://doi.org/10.1007/978-3-031-66159-4

This Springer imprint is published by the registered company Springer Nature Switzerland AG
The registered company address is: Gewerbestrasse 11, 6330 Cham, Switzerland

If disposing of this product, please recycle the paper.

Preface

The 28th International Conference on Developments in Language Theory (DLT 2024) was held from August 12–16 2024 in Göttingen, Germany, in the historic Alte Mensa building. It was organised by the Institute of Computer Science of the University of Göttingen, and was held in conjunction with the 14th International Workshop on Non-Classical Models of Automata and Applications (NCMA 2024).

The DLT conference series provides a forum for presenting current developments in formal languages and automata. Its scope is very general and includes, among others, the following topics and areas: grammars, acceptors and transducers for words, trees and graphs; algebraic theories of automata; algorithmic, combinatorial and algebraic properties of words and languages; relations between formal languages and artificial neural networks; variable length codes; symbolic dynamics; cellular automata; groups and semigroups generated by automata; polyominoes and multi-dimensional patterns; decidability questions; image manipulation and compression; efficient text algorithms; relationships to cryptography, concurrency, complexity theory and logic; bio-inspired computing; and quantum computing.

Following its establishment by Grzegorz Rozenberg and Arto Salomaa in Turku (1993), a DLT conference was held every other year in Magdeburg (1995), Thessaloniki (1997), Aachen (1999), and Vienna (2001). From 2001 onwards, DLT conferences have been organised yearly, typically alternating between locations within and outside of Europe. The locations of DLT conferences since 2002 were: Kyoto (2002), Szeged (2003), Auckland (2004), Palermo (2005), Santa Barbara (2006), Turku (2007), Kyoto (2008), Stuttgart (2009), London (2010), Milano (2011), Taipei (2012), Marne-la-Vallée (2013), Ekaterinburg (2014), Liverpool (2015), Montréal (2016), Liège (2017), Tokyo (2018), Warsaw (2019), Porto (2021), Tampa (2022), and Umeå (2023); since DLT 2001, the proceedings were published in the Lecture Notes in Computer Science series of Springer. DLT 2020 was initially planned to be held in Tampa, Florida, but was unfortunately cancelled due to the COVID-19 pandemic (and Tampa hosted DLT 2022). The accepted papers of DLT 2020 were nevertheless published as a volume of Lecture Notes in Computer Science, and the authors of these papers could present their work at DLT 2021 in Porto. DLT 2024 breaks from the rule of alternating European conferences with the ones organised outside Europe, but DLT 2025 will be again organised outside Europe.

In 2018, the DLT conference series instituted the Salomaa Prize to honour the work of Arto Salomaa, and to increase the visibility of research on automata and formal language theory. The prize is funded by the University of Turku. The ceremony for the Salomaa Prize 2024 took place in Göttingen on August 14, 2024, and the prize winner was Mikołaj Bojańczyk, from the University of Warsaw, who was presented with the prize by Jürgen Dassow, member of the Prize Committee of 2024.

This volume contains invited contributions as well as the accepted papers of DLT 2024. There were 26 submissions. Each submission was reviewed by at least three experts in the field and thoroughly discussed by the Program Committee. In the end, the Program Committee decided to accept 17 of these submissions for presentation at the conference. We would like to thank the members of the Program Committee, and all external reviewers, for their hard work in evaluating the papers and for valuable comments that led to the selection of the contributed papers.

There were four invited talks which were presented by:

- Laura Ciobanu (Edinburgh): Word Equations, Constraints and Formal Languages
- Paweł Gawrychowski (Wrocław): Algorithms and Combinatorics on Two-Dimensional Strings
- Sandra Kiefer (Oxford): Polyregular Functions – Characterisations and Refutations
- Martin Kutrib (Giessen): Cellular Automata: From Black-and-White to High Gloss Color (unifying talk with NCMA 2024)

DLT 2024 also hosted a series of highlight-talks, covering excellent papers in the area of language theory, which were recently presented at other conferences or published in scientific journals. The five highlight-talks, given in this track of the conference were:

- C. Aiswarya, Amaldev Manuel and Saina Sunny. Edit Distance of Finite State Transducers (based on a paper presented at ICALP 2024)
- Paul Gallot, Sebastian Maneth, Keisuke Nakano and Charles Peyrat. Deciding Linear Height and Linear Size-to-Height Increase of Macro Tree Transducers (based on a paper presented at ICALP 2024)
- Stefan Göller and Nathan Grosshans. The AC0-Complexity of Visibly Pushdown Languages (based on a paper presented at STACS 2024)
- Markus Lohrey, Markus L. Schmid: Enumeration for MSO-Queries on Compressed Trees (based on a paper presented at PODS 2024)
- Irmak Sağlam, Moses Ganardi and Georg Zetzsche. Directed Regular and Context-Free Languages (based on a paper presented at STACS 2024).

The conference program also contained, as a highlight, an evening lecture held by Anke Holler, on language theory from a linguistic perspective. The evening lecture was organised in the historic astronomical observatory, which served as the place of residence and work of Carl Friedrich Gauss.

We would like to express our sincere thanks to the four invited speakers, to Anke Holler for the evening lecture, to the speakers in the highlight-talks track, to the presenters of contributed talks, and to all authors of submitted papers.

The EasyChair conference system provided excellent support in the selection of the papers and the preparation of these proceedings. Special thanks are due to the Lecture Notes in Computer Science team at Springer for having granted us the opportunity to publish these proceedings in the series, and also for their help during the process. The registration of the conference was implemented using the conference management systems provided by Converia GmbH.

Most importantly, the DLT 2024 conference would not have been possible without the financial support of the German Research Foundation (Deutsche Forschungsgemeinschaft, DFG) through the grant MA 5725/6-1 (project number 542832826). We express our deepest thanks to the DFG for this support.

We are grateful, as well, to the members of the Organizing Committee: Patricia Nitzke, Tore Koß, Paul Sarnighausen-Cahn, Stefan Siemer, all members of the Theoretical Computer Science group of the University of Göttingen.

August 2024   Florin Manea
                                                           Joel D. Day

Organization

Program Committee

Marie-Pierre Béal	Université Gustave Eiffel, Marne-la-Vallée, France
Joel Day (Chair)	Loughborough University, UK
Dora Giammarresi	Università di Roma "Tot Vergata", Italy
Yo-Sub Han	Yonsei University, South Korea
Mika Hirvensalo	University of Turku, Finland
Markus Holzer	Univertsität Gießen, Germany
Tomohiro I	Kyushu Institute of Technology, Japan
Zsuzsanna Lipták	Università di Verona, Italy
Florin Manea (Chair)	Universität Göttingen, Germany
Sebastian Maneth	Universität Bremen, Germany
Ian McQuillan	University of Saskatchewan, Canada
Robert Mercaş	Loughborough University, UK
Cyril Nicaud	Université Gustave Eiffel, Marne-la-Vallée, France
Svetlana Puzynina	Saint Petersburg State University, Russia
Daniel Reidenbach	Keele University, UK
Arseny Shur	Bar Ilan University, Israel
Manon Stipulanti	Université de Liège, Belgium
Bianca Truthe	Universität Gießen, Germany
Mikhail Volkov	Yekaterinburg, Russia
Markus Whiteland	Université de Liège, Belgium
Georg Zetzsche	Max Planck Institute for Software Systems, Kaiserslautern, Germany

Additional Reviewers

Marcella Anselmo	Panagiotis Charalampopoulos
Hideo Bannai	Pamela Fleischmann
Pascal Bergsträßer	Rudolf Freund
Valérie Berthé	Daniel Gabrić
Laurent Bulteau	Philipp Gohlke
Péter Burcsi	Joonghyuk Hahn
Michaël Cadilhac	Benjamin Hellouin de Menibus

Ismaël Jecker
Makoto Kanazawa
Sungmin Kim
Dmitry Kosolobov
Dietrich Kuske
Martin Kutrib
Thierry Lecroq
Brennan Lockinger
Andreas Malcher
Richard Mandel
Tomáš Masopust
Théo Matricon
Friedrich Otto
Luca Prigioniero
Narad Rampersad

Priscilla Raucci
Christian Rauch
Antoine Renard
Dan Rust
Andrew Ryzhikov
Ville Salo
Joe Sawada
Luke Schaeffer
Markus L. Schmid
Wolfgang Steiner
Marek Szykuła
György Vaszil
Viet Anh Martin Vu
Aaron Williams

Algorithms and Combinatorics on Two-Dimensional Strings

Paweł Gawrychowski ⓘ

Institute of Computer Science, University of Wrocław, Poland
gawry@cs.uni.wroc.pl

Abstract. The primary objects considered in the areas of algorithms on strings and combinatorics on words are sequences of characters, called strings and words in the algorithmic and the combinatorial context, respectively. We could ask how to find one string in another, check how similar two given strings are, or search for the repetitive fragments of a given string. In some applications, it is natural to consider more general objects, though. For example, motivated by possible applications in image processing, we might consider two-dimensional strings, or 2D strings for short.

A 2D string is simply a two-dimensional array of characters. The fundamental algorithmic question concerning two-dimensional strings (2D strings for short) is to find a copy of a 2D string in another. This was already considered in the late 70s, but interestingly, it took till the mid-90s to design a linear-time algorithm without any assumption on the size of the alphabet. The linear-time algorithm is based on a non-trivial generalization of the well-known concept of periodicity to two dimensions, which becomes quite technical.

The notion of periodicity and the related concept of repetitions appear quite challenging to work with in two dimensions. In particular, it is unclear what the right definition should be for the latter. In this talk, I will survey recent results on the so-called tandems, quartics, and runs in 2D strings. All of those are attempts at defining a notion of repetition in two dimensions. I will also present some of the ongoing work on generalizing the notion of k-periods to 2D strings and using it to obtain efficient approximate pattern-matching algorithms. I will describe multiple open problems that might be interesting from both the algorithmic and the combinatorial point of view.

Keywords: 2D strings · periodicity · repetitions

Contents

Word Equations, Constraints, and Formal Languages

Laura Ciobanu$^{(\boxtimes)}$ (ID)

Heriot-Watt University and Maxwell Institute, Edinburgh EH14 4AS, Scotland
L.Ciobanu@hw.ac.uk

Abstract. In this short survey we describe recent advances on word equations with non-rational constraints in groups and monoids, highlighting the important role that formal languages play in this area.

Keywords: word equations · free groups and monoids · length and counting constraints · EDT0L and indexed languages · decidability

1 Introduction

The question of finding solutions to equations in algebraic structures such as semigroups, groups or rings is a fundamental topic in mathematics and theoretical computer science that finds itself at the frontier between decidability and undecidability. Beyond asking about the existence of solutions to an equation it is often of interest, for both practical and theoretical reasons, to ask that the solutions belong to a certain set or satisfy a certain property; in that case one asks about the decidability of solving *equations with constraints*.

In the context of monoids and groups there are broadly two types of constraints: *rational* and *non-rational*. The rational ones concern properties of the solutions that can be recognised by finite automata, e.g. if there are solutions of even length in a given monoid or group, with respect to a fixed alphabet (or generating set). These are well understood in most settings, and typically in the cases when the satisfiability of equations is decidable, adding rational constraints or versions thereof remains decidable [11,13].

The second class of constraints, the *non-rational* ones, refers to sets that cannot be recognised by finite automata; examples of such constraints are relations between the lengths of the solutions or their Parikh images, or membership in context-free sets. There is an extensive literature on word equations, that is, equations in free monoids, with various non-rational constraints [1,3,12,19,22] and, recently, equations in groups with similar (non-rational) constraints have received attention. We highlight here some of the new results in both free monoids and groups, and refer the reader to [4–6,9] for a more in-depth treatment.

2 Word Equations with Constraints

In this section we discuss equations in free monoids, which are classically referred to as 'word equations' in the literature and are defined as follows.

© The Author(s), under exclusive license to Springer Nature Switzerland AG 2024
J. D. Day and F. Manea (Eds.): DLT 2024, LNCS 14791, pp. 1–12, 2024.
https://doi.org/10.1007/978-3-031-66159-4_1

Let Ω be a finite collection of variables and A a finite set of letters, and consider the word equation $U = V$ over (A, Ω), where U, V are words in the free monoid $(A \cup \Omega)^*$. A *solution* to $U = V$ is a morphism $\sigma \colon (\Omega \cup A)^* \to A^*$ that fixes A point-wise and such that $\sigma(U) = \sigma(V)$. We will refer to the question of whether a given equation has solutions as (**WordEqn**). This problem is solvable by the groundbreaking work of Makanin [23] (and subsequent improvements).

A *word equation with rational constraints* is a word equation $U = V$ together with a regular language $R_X \subseteq A^*$ for each variable $X \in \Omega$. In this situation, we say that $\sigma \colon \Omega \to A^*$ is a *solution* if it is a solution to the word equation $U = V$ and also satisfies $\sigma(X) \in R_X$ for each $X \in \Omega$. We will refer to the question of whether a given equation has solutions satisfying rational constraints as (**WordEqn, RAT**). This is decidable by [13].

To describe the relevant non-rational constraints in this paper we need to introduce a few basic concepts. Let $|w|$ denote the word length of any $w \in A^*$. For any $a \in A$, let $|w|_a$ count the number of occurrences of a in w; for example, $|abab^2|_a = 2$. Denote by $\Psi(L) \subseteq \mathbb{N}^k$ (with $k = |\Sigma|$), the Parikh image of a set $L \subseteq A^*$, that is, the abelianisation of L.

The starting point of the study of word equations with non-rational constraints is Büchi and Senger's well-known paper [3], where they show that (positive) integer addition and multiplication can be encoded into word equations in a free monoid, when requiring that the solutions satisfy one of several counting constraints; by the undecidability of Hilbert's 10th problem, deciding the satisfiability of such enhanced word equations is also undecidable.

We present the problems concerning word equations with non-rational constraints via an easy example, where each question represents one of the types of constraints that we are interested in.

Example 1. Let $A = \{a, b\}$, $\Omega = \{X, Y, Z\}$ and consider the word equation

$$X^2 a Y = Y Z^2 a. \tag{1}$$

A possible solution is $\sigma_0 = (ab, a, ba)$, as $\sigma_0(X) = ab$, $\sigma_0(Y) = a$ and $\sigma_0(Z) = ba$ give $\sigma_0(X^2 a Y) = \sigma_0(Y Z^2 a)$.

1. (**WordEqn, LEN**) asks about the existence of solutions whose lengths satisfy some given linear system of equations over the integers.
   For example, is there a solution σ to (1) such that

   $$|\sigma(X)| = |\sigma(Z)| + 1?$$

   A simple argument comparing the lengths of solutions on the left and right hand side will show that no such solution exists.

2. (**WordEqn, EXP-SUM**) asks about the existence of solutions where the number of occurrences/exponent-sums of letters from A satisfy some given linear system of equations over the integers.

For example, is there a solution σ to (1) such that

$$|\sigma(X)|_a = 2|\sigma(Z)|_a?$$

A simple argument comparing the numbers of a's on the left and right hand side will show that no such solution exists.

3. (**WordEqn, PARIKH**) asks about the existence of solutions whose Parikh images satisfy some given linear system of equations over the integers. (These are called AbelianEq in [12]).

   For example, is there a solution σ such that

$$\Psi(\sigma(X)) = \Psi(\sigma(Z))?$$

   The answer is clearly 'yes', as solution σ_0 shows: $\Psi(ab) = \Psi(ba) = (1, 1)$.

4. (**WordEqn, INEQ**) asks about the existence of solutions for which length, exponent-sum or other counting identities never hold.

   (i) For example, is there a solution σ such that

$$|\sigma(X)|_a \neq |\sigma(Z)|_a?$$

   If we count the possible numbers of a's on the left and right hand sides of the equation, and consider the fact that both X and Z appear with the same exponent, we get $|\sigma(X)|_a = |\sigma(Z)|_a$ for every solution σ; thus the answer to the question above is negative.

   (ii) Another such problem would be: Is there a solution σ such that

$$|\sigma(X)| \neq |\sigma(Z)|?$$

   An immediate length argument shows that the answer is negative.

As Büchi and Senger showed [3, Corollary 4], problems such as (2) and (3) in full generality are undecidable. In fact, questions (2) and (3) are equivalent (see [5, Lemma 3.2]). However, it remains a well-known open problem whether the satisfiability of word equations with *length constraints*, as in (1), is decidable. Deciding algorithmically whether a word equation has solutions satisfying certain linear length constraints is a major open question, and it has deep implications, both theoretical (if undecidable, it would offer a new solution to Hilbert's 10th problem about the satisfiability of polynomial equations with integer coefficients) and practical, in the context of string solvers for security analysis. We refer the reader to the surveys [2, 18] for an overview of the area from several viewpoints, of both theoretical and applied nature.

One of the recent results about counting constraints concerns questions such as (4), where one is asking about certain properties of the solutions to *never* hold. The result below applies to more general counting functions than just length and exponent-sums of letters in the solutions.

Theorem 1 *[9, Theorem 5]. The problem of satisfiability of word equations with rational constraints and counting inequations is decidable.*

This result relies on two important ingredients, both of which revolve around formal languages:

1. Solution sets to word equations, as represented in (2), are EDT0L, and therefore indexed [7].
2. A certain semilinear set, called the *slice closure*, is effectively computable for any indexed language L. Moreover, this set, containing the Parikh image of L, records the counting properties (such as lengths of words, exponent-sums of letters etc.) of L to a large degree ([9, Section 2]).

We represent the set of solutions to word equations as follows. Given a map $\sigma\colon \Omega \to A^*$ with $\Omega = \{X_1, \ldots, X_k\}$, we define

$$\mathsf{enc}(\sigma) \;=\; \sigma(X_1)\# \cdots \#\sigma(X_k), \tag{2}$$

where $\#$ is a fresh letter not in A. Hence, $\mathsf{enc}(\sigma) \in (A\cup\{\#\})^*$. Let $\Sigma = A\cup\{\#\}$.

The counting constraints in Theorem 1 can be expressed using rational counting functions; these are functions $f\colon \Sigma^* \to \mathbb{Z}^n$ for which the graph of f is a rational subset of $\Sigma^* \times \mathbb{Z}^n$, that is, $\{(w, f(w)) \mid w \in \Sigma^*\} \subseteq \Sigma^* \times \mathbb{Z}^n$ can be defined by a rational transducer. These counting functions can then be applied to encodings of solutions, as in the following examples from [9].

1. Exponent-sums/letter occurrence counting:
   The function $f_{X,a}$ with $f_{X,a}(\mathsf{enc}(\sigma)) = |\sigma(X)|_a$ for some letter $a \in A$ and variable $X \in \Omega$. Here, the transducer increments the counter for each a between the i-th and $(i+1)$-st occurrence of $\#$, where $X = X_i$.
2. Counting positions with MSO properties:
   Consider a monadic second-order logic (MSO) formula $\varphi(x)$ with one free first-order variable x, evaluated over finite words. Then one can define the function f_φ such that $f_\varphi(\mathsf{enc}(\sigma))$ is the number of positions x in $\mathsf{enc}(\sigma)$ where $\varphi(x)$ is satisfied. Then f_φ is a counting function, which follows from the fact that MSO formulas define regular languages. For example, we could count the number of a's such that there is no c between the a and the closest even-distance b.
3. Linear combinations of counting functions:
   If $f_1, \ldots, f_m\colon \Sigma^* \to \mathbb{Z}^n$ are counting functions, then so is f with $f(w) = \lambda_1 f_1(w) + \cdots + \lambda_m f_m(w)$ for some $\lambda_1, \ldots, \lambda_m \in \mathbb{Z}$. This can be shown with a simple product construction.
4. Length functions:
   The function L_X with $L_X(\mathsf{enc}(\sigma)) = |\sigma(X)|$ for some $X \in \Omega$. For this, we can take a linear combination of letter counting functions.

Using rational counting functions one can express the problems (**WordEqn, LEN**), (**WordEq, EXP-SUM**) or (**WordEqn, PARIKH**) in the general framework:

Input. A word equation (U, V) with rational constraints and a counting function
  $f \colon (A \cup \{\#\})^* \to \mathbb{Z}^n$
Question. Is there a solution σ that satisfies $f(\mathrm{enc}(\sigma)) = 0$?

Then the question answered positively in Theorem 1, that is, the problem of solving *word equations with rational constraints and counting inequations*, can be phrased as follows.

Input. A word equation (U, V) with rational constraints and a counting function
  $f \colon (A \cup \{\#\})^* \to \mathbb{Z}^n$
Question. Is there a solution σ that satisfies $f(\mathrm{enc}(\sigma)) \neq \mathbf{0}$?

3   Groups Equations with Constraints

We find similarities to word equations with constraints, but also new territory, when entering the world of groups. The question of decidability of (systems of) equations is widely known as the *Diophantine Problem* in groups rather than solving *Word Equations*, but to have a uniform notation throughout the paper we will denote this question by (**WordEqnGp**) rather than the Diophantine Problem.

The main classes of groups (not necessarily disjoint, not an exhaustive list), for which (**WordEqnGp**) is decidable, are listed below:

1. virtually abelian groups [17],
2. hyperbolic groups, with subclasses treated in several stages, chronologically:
   (a) free groups [23],
   (b) torsion-free hyperbolic groups [25],
   (c) virtually free and (torsion) hyperbolic groups [11],
3. partially commutative groups (or right-angled Artin groups) [15],
4. graph products of groups with decidable (**WordEqnGp**) [14],
5. virtually direct products of hyperbolic groups (which include dihedral Artin groups, and therefore all Baumslag-Solitar groups $BS(m, m)$) [8],
6. central extensions of hyperbolic groups [20],
7. some torsion-free relatively hyperbolic groups [10].

For several additional classes of groups there are positive results for restricted types of equations, for example:

1. single equations in the Heisenberg group [16],
2. quadratic equations in the Baumslag-Solitar groups $BS(1, n)$ [24],
3. quadratic equations in the Grigorchuk group [21].

In many of the above cases the decidability of (**WordEqnGp**) with rational constraints, so (**WordEqnGp, RAT**), has also been established. Notable exceptions are the partially commutative groups and the hyperbolic groups, where (**WordEqnGp, RAT**) is undecidable in full generality, and one needs to weaken the type of constraints to get decidability.

To make this paper accessible to a wide audience we give some background before describing the non-rational constraints that are considered in groups.

3.1   Background on Group Theory

Let Σ be a finite set, and as before, let $|w|$ denote the word length of any $w \in \Sigma^*$. For any $a \in \Sigma$, let $|w|_a$ count the number of occurrences of a in w.

Free Groups. Define Σ^{-1} as the set of formal inverses of elements in Σ and denote the free group with generating set Σ by $F(\Sigma)$; $F(\Sigma)$ can be viewed as the set of all *freely reduced words* over $\Sigma^{\pm 1} = \Sigma \cup \Sigma^{-1}$, that is, words not containing xx^{-1} as subwords, $x \in \Sigma^{\pm 1}$, together with the operations of concatenation and free reduction (that is, the removal of any xx^{-1} that might occur when concatenating two words, where $(x^{-1})^{-1}$ for any $x \in \Sigma^{\pm 1}$.

Partially Commutative Groups (RAAGs). A group G generated by the finite set Σ, subject to a set of relations $R \subset F(\Sigma) \times F(\Sigma)$, is denoted as $G = \langle \Sigma \mid R \rangle$ and can be viewed as $F(\Sigma)$ modulo the relations in R: two elements are equal in G if there is a way of writing them as words which can be transformed into each other via the relations in R, together with deleting or inserting xx^{-1}, $x \in \Sigma^{\pm}$. For example, the free abelian group $(\mathbb{Z}^2, +)$ can be given by generators $\Sigma = \{a, b\}$ that satisfy the relation (ab, ba), which we write as $ab = ba$. One may replace $ab = ba$ by $aba^{-1}b^{-1} = 1$, and use the commutator notation $[a, b] = 1$.

A class of groups that lie between the free (non-abelian) and the free abelian groups, in terms of their presentations, are the *partially commutative groups*. They are the group theoretic counterpart to partially commutative monoids, or trace monoids. In geometric group theory these are called *right-angled Artin groups* (RAAGs), and we will use this short-hand notation to save space. The most common way of describing a RAAG is via a finite undirected graph Γ with no auto-adjacent vertices (i.e. no loops at any vertex) and no multiple edges between two vertices, and letting the vertices of Γ be the generators of the RAAG $G\Gamma$ based on Γ. The relations between generators correspond to the edges: for every edge (u, v) in Γ we introduce the commuting relation $uv = vu$.

One often writes G instead of $G\Gamma$ when the graph Γ is unambiguous.

Example 2. Let Γ be the graph below with vertices $\{a, b, c, d\}$ and 3 edges. The RAAG $G = G\Gamma$ based on Γ has the presentation $\langle a, b, c, d \mid [a, b] = [b, c] = [c, d] = 1 \rangle$; that is, G has generators $\{a, b, c, d\}$ which satisfy the relations $ab = ba$, $bc = cb$, and $cd = dc$.

The Abelianization of a Group. The abelianization of a group is the analogue of the Parikh map $\Psi : \Sigma^* \mapsto \mathbb{N}^{|\Sigma|}$ for free monoids, already defined, but where instead of Σ^* we consider a group defined by Σ.

For any group G, let $\mathsf{ab} : G \rightarrow G^{\mathsf{ab}}$ be the natural abelianization map. That is, G^{ab} is the group with the presentation of G, plus the additional relations that any two generators commute. In algebraic terms, $G^{\mathsf{ab}} = G/G'$, that is, the quotient of G by its commutator subgroup; G^{ab} is a commutative group which will decompose into an infinite part of the form $\mathbb{Z}^m$, for some $m \geq 1$, and a finite abelian group H. The integer m is called *the free rank* of G^{ab}. So for example, the abelianization of $F(a, b)$ is the free abelian group $(\mathbb{Z}^2, +)$ and is of free rank 2; this is because $F(a, b)$ has generators $\{a, b\}$ and no relations, and the abelianization has the same generators, but now they commute.

Length and Exponent-Sum. Every element g in a group G generated by Σ has a length $|g|_\Sigma$, which is the word length $|.|$ of a shortest word w representing g in G. For example, the length of aba^{-1} in $\mathbb{Z}^2$ is $|aba^{-1}|_\Sigma = |b| = 1$.

For any generator x of $F(\Sigma)$, the map $||.||_x : F(\Sigma) \rightarrow \mathbb{Z}$ represents the *exponent-sum* of x in a word w; that is, $||w||_x = |w|_x - |w|_{x^{-1}}$, so for example $||xyx^{-1}y^2||_x = 0$. One can define the exponent-sum of a generator x in an element g for certain (but not all) groups beyond free groups, and then we use the same notation $||g||_x$. The length and the image under abelianization are well-defined for any element in any finitely generated group. However, the situation regarding the exponent-sum (of a generator) is more complicated. For example, if H is a group and $x \in H$ a generator of order 5, then one may claim the exponent-sum of x in the element x^3 to be 3; but $x^3 = x^{-2}$ in H, and in x^{-2} the exponent-sum of x appears to be -2.

Formal Languages and Constraints. As before, suppose G is a group generated by $S(= \Sigma \cup \Sigma^{-1})$, and let $\pi : S^* \rightarrow G$ be the natural projection from the free monoid S^* generated by S to G, taking a word over the generators to the element it represents in the group. A language over S is *regular* if it is recognised by a finite state automaton, as is standard.

Definition 1.

(1) A subset L of G is recognisable *if the full preimage $\pi^{-1}(L)$ is a regular subset of S^*.*

(2) A subset L of G is rational *if $L = \pi(L')$, where L' is a regular subset of S^*.*

It follows immediately that recognisable subsets of G are rational.

3.2 Word Equations with Constraints in Groups

Word equations in groups are essentially the same as word equations in monoids, the only difference consisting in the fact that inverses, of both elements and variables, are allowed.

For a group G, a *finite system of equations* in G over the variables Ω is a finite subset $\mathcal{E}$ of the free product $G * F(\Omega)$, where $F(\Omega)$ is the free group on

Ω. If $\mathcal{E} = \{w_1, \ldots, w_n\}$, then a *solution* to the system $w_1 = \cdots = w_n = 1$ is a homomorphism $\phi\colon G * F(\Omega) \to G$, such that $\phi(w_1) = \cdots = \phi(w_n) = 1_G$, where $\phi(g) = g$ for all $g \in G$. If $\mathcal{E}$ has a solution, then it is *satisfiable*.

Example 3. Consider the system $\mathcal{E} = \{w_1, w_2\} \subset F(a, b) * F(X_1, X_2)$ over the free group $F(a, b)$, where $w_1 = X_1^2(abab)^{-1}$, $w_2 = X_2 X_1 X_2^{-1} X_1^{-1}$; we set $w_1 = w_2 = 1$, which can be written as $X_1^2 = abab$, $X_2 X_1 = X_2 X_1$. The solutions are $\phi(X_1) = ab$, $\phi(X_2) = (ab)^k$, $k \in \mathbb{Z}$.

For a group G, we say that systems of equations over G are *decidable* over G if there is an algorithm to determine whether any given system is satisfiable. The question of decidability of (systems of) equations is often called the *Diophantine Problem* for G, and will be denoted here by (**WordEqnGp**(G)).

The *Diophantine Problem with rational* constraints, denoted (**WordEqnGp, RAT**), or *recognisable constraints*, denoted (**WordEqnGp, REC**), asks about the existence of solutions to $\mathcal{E}$, with some of the variables restricted to taking values in specified rational or recognisable sets, respectively.

One can attach similar types of non-rational constraints in the group setting as for word equations in free monoids. We illustrate the kind of problems concerning group equations with non-rational constraints by an easy example, where each question represents one of the types of constraints that we are interested in. For each case the problem is about the existence of an algorithm to decide the satisfiability of an equation with constraints.

Example 4. Consider the equation

$$XaY^2bY^{-1} = 1 \tag{3}$$

over variables X, Y in the free group on two generators $F(a, b)$ with length function $|.| = |.|_{\{a,b\}}$ and abelianization $(\mathbb{Z}^2, +)$.

1. (**WordEqnGp, LEN**) asks about the existence of solutions whose lengths satisfy some given linear system of equations over the integers.
   An instance of this problem is: decide whether there are any solutions (x, y) to (3) such that $|x| = |y| + 2$; the answer is yes, since $(x, y) = (b^{-1}a^{-1}, 1)$ is a solution with $|x| = |y| + 2$.
2. (**WordEqnGp, EXP-SUM**) asks about the existence of solutions where the exponent-sums of generators satisfy some given linear system of equations over the integers.
   An instance of this problem is: decide whether there are any solutions (x, y) to (3) such that $|x|_a = 2|y|_a + |y|_b$ and $|x|_b = 3|y|_b$; the answer is 'no' by solving a basic linear system over the integers with variables $|x|_a, |x|_b, |y|_a, |y|_b$ and we leave this as an exercise.
3. (**WordEqnGp, ab**) asks about the existence of solutions in G whose abelianisation satisfy some given system of equations in G^{ab}.
   An instance of this problem is: decide whether there are any solutions (x, y) to (3) such that $\mathsf{ab}(x) = 3\mathsf{ab}(y)$ (we use additive notation for $\mathbb{Z}^2$); the answer is no, since $\mathsf{ab}(xax^2by^{-1}) = \mathsf{ab}(x) + \mathsf{ab}(y) + (1, 1) = (0, 0)$ together with $\mathsf{ab}(x) = 3\mathsf{ab}(y)$ lead to $4\mathsf{ab}(y) = (-1, -1)$, which is not possible in $\mathbb{Z}^2$.

Several other constraints can be imposed, such as requiring that the solutions belong to specified context-free sets, or that they can be compared via a lexicographic order. See [4] for details.

Positive results have been established for virtually abelian groups, that is, groups that have abelian subgroups of finite index.

Theorem 2 ([4]). *In any finitely generated virtually abelian group, it is effectively decidable whether a finite system of equations with the following kinds of constraints has solutions:*

(i) *linear length constraints (with respect to any weighted word metric),*
(ii) *abelianisation constraints,*
(iii) *context-free constraints,*
(iv) *lexicographic order constraints.*

The main undecidability results for groups have so far been obtained with respect to (**WordEqnGp, ab**):

Theorem A ([5]). *Let G be a partially commutative group (or right-angled Artin group) that is not abelian. Then (**WordEqnGp, ab**) is undecidable.*

Theorem B ([5]). *Let G be a hyperbolic group with abelianisation of torsion-free rank ≥ 2. Then (**WordEqnGp, ab**) is undecidable.*

The key approach for establishing Theorems A and B is the interleaving of algebra and model theory. The main tool used to prove undecidability results is *interpretability* using disjunctions of equations (i.e. positive existential formulas), or *PE-interpretability*, which allows one to translate one structure into another and to reduce the Diophantine Problem from one structure to the other. The main reductions are to the Diophantine Problem in the ring of integers, which is a classical undecidable problem (Hilbert's 10th problem).

As for free monoids, the starting point is Büchi and Senger's paper [3], where they show that (positive) integer addition and multiplication can be encoded into word equations in a free semigroup, when requiring that the solutions satisfy an abelian predicate; by the undecidability of Hilbert's 10th problem, such enhanced word equations are also undecidable. In order to encode the multiplication in the ring $(\mathbb{Z}, \oplus, \odot)$ within a (semi)group, one needs two 'independent' elements, which can be taken to be two of the free generators if the (semi)group is free. In non-free groups one requires the existence of two elements that play a similar role; however, it is not enough to pick two generators, these elements also need to satisfy additional properties with respect to the abelianization. Finding such a pair of elements, called *abelian-primitive* in [5], is difficult or impossible for arbitrary groups, but they can be found in (non-abelian) partially commutative groups and hyperbolic groups with 'large' abelianisation.

Finally, Theorem C provides an interesting contrast to the previous results, in that positive outcomes for (**WordEqnGp, ab**) hold when G has finite abelianization. In this case the (**WordEqnGp, ab**) can be reduced to the (**WordEqnGp**) with recognisable constraints, and we showcase groups for which these problems are decidable.

Theorem C ([5]). *Let G be a group where the (**WordEqnGp, REC**) is decidable. Then (**WordEqnGp, ab**) is decidable. In particular, this holds if:*

1. *G is a hyperbolic group with finite abelianization.*
2. *G be a graph product of finite groups, such as a right-angled Coxeter group.*

We finish this summary of results with some illustrative examples.

Example 5. 1. The graph product G of finite cyclic groups G_i based on

$$\underset{\bullet}{\overset{G_1}{}} \underline{\hspace{1cm}} \underset{\bullet}{\overset{G_2}{}} \underline{\hspace{1cm}} \underset{\bullet}{\overset{G_3}{}} \underline{\hspace{1cm}} \underset{\bullet}{\overset{G_4}{}}$$

is $\langle x_1, x_2, x_3, x_4 \mid [x_1, x_2] = [x_2, x_3] = [x_3, x_4] = 1, x_1^2 = x_2^4 = x_3^6 = x_4^8 = 1 \rangle$, has finite abelianisation G^{ab}, and so (**WordEqnGp, ab**) is decidable by Theorem C.

2. The surface group $G = \langle a, b, c, d \mid aba^{-1}b^{-1}cdc^{-1}d^{-1} = 1 \rangle$ is hyperbolic and the abelianisation G^{ab} of G is $\mathbb{Z}^4$, so (**WordEqnGp, ab**) is undecidable by Theorem B.

3. The free product $G = \langle x_1, x_2, x_3, x_4 \mid x_1^2 = x_2^4 = x_3^6 = x_4^8 = 1 \rangle$ of finite groups is hyperbolic and has finite abelianisation, so (**WordEqnGp, ab**) is decidable by Theorem C.

4   Conclusions

Word equations in free monoids are an established research direction in theoretical computer science, and imposing constraints on the solutions is of interest for both theoretical and practical reasons. Most results tend to lead to undecidability, and some outstanding open problems remain. In this survey we highlighted a rare positive result, Theorem 1, for a type of constraints that require none of the solutions to satisfy certain properties.

The same type of problems can be posed in the infinite groups for which the satisfiability of equations is known to be decidable, such as virtually abelian, partially commutative groups (the group counterparts of trace monoids) and more generally, graph products of groups, as well as hyperbolic groups. The motivation to study non-rational constraints is twofold: first, explore extensions of word equations that have not been systematically studied for groups before, and second, develop algebraic and model-theoretic tools that can complement the combinatorial techniques used for solving word equations. As for monoids, most results so far show undecidability, but for virtually abelian groups and certain groups with finite abelianisation several problems turn out to be decidable. Many interesting questions are yet to be explored.

The systematic crossover of non-rational constraints from computer science into algebra has begun only recently, and there are undoubtedly many more avenues to explore by translating questions on free monoids, where decades of literature exists, into groups. Once in the world of groups, there are numerous tools to tackle these problems which are not always available for free monoids, tools coming from geometry, topology and algebra. We expect that the progress and ideas from algebra can inform the work in computer science.

References

1. Abdulla, P.A., et al.: String Constraints for Verification. In: Biere, A., Bloem, R. (eds.) CAV 2014. LNCS, vol. 8559, pp. 150–166. Springer, Cham (2014). https://doi.org/10.1007/978-3-319-08867-9_10

2. Amadini, R.: A survey on string constraint solving. ACM Comput. Surv. **55**(1) (2021)

3. Büchi, J.R., Senger, S.: Definability in the existential theory of concatenation and undecidable extensions of this theory. Zeitschrift mathematische Logik Grundlagen Math. **34**(4), 337–342 (1988)

4. Ciobanu, L., Evetts, A., Levine, A.: Effective equation solving, constraints and growth in virtually abelian groups (2023). https://arxiv.org/abs/2309.00475

5. Ciobanu, L., Garretta, A.: Group equations with abelian predicates. Int. Math. Res. Not. **2024**(5), 4119–4159 (2024). https://arxiv.org/abs/2204.13946

6. Ciobanu, L., Levine, A.: Languages, Groups and Equations, Languages and Automata: Gagta Book 3 (2024). https://arxiv.org/abs/2303.07825

7. Ciobanu, L., Diekert, V., Elder, M.: Solution sets for equations over free groups are EDT0L languages. Internat. J. Algebra Comput. **26**, 843– 886 (2016). Conference Abstract in ICALP 2015. LNCS, vol. 9135 with full version on ArXiv e-prints: abs/1502.03426

8. Ciobanu, L., Holt, D., Rees, S.: Equations in groups that are virtually direct products. J. Algebra **545**, 88–99 (2020). MR4044690

9. Ciobanu, L., Zetzsche, G.: Slice closures of indexed languages and word equations with counting constraints. In: Proceedings of the Thirty-Ninth Annual ACM/IEEE Symposium on Logic in Computer Science (LICS 2024) (2024, to appear)

10. Dahmani, F.: Existential questions in (relatively) hyperbolic groups. Israel J. Math. **173**, 91–124 (2009). MR2570661

11. Dahmani, F., Guirardel, V.: Foliations for solving equations in groups: free, virtually free, and hyperbolic groups. J. Topol. **3**(2), 343–404 (2010)

12. Day, J.D., Ganesh, V., He, P., Manea, F., Nowotka, D.: The satisfiability of word equations: decidable and undecidable theories. In: International Conference on Reachability Problems, pp. 15–29 (2018)

13. Diekert, V., Gutiérrez, C., Hagenah, C.: The existential theory of equations with rational constraints in free groups is PSPACE-complete. In: Proceedings of the 18th Annual Symposium on Theoretical Aspects of Computer Science (STACS 2001), Dresden, Germany, pp. 170–182 (2001)

14. Diekert, V., Lohrey, M.: Word equations over graph products. Internat. J. Algebra Comput. **18**(3), 493–533 (2008). MR4113851

15. Diekert, V., Muscholl, A.: Solvability of equations in graph groups is decidable. Internat. J. Algebra Comput. **16**(6), 1047–1069 (2006). MR2286422

16. Duchin, M., Liang, H., Shapiro, M.: Equations in nilpotent groups. Proc. Amer. Math. Soc. **143**(11), 4723–4731 (2015). MR3391031

17. Evetts, A., Levine, A.: Equations in virtually abelian groups: languages and growth. Internat. J. Algebra Comput. **32**(3), 411–442 (2022). MR4417480

18. Ganesh, V., Minnes, M., Solar-Lezama, A., Rinard, M.: Word equations with length constraints: what's decidable? In: Biere, A., Nahir, A., Vos, T. (eds.) HVC 2012. LNCS, vol. 7857, pp. 209–226. Springer, Heidelberg (2013). https://doi.org/10.1007/978-3-642-39611-3_21

19. Garreta, A., Gray, R.: On equations and first-order theory of one-relator monoids. Inform. and Comput. **281** (2021)

20. Liang, H.: Equation problem over central extensions of hyperbolic groups. J. Topol. Anal. **6**(2), 167–192 (2014). MR3191648
21. Lysenok, I., Miasnikov, A., Ushakov, A.: Quadratic equations in the Grigorchuk group. Groups Geom. Dyn. **10**(1), 201–239 (2016)
22. Majumdar, R., Lin, A.W.: Quadratic word equations with length constraints, counter systems, and presburger arithmetic with divisibility. Log. Methods Comput. Sci. **17** (2021)
23. Makanin, G.S.: Equations in a free group. Izv. Akad. Nauk SSR Ser. Math. **46**, 1199–1273 (1983). English transl. in Math. USSR Izv. 21 (1983)
24. Mandel, R., Ushakov, A.: Quadratic equations in metabelian Baumslag-Solitar groups. Internat. J. Algebra Comput. **33**(06), 1195–1216 (2023)
25. Rips, E., Sela, Z.: Canonical representatives and equations in hyperbolic groups. Invent. Math. **120**(3), 489–512 (1995). MR1334482

Polyregular Functions: Characterisations and Refutations

Sandra Kiefer[(✉)]

Department of Computer Science, University of Oxford, Oxford, UK
sandra.kiefer@cs.ox.ac.uk

Abstract. Regular functions are a well-studied robust class of string-to-string functions, one of whose characterisations is that they are exactly the functions recognisable by deterministic two-way transducers, that is, finite automata with output. This implies that the growth of a regular function—the function describing the output length in terms of the input length—is always linear. To go beyond linear growth, one can equip the two-way transducers with multiple reading heads (pebbles), the number of which then still constitutes a bound on the degree of the polynomial describing the growth. The functions recognised by these pebble automata are called polyregular.

Over the past years, the properties of polyregular functions have been studied extensively. Just as for regular functions, various equivalent characterisations have been found, and variants of the corresponding models have been investigated. This paper gives an introduction to the realm of polyregular functions by discussing some of those characterisations, recent developments, and the parameters in the models that are linked to the growth. The second part presents simple constructions which show the asymmetry of the link between the growth degree and the number of heads. That is, in general, the growth degree of a polyregular function does not bound the minimum number of pebbles needed in an automaton to compute the function.

Keywords: polyregular functions · pebble transducers · MSO interpretations

1 Introduction

Regular languages are the formal languages that are accepted by finite automata, equivalently by deterministic or non-deterministic one-way or two-way ones. Various other equivalent characterisations have been found; among those are the expressibility of the language via a regular grammar and its definability in monadic second-order logic (MSO).

Every regular language L can also be studied via its Boolean characteristic function, that is, the function that assigns to a word the value 1 if it belongs to L, and 0 otherwise. When considering also non-Boolean values as output,

The original version of the chapter has been revised. The chapter title and subtitle have been corrected. A correction to this chapter can be found at https://doi.org/10.1007/978-3-031-66159-4_21

© The Author(s), under exclusive license to Springer Nature Switzerland AG 2024, corrected publication 2024
J. D. Day and F. Manea (Eds.): DLT 2024, LNCS 14791, pp. 13–21, 2024.
https://doi.org/10.1007/978-3-031-66159-4_2

we transition from languages to actual functions. There are several ways to extend the computational models for regular languages to non-Boolean outputs. For example, we can replace deterministic two-way automata with deterministic two-way transducers, i.e., automata that produce an output word from left to right while parsing the input word. The string-to-string functions that these transducers compute are the *regular functions*. Similarly to regular languages, regular functions are a well-studied and robust concept with various equivalent characterisations and useful properties (see, e.g., [16,23]), such as closure under composition [7].

However, also regular functions have limited expressivity in the sense that their *growth*, i.e., the function that describes the output length in terms of the input size, is at most linear. This asymptotic bound follows from the fact that, otherwise, the transducer would enter a loop of configurations. To make the model more expressive, one can modify it towards enabling superlinear growth. Polynomial growth can be achieved by allowing for more configurations of the transducer. This works, for example, by introducing multiple pairwise distinguishable reading heads instead of just a single one. The movement of the reading heads is subject to certain constraints, and the transition function is then defined based on the state of the transducer, the tuple of letters that the reading heads read, and the relative position of the heads. In this sense, the heads also serve as markers or "dropped pebbles". The resulting model is called a *pebble transducer* and the functions that pebble transducers compute are the *polyregular functions*.

By now, numerous equivalent characterisations of polyregular functions and variants of them have been found, which speaks for the robustness of the notion. Apart from this, polyregular functions appear as a natural concept in other areas that involve functions on strings, for example when studying reduplication in linguistics [26], the complexity of strings in string compression [19], or the use of transformers in natural language processing [28].

When the pebble transducer has k reading heads, the configuration contains k positions (and a state from a finite set) and thus, the bound on the output length is polynomial, namely $O(|w|^k)$ (and this bound is optimal), where w is the input string. A natural question to ask is whether the converse also holds: *if the output of a pebble transducer has length in $O(|w|^k)$ for some $k \in \mathbb{N}$, is there always a way to compute it with at most k reading heads?*

Outline. This paper surveys developments in the realm of polyregular functions. In the next section, we give an overview of the history of characterisations for polyregular functions. Section 3 then treats the growth and the refutation of pebble minimisation for polyregular functions, i.e., a proof that, in general, the growth degree is not an upper bound on the number of pebbles needed to compute the function with a pebble transducer. In Sect. 4, we conclude with a brief discussion of recently studied variants of polyregular functions.

2   History of Characterisations

This section aims at revisiting important elements in the history of the characterisation of polyregular functions. Since we intend to keep the formalism and technicalities light, we mainly provide pointers here and give intuition for selected characterisations via the running example of a specific simple polyregular function.

Polyregular functions are a certain class of functions that map strings to strings. The original motivation for this class was to introduce multiple reading heads into a transducer to enable a growth that is polynomial in the input size. For example, consider the function that maps a word to the concatenation of all its prefixes in descending length:

$$abcd \mapsto abcd\,abc\,ab\,a\,.$$

A transducer that produces the desired output could use one reading head (i) to mark the position of the end of the current prefix, and a second reading head (j) to produce the current prefix letter by letter from left to right. So the positions of the reading heads in the output $abcd\,abc\,ab\,a$ for the example input $abcd$ are

$$\binom{1}{4}\binom{2}{4}\binom{3}{4}\binom{4}{4}\ \binom{1}{3}\binom{2}{3}\binom{3}{3}\ \binom{1}{2}\binom{2}{2}\ \binom{1}{1}.$$

This way, positions in the output string correspond to tuples of positions (namely, of the reading heads) in the input string. However, without any further restrictions on the movement and power of these reading heads, the expressive power of the computation model captures LOGSPACE and many fundamental problems become undecidable [17,18][1]. Therefore, a natural approach to control the interaction between the reading heads was to impose a so-called *stack discipline* or *stack hierarchy*. The reading heads are ordered and they cannot move absolutely independently. They can be placed on the word (*push* to the stack) one by one, but only the top-most reading head is allowed to move. When a lower reading head is to be moved, the ones higher up in the stack first have to be removed/forgotten (*pop* from the stack).

Note that the prefix example can be computed with reading heads that follow the stack hierarchy. Indeed, the reading head i is first pushed to the stack and remains on the stack until the entire output is produced. It moves to position 4, the end of the input, without producing any output letters. Then the reading head j is pushed and moves from position 1 through until it finds i. In each of these steps where both i and j are on the stack, an output letter that is identical to the letter that j currently reads is produced. The reading head j is popped every time it reaches the same position as i, then i is moved one position to the left (to produce the next-shortest prefix), and j is pushed again starting at position 1. The computation ends when both i and j have reached the start of the input.

The first occurrence of this model in the literature was in the pebble tree transducers introduced by Milo, Suciu, and Vianu in [22] for the purpose of processing XML documents. Actual string-to-string pebble transducers first appeared in the work [14] by Engelfriet and Maneth. In the same article, the

---

[1] In [18], the result is attributed to Alan Cobham.

authors also proved that the class of functions computed by these transducers is closed under composition, and in [11], an optimal bound on the number of pebbles in the transducer for the composition was given.

About 15 years later, Bojańczyk moved the focus from the computational model towards the class of functions that it computes and introduced the term *polyregular functions* for them [2]. In this and follow-up work (see also [5]), he showed that polyregular functions are equivalently characterised as the ones computed by string-to-string pebble transducers, a simple imperative programming language, and certain functional programming languages, and that they form the closure of a class of string-to-string functions under a restricted set of simple operations.

In [6], we added the missing logical characterisation.

Theorem 1 ([6]). *Polyregular functions are precisely the functions that are definable via string-to-string MSO interpretations.*

MSO interpretations consist of three types of MSO formulas, namely

- a *domain formula* $\varphi_{\text{dom}}(x_1, \ldots, x_k)$ that expresses which tuples of input positions in the input string are represented in the output string,
- an *order formula* $\varphi_<(x_1, \ldots, x_k, x'_1, \ldots, x'_k)$ that determines the order of two k-tuples of input positions in the output,
- *label formulas* $\varphi_a(x_1, \ldots, x_k)$ that describe for each letter a in the output alphabet which tuples of input positions are labelled with a in the output string.

The *dimension* of the interpretation is k.

Next, we provide a 2-dimensional MSO interpretation for our running example. A tuple (i, j) of reading-head positions in the input string is represented in the output whenever $i \leq j$. Therefore,

$$\varphi_{\text{dom}}(i, j) = i \leq j.$$

Moreover, a position tuple (i, j) appears earlier than a tuple (i', j') in the output when (i, j) belongs to the parsing of a longer prefix than (i', j'), or when they belong to the parsing of the same prefix but j is a position left of j'. Hence,

$$\varphi_<(i, j, i', j') = (i \geq i') \vee ((i = i') \wedge (j \leq j')).$$

The label is inherited from the topmost reading head. Thus, for every output letter a, we have

$$\varphi_a(i, j) = a(j).$$

As a further connection of polyregular functions to logic, in the recent article [3], a new characterisation via a type system inspired by linear logic was found.

Besides the pebble-minimisation question treated in the next section, a fundamental open question concerning polyregular functions is that of equivalence, i.e., the task to decide whether two transducers or interpretations compute the same polyregular function. While the problem is still open for general string-to-string functions, it was shown in [4] that it is decidable on unordered trees of bounded height.

3   Refutation of Pebble Minimisation

In the previous section, we have discussed characterisations of polyregular functions, with a focus on the original one via pebble transducers and the logical one via MSO interpretations. We now take a closer look at the parameters of the models in these characterisations.

There is a precise correspondence between the growth degree and the dimension of the MSO interpretation for a polyregular function.

Theorem 2 ([1]). *Let $k \in \mathbb{N}$. A polyregular function has growth in $O(n^k)$ if and only if the function can be defined via a k-dimensional MSO interpretation, where n is the input length.*

For example, the growth of the prefix-concatenation function is at most quadratic in the input size, and it can be defined via a 2-dimensional MSO interpretation, as we have seen. In this section, we use the example of another polyregular function to prove that the relation between the minimal number of pebbles needed to produce the output of a polyregular function and the growth degree is *not* symmetric: even though it clearly holds that, on input a string w, a pebble transducer with k pebbles only produces an output of length $O(|w|^k)$, the converse is not true. So the number of pebbles or reading heads cannot always be minimised to correspond to the growth degree of the polyregular function.

This came as a surprise given that, a paper published at the conference LICS 2020 had suggested a pebble minimisation procedure. However, by now, there are three independently developed refutation proof techniques for the pebble minimisation. The hypothesis was first shown to be wrong in [1]. In fact, Bojańczyk proved that the growth degree, i.e., the degree of the polynomial that describes the growth does not even provide any bound on the number of pebbles.

Theorem 3 ([1]). *No constant number of pebbles suffices to compute every polyregular function with quadratic growth.*

Bojańczyk's proof proceeds via inner squaring of words formed with an infinite alphabet and uses a concept called *pebble transducers with atoms.* (The reduction from finite to infinite alphabets is not straightforward.) In [20], we provide two different and simpler refutation proofs, one of which shows a slightly stronger version of Bojańczyk's inapproximability result. That proof exploits the connection of polyregular functions with macro tree transducers studied in [15] as well as [13, Theorem 18].

The other proof in [20] is a technique to show the following weaker statement with much less involved tools than in [1].

Theorem 4 ([5,20]). *The inner-squaring function is a polyregular function which has quadratic growth and cannot be computed with a 2-pebble transducer.*

The inner-squaring function innsq is defined on strings over the 3-letter alphabet $\{a, b, \#\}$ and maps, for $n \in \mathbb{N}$ and $w_0, \ldots, w_n \in \{a, b\}^*$, strings as follows

$$\text{innsq}: w_0\# \ldots \#w_n \;\mapsto\; w_0^n\# \ldots \#w_n^n,$$

where w_i^n means the n-fold concatenation of w_i.

For every input string $w = w_0 \# \dots \# w_n$, we have that

$$|\texttt{innsq}(w)| \le |w| \cdot n \le |w|^2,$$

and thus, $\texttt{innsq}$ has growth degree at most.

A straightforward 3-pebble transducer to compute this function uses one reading head to mark the start of the word w_i that is currently copied, one reading head to track the number of copies of w_i that have been output (by parsing one $\#$ occurrence for each copy), and one reading head to actually copy w_i letter by letter. We show in [20] that there is no way to save one of these reading heads and get by with just two.

The proof falls back on very elementary properties of regular functions and a pumping statement about the shape of their outputs [27]. We consider the function on inputs of the form $(a^*b\#)^*\#^*$ and prove by contradiction that no function computable with a 2-pebble transducer produces the required outputs on all such inputs. Assuming that there is such a function f, each subcomputation between consecutive *push* and *pop* of the same reading head in the 2-pebble transducer can be simulated via a two-way transducer (with just one reading head), i.e., the image of a regular function. This implies that the output of f can be obtained by adequately nesting regular functions (with nesting depth 2). Since each of the regular functions has linear growth, each of the function calls can only produce a linearly long infix. Via this observation, we can prove that there must be a language $L \subseteq b\{a,b\}^*b$ that

1. is the image of a regular function,
2. consists of infixes of outputs of $\texttt{innsq}$ on words of the form $(a^*b\#)^*\#^*$, and
3. contains for every $N \in \mathbb{N}$ a word

$$b\underbrace{a\cdots a}_{a\text{-block}}b\dots b\underbrace{a\cdots a}_{a\text{-block}}b$$

with at least N occurrences of the letter b and where all a-blocks have the same length $n \ge N$.

We can then apply a pumping argument due to Rozoy [27] to the words in the third item to show that L cannot be the image of a regular function, a contradiction. Therefore, L does not exist, and hence there is no function that coincides with the inner squaring function on $(a^*b\#)^*\#^*$, which refutes pebble minimisation.

4   Conclusion

Polyregular functions are a class of string-to-string functions with polynomial growth. They have various equivalent characterisations, one of which is via MSO interpretations. This characterisation is arguably particularly natural because the growth degree equals the dimension of the logical interpretation. For the (original) characterisation via pebble transducers, the link between the parameters of the models is asymmetric: the number of pebbles is an upper bound on the growth degree, but not vice versa.

In the classical pebble transducers, the state transitions may depend on the order of the positions of the reading heads. When disallowing comparisons of positions, we obtain a computation model for the *polyblind functions*, comparison-free polyregular functions [25]. These have a second (and their original) motivation in a connection to λ-calculus [24]. As it turns out, for this class of functions, the pebble-minimisation attempt works: for polyblind functions, growth degree k does imply that there is a pebble transducer with k reading heads that computes the function (see also [10]). Similarly, pebble minimisation, i.e., getting by with only as many pebbles as the growth degree indicates, works in *last pebble transducers* [10], which can only compare the topmost two heads of the stack and are a variant of the invisible pebble transducers from [12].

The concept of polyregular functions has been carried further to develop and study concepts such as $\mathbb{N}$-polyregular functions [9,21] and $\mathbb{Z}$-polyregular functions [8,10]. The first of these classes is the class of polyregular functions that have unary output, the second is the closure of $\mathbb{N}$-polyregular functions under the subtraction operation.

Future perspectives to advance the understanding of polyregular functions include generalising them to trees (or even graphs of bounded treewidth) and extending decidability results for fundamental problems such as equivalence and subclass membership from $\mathbb{N}$-polyregular and $\mathbb{Z}$-polyregular functions to the general class.

Acknowledgments. The author would like to thank the organisers of DLT 2024, Joel D. Day and Florin Manea, for the opportunity for this contribution. She is also grateful to Lê Thành Dũng (Tito) Nguyễn for very helpful discussions about the literature mentioned in this paper.

Disclosure of Interests. The author has no competing interests to declare that are relevant to the content of this article

References

1. Bojanczyk, M.: On the growth rates of polyregular functions. In: 38th Annual ACM/IEEE Symposium on Logic in Computer Science, LICS 2023, Boston, MA, USA, 26–29 June 2023, pp. 1–13. IEEE (2023). https://doi.org/10.1109/LICS56636.2023.10175808
2. Bojańczyk, M.: Polyregular functions. CoRR abs/1810.08760 (2018). http://arxiv.org/abs/1810.08760
3. Bojańczyk, M.: Folding interpretations. In: 38th Annual ACM/IEEE Symposium on Logic in Computer Science (LICS), pp. 1–13 (2023). https://doi.org/10.1109/LICS56636.2023.10175796
4. Bojańczyk, M., Klin, B.: Polyregular functions on unordered trees of bounded height. Proc. ACM Program. Lang. **8**(POPL), 1326–1351 (2024). https://doi.org/10.1145/3632887

5. Bojańczyk, M.: Transducers of polynomial growth. In: Baier, C., Fisman, D. (eds.) LICS 2022: 37th Annual ACM/IEEE Symposium on Logic in Computer Science, Haifa, Israel, 2–5 August 2022, pp. 1:1–1:27. ACM (2022). https://doi.org/10.1145/3531130.3533326

6. Bojańczyk, M., Kiefer, S., Lhote, N.: String-to-string interpretations with polynomial-size output. In: 46th International Colloquium on Automata, Languages, and Programming, ICALP 2019, 9–12 July 2019, Patras, Greece, pp. 106:1–106:14 (2019). https://doi.org/10.4230/LIPIcs.ICALP.2019.106

7. Chytil, M.P., Jákl, V.: Serial composition of 2-way finite-state transducers and simple programs on strings. In: Salomaa, A., Steinby, M. (eds.) ICALP 1977. LNCS, vol. 52, pp. 135–147. Springer, Heidelberg (1977). https://doi.org/10.1007/3-540-08342-1_11

8. Colcombet, T., Douéneau-Tabot, G., Lopez, A.: $\mathbb{Z}$-polyregular functions. In: 38th Annual ACM/IEEE Symposium on Logic in Computer Science, LICS 2023, Boston, MA, USA, 26–29 June 2023, pp. 1–13. IEEE (2023). https://doi.org/10.1109/LICS56636.2023.10175685

9. Douéneau-Tabot, G.: Hiding pebbles when the output alphabet is unary. In: Bojańczyk, M., Merelli, E., Woodruff, D.P. (eds.) 49th International Colloquium on Automata, Languages, and Programming, ICALP 2022, 4–8 July 2022, Paris, France. LIPIcs, vol. 229, pp. 120:1–120:17. Schloss Dagstuhl - Leibniz-Zentrum für Informatik (2022). https://doi.org/10.4230/LIPICS.ICALP.2022.120

10. Douéneau-Tabot, G.: Optimization of string transducers. Ph.D. thesis, Université Paris Cité (2023). https://gdoueneau.github.io/pages/phd.html

11. Engelfriet, J.: Two-way pebble transducers for partial functions and their composition. Acta Inform. **52**(7–8), 559–571 (2015). https://doi.org/10.1007/s00236-015-0224-3

12. Engelfriet, J., Hoogeboom, H.J., Samwel, B.: XML navigation and transformation by tree-walking automata and transducers with visible and invisible pebbles. Theor. Comput. Sci. **850**, 40–97 (2021). https://doi.org/10.1016/j.tcs.2020.10.030. Extended version of a PODS 2007 paper

13. Engelfriet, J., Maneth, S.: Output string languages of compositions of deterministic macro tree transducers. J. Comput. Syst. Sci. **64**(2), 350–395 (2002). https://doi.org/10.1006/jcss.2001.1816

14. Engelfriet, J., Maneth, S.: Two-way finite state transducers with nested pebbles. In: Diks, K., Rytter, W. (eds.) MFCS 2002. LNCS, vol. 2420, pp. 234–244. Springer, Heidelberg (2002). https://doi.org/10.1007/3-540-45687-2_19

15. Engelfriet, J., Maneth, S.: A comparison of pebble tree transducers with macro tree transducers. Acta Inform. **39**(9), 613–698 (2003). https://doi.org/10.1007/s00236-003-0120-0

16. Filiot, E., Reynier, P.A.: Transducers, logic and algebra for functions of finite words. ACM SIGLOG News **3**(3), 4–19 (2016). https://doi.org/10.1145/2984450.2984453

17. Hartmanis, J.: On non-determinancy in simple computing devices. Acta Inform. **1**, 336–344 (1972). https://doi.org/10.1007/BF00289513

18. Ibarra, O.H.: Characterizations of some tape and time complexity classes of turing machines in terms of multihead and auxiliary stack automata. J. Comput. Syst. Sci. **5**(2), 88–117 (1971). https://doi.org/10.1016/S0022-0000(71)80029-6

19. Jordon, L.: An investigation of feasible logical depth and complexity measures via automata and compression algorithms. Ph.D. thesis, National University of Ireland Maynooth (2022). https://mural.maynoothuniversity.ie/16566/

20. Kiefer, S., Nguyên, L.T.D., Pradic, C.: Refutations of pebble minimization via output languages (2023, submitted). https://arxiv.org/abs/2301.09234
21. Lopez, A.: Commutative N-polyregular functions (2024). https://arxiv.org/abs/2404.02232
22. Milo, T., Suciu, D., Vianu, V.: Typechecking for XML transformers. J. Comput. Syst. Sci. **66**(1), 66–97 (2003). https://doi.org/10.1016/S0022-0000(02)00030-2. Journal version of a PODS 2000 paper
23. Muscholl, A., Puppis, G.: The many facets of string transducers. In: Niedermeier, R., Paul, C. (eds.) 36th International Symposium on Theoretical Aspects of Computer Science (STACS 2019). Leibniz International Proceedings in Informatics (LIPIcs), vol. 126, pp. 2:1–2:21. Schloss Dagstuhl-Leibniz-Zentrum fuer Informatik (2019). https://doi.org/10.4230/LIPIcs.STACS.2019.2
24. Nguyên, L.T.D.: Implicit automata in linear logic and categorical transducer theory. Ph.D. thesis, Université Paris XIII (Sorbonne Paris Nord) (2021). https://theses.hal.science/tel-04132636
25. Nguyên, L.T.D., Noûs, C., Pradic, C.: Comparison-free polyregular functions. In: Bansal, N., Merelli, E., Worrell, J. (eds.) 48th International Colloquium on Automata, Languages, and Programming, ICALP 2021, 12–16 July 2021, Glasgow, Scotland (Virtual Conference). LIPIcs, vol. 198, pp. 139:1–139:20. Schloss Dagstuhl - Leibniz-Zentrum für Informatik (2021). https://doi.org/10.4230/LIPICS.ICALP.2021.139
26. Rawski, J., Dolatian, H., Heinz, J., Raimy, E.: Regular and polyregular theories of reduplication. Glossa: J. Gener. Linguist. **8**(1) (2023). https://doi.org/10.16995/glossa.8885
27. Rozoy, B.: Outils et résultats pour les transducteurs boustrophedon. RAIRO Theor. Inform. Appl. **20**(3), 221–249 (1986)
28. Strobl, L., Angluin, D., Chiang, D., Rawski, J., Sabharwal, A.: Transformers as transducers. CoRR abs/2404.02040 (2024). https://doi.org/10.48550/ARXIV.2404.02040

Cellular Automata: From Black-and-White to High Gloss Color

Martin Kutrib$^{(\boxtimes)}$ ⓘ and Andreas Malcher ⓘ

Institut für Informatik, Universität Giessen, Arndtstr. 2, 35392 Giessen, Germany
{kutrib,andreas.malcher}@informatik.uni-giessen.de

Abstract. We provide an overview of three different operating modes of cellular automata. A classical point of view is to consider cellular automata as acceptors for formal languages. As usual, these systems then act as deciders whose output is yes or no. A second point of view takes the perspective of more complex outputs. Now cellular automata not only compute a binary output, but they transform inputs into outputs, that is, they act as transducers. In the third case, cellular automata are considered that generate outputs from almost nothing, only the length of the pattern to be generated is given. A brief comparative overview of the basic capabilities in all three cases is given.

1 Introduction

Since the forties of the last century cellular automata have been studied and investigated as an important area of computer science. In particular, there are numerous results concerning theoretical properties of cellular automata. Surveys on such aspects are given in [4,9,23]. In particular, we refer to surveys concerning reversible cellular automata [21], computational aspects [12], formal language aspects [13], and the descriptional complexity of cellular automata [16].

Here, we complement these surveys by summarizing and comparing results on cellular automata that work in three different modes. Classically, in the literature, the computational capacity of cellular automata is often measured by their ability to accept formal languages. In this context, first the input is given to the cells in a pre-initial step where an input of length n requires n cells. Second, the cellular automaton processes this input by iteratively and synchronously applying its local transition function to the cells. Finally, the input is accepted if at some time step the leftmost cell enters an accepting state. In addition, the available time is often restricted to real-time, which means that at most n time steps are allowed to accept an input of length n. The resulting family of formal languages accepted by cellular automata working in real time is then compared with language classes of the Chomsky hierarchy or other cellular automaton language classes. Results on this point of view on cellular automata such as the computational capacity, closure properties, and decidability questions of such devices can be found, for example, in the surveys [12,13].

J. D. Day and F. Manea (Eds.): DLT 2024, LNCS 14791, pp. 22–36, 2024.
https://doi.org/10.1007/978-3-031-66159-4_3

Rather than a device that accepts formal languages, cellular automata that transform or transduce formal languages are considered in [7,15]. As a language accepting device a cellular automaton answers to a given input with yes or no within certain time constraints. As a language transducing device a cellular automaton transforms a given input into a corresponding output within certain time constraints. In [7] cellular transducers obeying several time constraints and inclusion relationships based on these constraints are investigated. In addition, closure properties and the relation to language accepting devices are studied. The paper [15] also considers cellular automata with sequential input mode and compares their language transducing ability with cellular automata with parallel input mode. Moreover, also the relationship to classical sequential transducing devices such as finite state transducers or pushdown transducers is established.

A third point of view on cellular automata is to consider them as devices that generate outputs from almost nothing, only the length of the pattern to be generated is given. Instead of starting with an input on n cells and answering yes or no or transforming an input into an output, we start here with n cells that are all in the quiescent state. Thus, the information given to the automaton is just the number of cells. Then, the cellular automaton processes this "empty input" by iteratively and synchronously applying its local transition function to the cells. Finally, if at some time step the configurations reach a fixed point, we will call the configuration reached the pattern generated. In this way, cellular automata can compute a (partial) function that maps an integer n to a pattern of length n over the underlying alphabet of the automaton. This variant of cellular automata has been introduced and investigated in [17,18].

2   Preliminaries and Definitions

We denote the non-negative integers $\{0, 1, 2, \dots\}$ by $\mathbb{N}$. We write A^* for the *set of all words* over the finite alphabet A. The *empty word* is denoted by λ, and we set $A^+ = A^* \setminus \{\lambda\}$. The *reversal* of a word w is denoted by w^R, and for the *length* of w we write $|w|$. The number of *occurrences* of a symbol $a \in A$ in $w \in A^*$ is written as $|w|_a$. We use $\subseteq$ for *inclusions* and $\subset$ for *strict inclusions*. We say that two language families $\mathscr{L}_1$ and $\mathscr{L}_2$ are *incomparable* if $\mathscr{L}_1$ is not a subset of $\mathscr{L}_2$ and vice versa. In order to avoid technical overloading in writing, two languages L and L' are considered to be equal, if they differ at most by the empty word, that is, $L \setminus \{\lambda\} = L' \setminus \{\lambda\}$.

A (one-dimensional) cellular automaton is a linear array of identical deterministic finite automata, called cells, numbered $1, 2, \dots, n$. Each cell except the two outermost ones is connected to its both nearest neighbors. The total number of cells in the array is determined by the input data. One can suppose that all cells fetch their input symbol during a pre-initial step. The state transition depends on the current state of a cell itself and the current states of its two neighbors, where the outermost cells receive information associated with a boundary symbol on their free input lines. The cells work synchronously at discrete time steps (Fig. 1).

Fig. 1. A (two-way) cellular automaton.

More precisely, a *cellular automaton* (CA) is a system $M = \langle S, F, A, \#, \delta \rangle$, where S is the finite, nonempty set of *cell states*, $F \subseteq S$ is the set of *accepting states*, $A \subseteq S$ is the nonempty set of *input symbols*, $\# \notin S$ is the permanent *boundary symbol*, and $\delta \colon (S \cup \{\#\}) \times S \times (S \cup \{\#\}) \to S$ is the *local transition function*.

A *configuration* c_t of M at time $t \geq 0$ is a description of its global state, which is formally a mapping $c_t \colon \{1, 2, \ldots, n\} \to S$, for $n \geq 1$. The configuration at time 0 is defined by the given input $w = a_1 a_2 \cdots a_n \in A^+$. We set $c_0(i) = a_i$, for $1 \leq i \leq n$. Configurations may be represented as words over the set of cell states in their natural ordering. For example, the initial configuration for w is represented by $\#a_1 a_2 \cdots a_n \#$. Successor configurations are computed according to the global transition function Δ, that is, $c_{t+1} = \Delta(c_t)$, as follows. For $2 \leq i \leq n - 1$, $c_{t+1}(i) = \delta(c_t(i - 1)), c_t(i), c_t(i + 1))$, and for the outermost cells we set $c_{t+1}(1) = \delta(\#, c_t(1), c_t(2))$ and $c_{t+1}(n) = \delta(c_t(n-1), c_t(n), \#)$. For $n = 1$, the next state of the sole cell is $\delta(\#, c_t(1), \#)$. Thus, the global transition function Δ is induced by δ.

3   Cellular Automaton Acceptors

In this section, we consider cellular automata as acceptors for formal languages, that is, due to their space and time bounds they can be seen as deciders. They decide whether a given input word belongs to a formal language or not. In other words, this is the black-and-white outcome of computations.

A cellular automaton M *accepts* a word $a_1 a_2 \cdots a_n \in A^+$, if at some time step during the course of its computation the leftmost cell enters an accepting state. The *language accepted by* M is $L(M) = \{\, w \in A^+ \mid w \text{ is accepted by } M \,\}$. Let $t \colon \mathbb{N} \to \mathbb{N}$ be a mapping. If all $w \in L(M)$ are accepted with at most $t(|w|)$ time steps, then M is said to be of time complexity t. If $t(n) = n$ then M operates in *real time*. If $t(n) = k \cdot n$ for a rational number $k \geq 1$ then M operates in *linear time*.

The family of languages accepted by CAs with time complexity t is denoted by $\mathscr{L}_t(\mathsf{CA})$. The index is omitted for arbitrary time. Actually, arbitrary time is exponential time due to the space bound. For real-time (linear-time) we write $\mathscr{L}_{rt}(\mathsf{CA})$ ($\mathscr{L}_{lt}(\mathsf{CA})$).

The Examples 1 and 2 are intended to illustrate the notation.

The possibility to send signals through cellular automata is a widely used tool to solve problems. Examples are basic signals with a certain speed or complex signals that allow to generate prime numbers. In general, signals are used to transmit or encode information. They have been used for a long time, but the systematic study originated from [20]. A related concept are *time-constructible functions* that originates from [6], where a cellular automaton is constructed whose leftmost cell distinguishes exactly the time steps that are prime numbers. Initially, all cells are in the same state. In [3] a time-constructor for the function $i \mapsto 2^i$ is given.

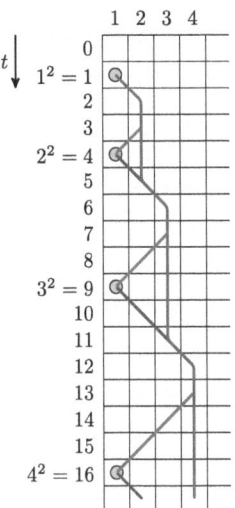

Fig. 2. Time construction of the function i^2. Signal α is depicted in blue and green, signal β is depicted in red. (Color figure online)

As a simple example, in Fig. 2 the time-construction of the function $i \mapsto i^2$ is given. Basically, the necessary signals can be derived from $(i+1)^2 = i^2 + 2i + 1$. In particular, after being designated at time i^2, the leftmost cell has to wait for $2i$ time steps before it is designated again at time $i^2 + 2i + 1$. This delay is exactly the time needed by an auxiliary signal α that moves from the leftmost cell 1 to cell $i+1$, stays there for one time step, and moves back to the leftmost cell. To this end, another auxiliary signal β is used that stays in cell i until it is hit by α and moves to cell $i+1$.

Example 1. Let φ be a time-constructible function. Then the unary language $\{\, a^{\varphi(n)} \mid n \geq 1 \,\}$ is accepted by some CA in real time.

At initial time the time construction of φ is started. In addition, a signal with maximal speed is sent to the left by the rightmost cell. This signal arrives at the leftmost cell at the time step which corresponds to the length of the input. If this time step is distinguished by the time construction, the input is accepted. ∎

t	1	2	3	4	5	6	7	8	9	10	11	12	
0 #	a	a	a	a	a	a	b	b	b	b	b	b	#
1 #	a	a	a	a	a	a	b	b	b	b	b	s	#
2 #	a	a	a	a	a	a	c	b	b	b	b	s	#
3 #	a	a	a	a	a	a	b	b	b	b	s		#
4 #	a	a	a	a	a	c	b	b	b	s			#
5 #	a	a	a	a	a	b	b	b	s				#
6 #	a	a	a	a	c	b	b	s					#
7 #	a	a	a	a	b	b	s						#
8 #	a	a	a	c	b	s							#
9 #	a	a	a	b	s								#
10 #	a	a	c	s									#
11 #	a	a	s										#
12 #	+												#

Fig. 3. Space-time diagram of a cellular automaton accepting an input from the language $\{\, a^n b^n \mid n \geq 1 \,\}$ in real time.

Example 2. The language $\{\, a^n b^n \mid n \geq 1 \,\}$ is accepted by some CA in real time. The idea of the construction is to establish two left-moving signals. See Fig. 3. The rightmost cell can identify itself by receiving a message from the border cell. It sends a signal s with maximum speed to the left. The unique cell which has an a in its input and has a right neighbor with a b in its input can identify itself as well if any cell initially communicates its input to the left. It sends a signal c with speed $1/2$ to the left. When both signals meet in a cell, an accepting state is entered. Clearly, each cell communicates only finitely often. The construction is easily modified to reject inputs having a wrong format. ∎

It is worth mentioning that the family of time-constructible functions is very rich. See [20] and [12] for details.

The next goal of this section is to establish a basic hierarchy of cellular language families, and to compare the levels with well-known families of the Chomsky hierarchy. The properness of some inclusions are long-standing open problems with deep relations to sequential complexity problems. In order to establish the hierarchy we start at the upper end. Straightforward constructions of linearly space-bounded Turing machines from CAs and vice versa show the following lemma [26].

Lemma 3. *The family $\mathscr{L}(CA)$ coincides with the family of deterministic context-sensitive languages, that is, with the complexity class* DSPACE(n).

Diving into the Chomsky hierarchy, we next see the family of context-free languages (CFL). However, whether or not CFL is included in the family $\mathscr{L}_{rt}(CA)$ is an open question raised in [26]. It is related to the open question whether or not sequential one-tape Turing machines are able to accept the context-free languages in square-time. A proof for the inclusion would imply the existence of square-time Turing machines. In fact, also the problem whether or not the context-free languages are included in $\mathscr{L}_{lt}(CA)$ is open. But for the important metalinear and deterministic context-free languages we can answer the inclusion problem in the affirmative [11].

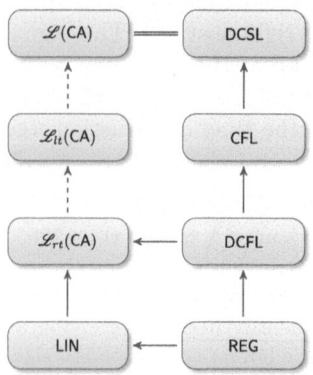

Moreover, the family $\mathscr{L}_{rt}(CA)$ contains further important subfamilies, for example, the linear context-free languages [25], the Dyck languages [24], and the bracketed context-free languages [5]. Furthermore, the non-semilinear language $\{(a^i b)^* \mid i \geq 0\}$ [24] and the inherently ambiguous language $\{a^i b^j c^k \mid i = j$ or $j = k$ for $i,j,k \geq 1\}$ [26] belong to $\mathscr{L}_{rt}(CA)$ as well.

At the lower end we consider the regular languages. Since already the leftmost cell is a deterministic finite automaton, clearly, the regular languages are a subset of $\mathscr{L}_{rt}(CA)$.

Fig. 4. Basic hierarchy of language families. A solid arrow indicates a proper inclusion, a dashed arrow an inclusion, and a double arrow an equality. In addition, the linear languages (LIN) are properly included in the context-free languages (CFL). Deterministic context-free languages are denoted by DCFL, regular languages by REG, and deterministic context-sensitive languages by DCSL.

Altogether we obtain the hierarchy depicted in Fig. 4. We emphasize that the inclusions between real-time CAs and arbitrary time CAs, that is exponential-time CAs, is not known to be proper. For further reading on cellular automata and language theory we refer to [13].

4   Cellular Automaton Transducers

In this section, we will consider CAs not only as language recognizing devices but as language transforming devices. In other words, the outcome of computations is more than yes or no, it is a word computed from the input. To some extent, this adds colors to the outcome.

In particular, we are interested in results that compare cellular automaton transducers with the conventional transducer models. To this end, we enhance the definition suitably. Now, every cell of a CA is equipped with an output register which is initially empty and can be filled once with some symbols of an output alphabet. An output is computed, if the input is accepted and all cells have filled their output register. Since CAs are deterministic devices every input accepted corresponds to exactly one output.

So, a *cellular automaton transducer* (CAT) is a system $M = \langle S, F, A, B, \#, \delta \rangle$, where the components S, F, A, and $\#$ are defined as for cellular automaton acceptors, B is the finite set of *output symbols* not including the special symbol $\perp$, and $\delta \colon (S \cup \{\#\}) \times S \times (S \cup \{\#\}) \to S \times (B^* \cup \{\perp\})$ is the *local transition function*.

A *configuration* of a cellular automaton transducer $M = \langle S, F, A, B, \#, \delta \rangle$ is a pair of mappings $c_t \colon \{1, 2, \ldots, n\} \to S$ and $o_t \colon \{1, 2, \ldots, n\} \to B^* \cup \{\perp\}$, for $n \geq 1$, which map the single cells to their current states and to their output emitted, where $\perp$ means no output so far. The *initial configuration* is defined by the given input $w = a_1 a_2 \cdots a_n \in A^+$, and no outputs. We set $c_0(i) = a_i$ and $o_0(i) = \perp$, for $1 \leq i \leq n$. As for acceptors, successor configurations are computed according to a global transition function Δ. For convenience, we write δ_s for the projection on the first component of δ, that is, the successor state, and δ_o for the projection on the second component, that is, the output emitted. Let (c_t, o_t), $t \geq 0$, be a configuration with $n \geq 2$, then the first component of its successor (c_{t+1}, o_{t+1}) is as for CA and the second component is as follows.

$$o_{t+1}(i) = \begin{cases} o_t(i) & \text{if } o_t(i) \neq \perp \\ \delta_o(c_t(i-1), c_t(i), c_t(i+1)) & \text{if } o_t(i) = \perp \text{ and } i \in \{2, 3, \ldots, n-1\} \\ \delta_o(\#, c_t(1), c_t(2)) & \text{if } o_t(i) = \perp \text{ and } i = 1 \\ \delta_o(c_t(n-1), c_t(n), \#) & \text{if } o_t(i) = \perp \text{ and } i = n \end{cases}$$

and, again, Δ is induced by δ.

Transducer M *transforms* input words $w \in A^+$ into output words $v \in B^*$. For a successful transformation M has to accept the input, otherwise the output is not recorded. Moreover, each cell has to emit an output: $M(w) = v$, if w is accepted, at some time t automaton M reaches a configuration so that $o_t(i) \neq \perp$, $1 \leq i \leq |w|$, and $v = o_t(1) o_t(2) \cdots o_t(|w|)$. The *transduction realized by M*, denoted by $T(M)$, is the set of pairs $(w, v) \in A^+ \times B^*$ such that $M(w) = v$.

Here, we are mainly interested in fast transductions and consider the time complexities of real-time and linear-time. Additionally, we consider the time complexities of accepting the input and computing the output separately. Let $t_i, t_o \colon \mathbb{N} \to \mathbb{N}$ be two mappings. If for all $(w, v) \in T(M)$, the input w is accepted after at most $t_i(|w|)$ time steps, and $o_t(i) \neq \perp$, $1 \leq i \leq |w|$, after at most $t_o(|w|)$ time steps, then M is said to be of time complexity (t_i, t_o) and we write CAT_{t_i,t_o}. The family of transductions realized by CAT_{t_i,t_o} is denoted by $\mathscr{T}(\mathsf{CAT}_{t_i,t_o})$.

If we build the projection on the first components of $T(M)$, then the cellular automaton transducer degenerates to a cellular automaton acceptor.

In order to clarify the notation we give two examples. The first example shows that CAT can copy their input in real time. The second example shows that CAT can sort a binary input in real time.

Example 4. The transduction $\{\,(w, ww) \mid w \in \{a,b\}^+\,\}$ belongs to the family $\mathscr{T}(\mathsf{CAT}_{rt,rt})$.

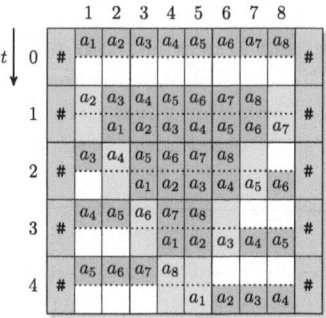

Fig. 5. Schematic computation of a $\mathsf{CAT}_{rt,rt}$ copying input $a_1 a_2 \cdots a_8$. The signals that cause cells to emit their output are highlighted (in yellow). (Color figure online)

The basic idea is to use two tracks. On one track the input is successively shifted to the left, while on the other track a copy of the input is successively shifted to the right. In addition, at the beginning of the computation one right-moving signal is started at the left end, and one left-moving signal is started at the right end. When a cell receives the signal from the left first, it emits two symbols from the input that is shifted to the left. Accordingly, when a cell receives the signal from the right first, it emits two symbols from the input that is shifted to the right (see Fig. 5).

More detailed, let $w = a_1 a_2 \cdots a_n$ be the input and n be even. In the first time step, cell 1 emits $a_1 a_2$ and cell n emits $a_{n-1} a_n$. In the next time step, cell 2 emits $a_3 a_4$ and cell $n-1$ emits $a_{n-3} a_{n-2}$. This is possible since the necessary information has been shifted to the corresponding cells and their neighborhood. Generalizing this observation to $1 \leq i \leq \frac{n}{2}$, we obtain that cell i emits $a_{2i-1} a_{2i}$ and cell $n-i+1$ emits $a_{n-2i+1} a_{n-2i+2}$ at time i. If n is odd, the construction is similar. Now cells i and $n-i+1$ emit $a_{2i-1} a_{2i}$ and $a_{n-2i+1} a_{n-2i+2}$ at time i, for $1 \leq i \leq \lfloor \frac{n}{2} \rfloor$. Finally, cell $\lceil \frac{n}{2} \rceil$ emits $a_n a_1$ at time step $\lceil \frac{n}{2} \rceil$.   ∎

Example 5. Here, we consider the transduction $\{\,(w, a^{|w|_a} b^{|w|_b}) \mid w \in \{a,b\}^+\,\}$ and show that it is computed by a $\mathsf{CAT}_{rt,rt}$. The basic idea is that any two neighboring cells where the left cell carries a b and the right cell carries an a switch their contents. By this local transpositions, all a's will eventually be in the left part of the array whereas all b's will be in the right part (see Fig. 6).

Clearly, on any input of length n the correct sorting has been achieved after at most n time steps. There is one problem to be solved: since the transpositions are only local, a cell cannot know whether its current content is final and has to be emitted. We can cope with the problem by synchronizing all cells at time n. The synchronization is realized by the well-known Firing Squad Synchronization Problem (FSSP), that is implemented on an additional track. In the first time step, two instances of the FSSP are started, one in the leftmost cell and the other one in the rightmost cell. Since both FSSP work in parallel, we obtain that all cells are synchronized at time step n.   ∎

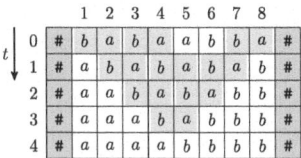

	1	2	3	4	5	6	7	8	
0	#	b	a	b	a	a	b	b	a	#
1	#	a	b	a	b	a	b	a	b	#
2	#	a	a	b	a	b	a	b	b	#
3	#	a	a	a	b	a	b	b	b	#
4	#	a	a	a	a	b	b	b	b	#

Fig. 6. Schematic computation of a $\mathsf{CAT}_{rt,rt}$ sorting an input *babaabba* by local transpositions. Neighboring cells that switch their input are highlighted (in yellow). The synchronization is not depicted. (Color figure online)

Any transduction computed by a cellular automaton can be divided into two tasks. One is the acceptance of the input, the other one the transformation of the input into the output. Both tasks have to end successfully in order to obtain a valid computation. On the one hand, this allows to modularize constructions of cellular automaton transducers as both parts can be implemented independently on different tracks. On the other hand, this implies that a language, which is not accepted by any cellular automaton in time t_i, cannot be the projection on the first components of any transduction belonging to any class $\mathscr{T}(\mathsf{CAT}_{t_i,t_o})$. Unfortunately, as mentioned above, it is a long-standing open problem whether there are languages accepted by two-way cellular automata in arbitrary, that is, exponential time but cannot be accepted in real time (see, for example, [13]).

4.1 Comparison with Finite State Transducers and Pushdown Transducers

Let us turn to comparing the computational capacities of massively parallel cellular automaton transducers with sequential finite state transducers (FST) and pushdown transducers (PDT). These devices are in essence finite automata and pushdown automata, where each transition is associated with a possibly empty output word (see [1]). In their most general form, FST and PDT are nondeterministic devices, that is, the partial transition function of an FST maps from $S \times (A \cup \{\lambda\})$ into the finite subsets of $S \times B^*$. As above, S denotes the state set and A the input alphabet. The partial transition function of a PDT maps from $S \times (A \cup \{\lambda\}) \times G$ into the finite subsets of $S \times B^* \times G^*$, where G denotes the pushdown alphabet. Since a nondeterministic transducer may transform an input into different outputs, which is impossible for deterministic CAT, in the sequel we only study deterministic, unambiguous, and single valued devices.

A nondeterministic FST M is called *single valued* (SFST) if for all pairs $(w_1, v_1), (w_2, v_2) \in T(M)$ either $(w_1, v_1) = (w_2, v_2)$ or $w_1 \neq w_2$. An SFST is said to be *unambiguous* (UFST) if for all pairs $(w, v) \in T(M)$ there is a unique computation transforming w into v. Finally, a UFST is *deterministic* (DFST) if any computation is deterministic. It has been shown in [27] that every single-valued finite state transducer can be simulated by an unambiguous one. Furthermore, it is known (see, for example, [27]) that $\mathscr{T}(\mathsf{DFST}) \subset \mathscr{T}(\mathsf{UFST}) = \mathscr{T}(\mathsf{SFST})$.

The properness of the next inclusions is witnessed by the transduction $\{ (w, w^R) \mid w \in \{a, b\}^* \}$, while the inclusions themselves are shown by a construction that shows that any SFST can be simulated by some $\mathsf{CAT}_{rt,rt}$ [15].

Lemma 6. *The families $\mathscr{T}(\mathsf{DFST})$ and $\mathscr{T}(\mathsf{SFST})$ are strictly included in $\mathscr{T}(\mathsf{CAT}_{rt,rt})$.*

The notions of single-valued PDT (SPDT), unambiguous PDT (UPDT), and deterministic PDT (DPDT) are defined in the same way as for FST. Additionally, a DPDT is called *real-time deterministic* (DPDT_λ) if it is not allowed to move on empty input. As opposed to finite state transducers, single-valued pushdown transducers have more computational power than unambiguous pushdown transducers. For example, the transduction $T = \{\, (a^n b^n a^m b^m, a^{2m+2n}) \mid m, n \geq 1 \,\} \cup \{\, (a^n b^m a^m b^n, a^{2m+2n}) \mid m, n \geq 1 \,\}$ belongs to $\mathscr{T}(\mathsf{SPDT})$ but not to $\mathscr{T}(\mathsf{UPDT})$ because the projection on the first components is known to be an inherently ambiguous context-free language [8]. The following proper hierarchy is known (see, for example, [14]): $\mathscr{T}(\mathsf{DPDT}_\lambda) \subset \mathscr{T}(\mathsf{DPDT}) \subset \mathscr{T}(\mathsf{UPDT}) \subset \mathscr{T}(\mathsf{SPDT})$.

The computational power of the resource pushdown store equipped to a deterministic finite state device cannot compensate the presence of a little bit of nondeterminism [15].

Lemma 7. *The family $\mathscr{T}(\mathsf{SFST})$ is incomparable with both families $\mathscr{T}(\mathsf{DPDT}_\lambda)$ and $\mathscr{T}(\mathsf{DPDT})$.*

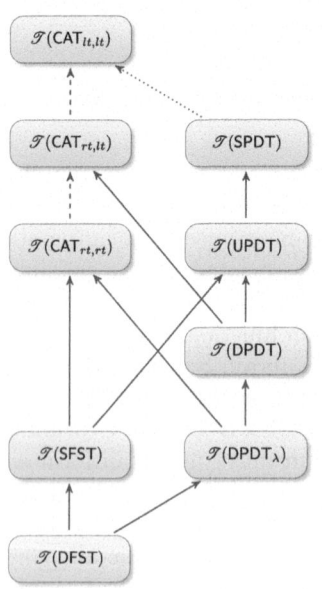

By Lemma 7, the families $\mathscr{T}(\mathsf{SFST})$ and $\mathscr{T}(\mathsf{DPDT}_\lambda)$ are incomparable, and by Lemma 6, $\mathscr{T}(\mathsf{SFST})$ is strictly included in $\mathscr{T}(\mathsf{CAT}_{rt,rt})$. For the sake of completeness, here, we mention that $\mathscr{T}(\mathsf{DPDT}_\lambda)$ is strictly included in $\mathscr{T}(\mathsf{CAT}_{rt,rt})$ as well.

Lemma 8. *The family $\mathscr{T}(\mathsf{DPDT}_\lambda)$ is strictly included in $\mathscr{T}(\mathsf{CAT}_{rt,rt})$.*

Finally, the transduction $T = \{\, (ww, wc^{|w|}) \mid w \in \{a,b\}^+ \,\}$ cannot be computed by any pushdown transducer, since the language $\{\, ww \mid w \in \{a,b\}^+ \,\}$ is not context free. On the other hand, transduction T can be computed by a $\mathsf{CAT}_{rt,rt}$ and, thus, by a $\mathsf{CAT}_{rt,lt}$. Moreover, cellular automata can simulate data structures as pushdown stores, queues, and rings in real time [12]. From these ingredients one can prove the following inclusion (Fig. 7).

Fig. 7. Summary of inclusions. Solid lines are proper inclusions, dashed lines are inclusions, and dotted lines are conjectured inclusions.

Lemma 9. *The family $\mathscr{T}(\mathsf{DPDT})$ is strictly included in $\mathscr{T}(\mathsf{CAT}_{rt,lt})$.*

We note that cellular automaton transducers have been investigated also in [7]. In particular, a speed-up theorem for cellular automaton transducers is shown there and a proper subclass of real-time transductions is established.

5 Cellular Automaton Pattern Generators

In this section, we turn to cellular automata that, in contrast to the point of view taken so far, are used to *generate* formal languages which are here called *patterns*. Instead of starting with an input in n cells and answering yes or no or transforming an input into an output, here we start with n cells that are all in a so-called quiescent state. Thus, the information given to the automaton is just the number of cells. Then, the cellular automaton processes this "empty input" by iteratively and synchronously applying its local transition function to the cells. Finally, if at some time step the configurations reach a fixed point, we will call the configuration reached the pattern generated. In this way, cellular automata can compute a (partial) function that maps an integer n to a pattern of length n over the underlying alphabet of the automaton. This variant of cellular automata has been introduced and investigated in [17,18]. So, to some extent we now add high gloss color to the outcome that is generated from almost nothing.

In order to obtain a cellular pattern generator, we take the definition of cellular language acceptors and add a quiescent state that is used to prevent from creating information spontaneously somewhere in the cells. Since the set of input symbols and the set of accepting states are not used when a cellular automaton operates as generator, we may safely omit them from its definition. So, a *cellular automaton pattern generator* (CAG) is a system $M = \langle S, s_0, \#, \delta \rangle$, where the components S, $\#$, and δ are defined as for cellular automaton acceptors, $s_0 \in S$ is the *quiescent state*, and δ satisfies $\delta(s_0, s_0, s_0) = s_0$.

A word $a_1 a_2 \cdots a_n$ is *generated* by M, if at some time step t during the computation on the initial configuration $c_0(i) = s_0$, $1 \leq i \leq n$, (i) the word appears as configuration (that is, $c_t(i) = a_i$, $1 \leq i \leq n$) and (ii) configuration c_t is a fixed point of the global transition function Δ (that is, the configuration is stable from time t on). The *pattern generated by* M is $P(M) = \{ w \in S^+ \mid w$ is generated by $M \}$. The family of patterns generated by CAGs with time complexity t is denoted by $\mathscr{P}_{rt}(\mathsf{CAG})$.

Example 10. We consider the *Oldenburger-Kolakoski sequence* [10,22] that is an infinite sequence over the alphabet $\{1, 2\}$. Basically, it is the sequence of run lengths in its own run-length encoding, where the sequence starts with 1 and consists of alternating blocks of 1's and 2's.

The sequence is described as follows: The first symbol is 1. For $n \geq 1$, the nth symbol is the length of the nth block. So, the Oldenburger-Kolakoski sequence can be constructed by setting the first symbol to 1. The 1 means that the first block has length 1, that is the 1 itself. The next block starts with 2 which means that the second block has length 2. This gives 122 Next, the 2 at position three means that a block of length 2 has to be appended to the sequence, which yields 12211. Now the 1 at the fourth position means to append 2 and the 1 at the fifth position to append a further 1 yielding 1221121, and so on.

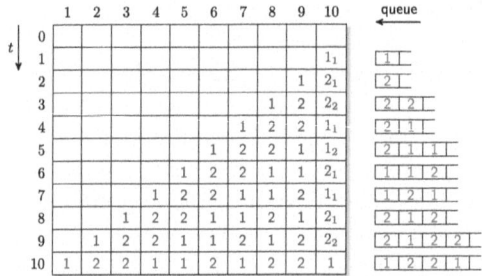

Fig. 8. Generation of the prefix 1221121221 of the Oldenburger-Kolakoski sequence. The prefix is fed from the right and shifted to the left. The queue is simulated at the rightmost cell. For the sake of readability, it is depicted separately. The boundary symbol is not depicted. The cells enter simultaneously a permanent state at time n, which can be achieved by an FSSP synchronization.

Since the nth run of the sequence is generated by its nth symbol, and the nth run generates the nth symbol, the Oldenburger-Kolakoski sequence is self-generating or self-describing. In fact, not much is known about the Oldenburger-Kolakoski sequence and several of its properties are non-trivial and open problems [2]. However, the pattern consisting of all prefixes of the sequence can be generated by a CAG in real time. Apart from other ingredients, the CAG utilizes the possibility of cellular automata to simulate the data structure *queue* in real time [12]. See Fig. 8 for the principle of the construction and [18] for details of the construction. ∎

5.1 Unary Patterns

Next, we consider the special case of unary patterns. Let, for example, $\varphi \colon \mathbb{N} \to \mathbb{N}$ be some function. Then pattern P_φ is defined to be a^n if there is some m such that $n = \varphi(m)$, and undefined otherwise. Assume now, that φ is a time-constructible function. Since the pattern is undefined for lengths n that are not in the range of φ, the pattern is easily generated by a CAG. The simple construction of the generator of P_φ works fine since, in principle, only the leftmost cell decides whether a pattern has to be generated or not. Moreover, the other cells can safely enter the pattern states in advance without violating the overall result. Now, we define the pattern to be generated as a total function on n and impose a condition on n such that the pattern a^n is to be generated if n meets the condition, but the pattern b^n otherwise. In this case, the leftmost cell can still decide which pattern has to be generated, but all the other cells cannot enter the pattern state in advance. Instead, they have to know whether the condition is met or not. For a time-constructible function $\varphi \colon \mathbb{N} \to \mathbb{N}$, we define the pattern $\hat{P}_\varphi$ to be a^n if there is some m such that $n = \varphi(m)$, and b^n otherwise. However, even these unary patterns are generated in real time.

Theorem 11. *Let $\varphi \colon \mathbb{N} \to \mathbb{N}$ be a time-constructible function. Then the pattern $\hat{P}_\varphi$ is generated by some real-time CAG.*

Moreover, in [20] the following relation between time-constructible functions and unary languages is shown: A function $\varphi \colon \mathbb{N} \to \mathbb{N}$ is time constructible if and only if the language $L_\varphi = \{\, a^{\varphi(m)} \mid m \geq 1 \,\}$ is accepted by a real-time cellular automaton.

Proposition 12. *Let* $\varphi\colon \mathbb{N} \to \mathbb{N}$ *be a function and* $\hat{P}_\varphi$ *be generated by some cellular automaton in real time. Then language* L_φ *is accepted by a real-time cellular automaton.*

Now by Theorem 11, Proposition 12, and the result from [20], we derive that for unary languages/patterns the three different notions of language acceptance, time-constructibility, and pattern generation, in fact, coincide.

Theorem 13. *A function* $\varphi\colon \mathbb{N} \to \mathbb{N}$ *is time constructible if and only if the language* L_φ *is accepted by a real-time cellular automaton if and only if the pattern* $\hat{P}_\varphi$ *can be generated by a cellular automaton in real time.*

5.2   Automatic Sequences and Morphic Words

Here, we turn to patterns that are defined as set of all prefixes of some infinite string of symbols. Example 10 already showed how the prefixes of the infinite Oldenburger-Kolakoski sequence are generated. We are going to consider wider classes and start with so-called *automatic sequences* [2].

So, what is an automatic sequence? It may be defined in a number of ways. For our purposes, the automata-theoretic definition is well suited.

Let $k \geq 1$ be a positive integer and M be a **DFA** with output (Moore automaton). The automaton defines a *k-automatic sequence* $(u_m)_{m\geq0}$ as follows. For each $m \geq 0$, the automaton is fed with the k-ary expansion of m (the unary expansion is 1^m). After processing this input word, the automaton reaches one of its states, whose image under some output function gives the mth element of the sequence.

Many well-known infinite sequences with applications in numerous fields of mathematics and non-trivial properties are automatic. The next example mentions three of them (see [2] for further details).

Example 14. The *Thue-Morse sequence* is a 2-automatic sequence over the alphabet $\{0,1\}$. There a several ways of generating the Thue-Morse sequence one of which is given by a Lindenmayer system with axiom 0 and rewriting rules $0 \to 01$ and $1 \to 10$. The generation of strings can be described as follows: starting with the axiom every symbol 0 (symbol 1) is in parallel replaced by the string 01 (10). This procedure is iteratively applied to the resulting strings. The first steps of this procedure yield the strings 0, 01, 0110, 01101001, and so on.

The *Rudin-Shapiro sequence* is a 2-automatic sequence over $\{1, -1\}$. Let $e(m)$ be the number of (possibly overlapping) occurrences of 11 in the binary expansion of m. Then the Rudin-Shapiro sequence $(u_m)_{m\geq0}$ can be defined by the formula $u_m = (-1)^{e(m)}$.

The *Regular Paperfolding sequence* takes its name from the following fact. Assume that a strip of paper is folded repeatedly in half in the same direction. If each fold is now opened out to create a right-angled corner, then a right turn is represented by 1 and a left turn by a 0. The *Regular Paperfolding sequence* is 2-automatic. ∎

It turned out that the prefixes of automatic sequences are patterns generated by cellular automata in real time.

Theorem 15. *Let $k \geq 1$ be a positive integer and $u = (u_m)_{m \geq 0}$ be a k-automatic sequence. Then the pattern $P_u = \{\, p \mid p \text{ is prefix of } u \,\}$ can be generated by a cellular automaton in real time.*

Finally, we turn to *morphic words*. To some extent, they are generalizations of automatic sequences. The idea is to define an infinite string by using an infinite sequence of finite strings such that each string at position x is a prefix of the string at position $x + 1$. To construct a finite string from its successor, homomorphisms are used (which justifies the notion morphic words).

To introduce morphic words, let A be an alphabet and $\mu\colon A^* \to A^*$ be a homomorphism. We say that a symbol $a \in A$ is *mortal* if there exists some integer $i \geq 1$ such that $\mu^i(a) = \lambda$. If there is some $a \in A$ such that $\mu(a) = as$, where s includes at least one letter that is not mortal, then μ is said to be *prolongable* on a. Now we have all ingredients to construct the infinite sequence of finite strings.

Let $\mu\colon A^* \to A^*$ be a homomorphism prolongable on the symbol $a \in A$ with $\mu(a) = as$. Then the sequence of strings a, $\mu(a)$, $\mu^2(a)$, ... converges in the limit to the infinite string $w = as\mu(s)\mu^2(s) \cdots$ In this case, w is called a *pure morphic word*. The image of a pure morphic word under an encoding, that is, a letter-to-letter homomorphism $\tau\colon A \to U$ with some alphabet U, is a *morphic word*.

What are the relations between automatic sequences and morphic words? It is known that any automatic sequence is a morphic word. Conversely, if a morphic word is defined via a so-called k-uniform homomorphism prolongable on some symbol, then the word is k-automatic [19].

The crucial point for a difference between automatic sequences and morphic words seems to be the uniformity of the homomorphisms for the former ones. A well-known and famous morphic word is defined via the following non-uniform homomorphism $f\colon \{0,1\}^* \to \{0,1\}^*$, where $f(0) = 01$ and $f(1) = 0$. Here the encoding is the identity. Clearly, f is prolongable on 0. The first few symbols of the word are $0100101001001 \cdots$ Since the lengths of the finite words 0, $f(0)$, $f^2(0)$, $f^3(0)$, that converge in the limit to the morphic word, are equal to the Fibonacci numbers $1, 2, 3, 5, 8, 13, \ldots$, the morphic word is called the *Fibonacci word*.

Not only the pattern of all prefixes of the Fibonacci word but all patterns of all prefixes of *every* morphic word are generated by some cellular automaton in real time.

Theorem 16. *Let $u = (u_m)_{m \geq 0}$ be a morphic word. Then the pattern of all of its prefixes $P_u = \{\, p \mid p \text{ is prefix of } u \,\}$ can be generated by a cellular automaton in real time.*

References

1. Aho, A.V., Ullman, J.D.: The Theory of Parsing, Translation, and Compiling, vol. I: Parsing. Prentice-Hall Inc. (1972)
2. Allouche, J., Shallit, J.O.: Automatic Sequences – Theory, Applications, Generalizations. Cambridge University Press, Cambridge (2003)
3. Choffrut, C., Čulik, K., II.: On real-time cellular automata and trellis automata. Acta Inform. **21**, 393–407 (1984)
4. Delorme, M., Mazoyer, J. (eds.): Cellular Automata – A Parallel Model. Kluwer Academic Publishers (1999)
5. Dyer, C.R.: One-way bounded cellular automata. Inf. Control **44**, 261–281 (1980)
6. Fischer, P.C.: Generation of primes by a one-dimensional real-time iterative array. J. ACM **12**, 388–394 (1965)
7. Grandjean, A., Richard, G., Terrier, V.: Linear functional classes over cellular automata. In: Formenti, E. (ed.) International workshop on Cellular Automata and Discrete Complex Systems and Journées Automates Cellulaires (AUTOMATA & JAC 2012). EPTCS, vol. 90, pp. 177–193 (2012)
8. Harrison, M.A.: Introduction to Formal Language Theory. Addison-Wesley (1978)
9. Kari, J.: Theory of cellular automata: a survey. Theor. Comput. Sci. **334**, 3–33 (2005)
10. Kolakoski, W.: Problem 5304: self generating runs. Am. Math. Mon. **72**, 674 (1965)
11. Kutrib, M.: Automata arrays and context-free languages. In: Martin-Vide, C., Mitrana, V. (eds.) Where Mathematics, Computer Science and Biology Meet, pp. 139–148. Kluwer Academic Publishers (2001)
12. Kutrib, M.: Cellular automata – a computational point of view. In: Bel-Enguix, G., Jiménez-López, M.D., Martín-Vide, C. (eds.) New Developments in Formal Languages and Applications, pp. 183–227. Springer, Cham (2008). https://doi.org/10.1007/978-3-540-78291-9_6
13. Kutrib, M.: Cellular automata and language theory. In: Meyers, R. (ed.) Encyclopedia of Complexity and System Science, pp. 800–823. Springer, Cham (2009). https://doi.org/10.1007/978-0-387-30440-3_54
14. Kutrib, M., Malcher, A.: Transductions computed by iterative arrays. In: Kari, J. (ed.) Symposium on Cellular Automata – Journées Automates Cellulaires (JAC 2010). TUCS Lecture Notes, vol. 13 pp. 156–167. Turku Center for Computer Science (2010)
15. Kutrib, M., Malcher, A.: One-dimensional cellular automaton transducers. Fund. Inform. **126**, 201–224 (2013)
16. Kutrib, M., Malcher, A.: Cellular automata: descriptional complexity and decidability. In: Adamatzky, A. (ed.) Reversibility and Universality. ECC, vol. 30, pp. 129–168. Springer, Cham (2018). https://doi.org/10.1007/978-3-319-73216-9_6
17. Kutrib, M., Malcher, A.: String generation by cellular automata. Complex Syst. **30**, 111–132 (2021)
18. Kutrib, M., Malcher, A.: One-dimensional pattern generation by cellular automata. Nat. Comput. **21**, 361–375 (2022)
19. Lothaire, M.: Applied Combinatorics on Words. Encyclopedia of Mathematics and its Applications. Cambridge University Press, Cambridge (2005)
20. Mazoyer, J., Terrier, V.: Signals in one-dimensional cellular automata. Theor. Comput. Sci. **217**, 53–80 (1999)
21. Morita, K.: Reversible computing and cellular automata – a survey. Theor. Comput. Sci. **395**, 101–131 (2008)

22. Oldenburger, R.: Exponent trajectories in symbolic dynamics. Trans. Am. Math. Soc. **46**, 453–466 (1939)
23. Rozenberg, G., Bäck, T., Kok, J.N. (eds.): Handbook of Natural Computing. Springer, Cham (2012). https://doi.org/10.1007/978-3-540-92910-9
24. Seidel, S.R.: Language recognition and the synchronization of cellular automata. Technical report. 79-02, Department of Computer Science, University of Iowa, Iowa City (1979)
25. Smith, A.R., III.: Cellular automata and formal languages. In: Symposium on Switching and Automata Theory (SWAT 1970), pp. 216–224. IEEE (1970)
26. Smith, A.R., III.: Real-time language recognition by one-dimensional cellular automata. J. Comput. Syst. Sci. **6**, 233–253 (1972)
27. Weber, A., Klemm, R.: Economy of description for single-valued transducers. Inf. Comput. **118**, 327–340 (1995)

Deciding Conjugacy of a Rational Relation
(Extended Abstract)

C. Aiswarya[1,2]($\boxtimes$) (iD), Amaldev Manuel[3]($\boxtimes$) (iD), and Saina Sunny[3]($\boxtimes$) (iD)

[1] Chennai Mathematical Institute, Siruseri, India
aiswarya@cmi.ac.in
[2] CNRS, ReLaX, IRL 2000, Siruseri, India
[3] Indian Institute of Technology Goa, Ponda, India
{amal,saina19231102}@iitgoa.ac.in

Abstract. A relation on the free monoid is conjugate if each pair of words in the relation is conjugate, i.e., cyclic shifts of each other. We show that checking whether a rational relation is conjugate is decidable. This extended abstract outlines the proof of this fact. A result of independent interest is a generalisation of the classical Lyndon-Schützenberger theorem from word combinatorics that equates conjugacy of a pair of words (u, v) and the existence of a word z (called a *witness*) such that $uz = zv$.

A full version of the paper, with details of the proof, can be found on arXiv [1].

Keywords: Rational relations · Finite state transducers · Conjugacy of words · Combinatorics of words

1 Conjugacy of a Relation

Conjugacy of two elements u and v in a group can be defined as any of the following equivalent cases:

1. $uz = zv$ for some z,
2. $u = xy$ and $v = yx$ for some x, y.

The conjugacy problem asks if a given pair of elements in a finitely presented group (typically infinite) is conjugate. It along with the word and isomorphism problems constitute the classical triad of decision problems on groups identified by Dehn in 1912 [15]. Dehn's prescient choice turned out to be instrumental not only in mathematics, but also to the theory of semigroups/monoids and automata in computer science. It turns out that the above conditions are equivalent for free monoids as well (i.e., when u, v, z, x, y are taken to be words over some finite alphabet). This is the well-known second theorem of Lyndon-Schützenberger. But unlike in the case of groups where condition (1) is taken to be the definition of conjugacy, in the case of monoids condition (2) is taken as the definition of conjugacy. Hence the statement reads the following way.

A. Manuel—Supported by the DST SERB MATRICS grant for the project *Deciding closeness of finite state transducers* [MTR/2022/000628].

J. D. Day and F. Manea (Eds.): DLT 2024, LNCS 14791, pp. 37–50, 2024.
https://doi.org/10.1007/978-3-031-66159-4_4

Theorem 1 (Proposition 1.3.4 of [14]). *A pair of nonempty words (u, v) is conjugate iff there exists a word z such that $uz = zv$. Moreover, $z \in (xy)^*x$ where x and y are such that $u = xy$ and $v = yx$.*

The conjugacy problem is solvable in polynomial time over free monoids and free groups. We consider a generalisation of the problem to a finitely-presented possibly infinite set of pairs. Let A be a finite alphabet. A relation $R \subseteq A^* \times A^*$ over the free monoid A^* is conjugate if each pair $(u, v) \in R$ is conjugate. Consider the following decision version: *Given a relation R over A^*, is it conjugate?*. Of particular interest is when R is automata-definable because of motivations detailed later. First, we recall the class of rational relations. The family of *rational subsets* of a monoid M is the smallest class containing $\emptyset$, all singleton subsets of M and closed under union, product and Kleene closure. A natural way to present a rational subset of M is as a rational expression: $\emptyset, m \in M$ are rational expressions, and if E_1, E_2 are rational expressions then $E_1 \cdot E_2$, $E_1 + E_2$, and E_1^* are also rational expressions. A rational relation over A^* is a rational subset of the product monoid $A^* \times A^*$. Coincidentally, rational relations are precisely those that are defined by nondeterministic finite state transducers.

Example 1. The rational expression $E_1 = (\epsilon, a)(ab, ba)^*(a, \epsilon)$ denotes the set of pairs $\{((ab)^n a, a(ba)^n) \mid n \geq 0\}$. The expression $E_2 = ((a, aa) + (b, \epsilon))^*$ represents $\{(u, v) \mid v$ is obtained from u by duplicating a's and discarding b's$\}$. The expression E_1 is conjugate, and in fact is a subset of identity relation. However, E_2 is not conjugate.

A strong justification for the above problem comes from the theory of word transducers. Checking a number of properties of word transducers, for instance sequentiality (can the given transducer be determinised?) or finite sequentiality (is the given transducer equivalent to a disjoint union of deterministic transducers?), bounded edit-distance [2] (is the edit-distance between the respective outputs of the given tranducers bounded?) etc. amounts to checking conjugacy of the rational relations defined by the strongly connected components of the transducer and certain specific properties of the underlying acyclic graph of strongly connected components. Loosely speaking, conjugacy of the relations defined by the strongly connected components imply that the loops of the transducer are pumpable. Historically, decidability of these properties were shown by tailor-made procedures, for instance *twinning property* of Choffrut for sequentiality [5], *weak twinning* for finite sequentiality [7,11,13]. However, there is no general procedure to decide conjugacy of rational relations.

Our main result is summarised by the following theorem.

Theorem 2. *Conjugacy of rational relations is decidable.*

The decidability hinges on a couple of crucial definitions. The first is that of a sumfree expression: a rational expression is sumfree if it does not use sum (i.e., $+$). Formally, they can be defined as a hierarchy. Given a class $\mathcal{C}$ of expressions over the monoid M, the *Kleene closure* of $\mathcal{C}$, denoted as $\mathcal{KC}$, is the class of expressions $\mathcal{KC} = \mathcal{C} \cup \{E^* \mid E \in \mathcal{C}\}$. Similarly, the *monoid closure* of $\mathcal{C}$, denoted as

$\mathcal{MC}$, is the class of expressions $\mathcal{MC} = \mathcal{C} \cup \{E_1 \cdots E_k \mid E_i \in \mathcal{C}, i \in \{1, \ldots, k\}, k \in \mathbb{N}\}$. The family $\mathcal{F}$ of sumfree expressions is given by: $\mathcal{F}_0 = M \cup \{\emptyset\}$ and $\mathcal{F}_{i+1} = \mathcal{MKF}_i$ for each $i \geq 0$, and

$$\mathcal{F} = \bigcup_{i \geq 0} \mathcal{F}_i \ .$$

The *star height* of an expression E is the smallest $k \in \mathbb{N}$ such that E belongs to $\mathcal{F}_k$.

Over the free monoid A^*, the set of expressions $\mathcal{F}_0$ is $A^* \cup \{\emptyset\}$ and $\mathcal{KF}_0$ is the set of expressions $\mathcal{F}_0 \cup \{w^* \mid w \in A^*\}$ (for convenience we assume that $\emptyset$ is not used in any other expression other than $\emptyset$ itself). It is not difficult to see that $\mathcal{MKF}_0$ is the set of expressions $\mathcal{KF}_0 \cup \{u_1 v_1^* u_2 v_2^* \cdots u_k v_k^* u_{k+1} \mid u_i, v_i \in A^*, k \in \mathbb{N}\}$.

Every rational expression is effectively equivalent to a sum of sumfree expressions (called sumfree normal form (SNF)), by inductively rewriting the expression using the identities $(a+b)^* = (a^*b^*)^*$ and $(a+b) \cdot (c+d) = ac + ad + bc + bd$. This fact is the rational-expression analogue of the factorisation forest theorem of Simon, a deep result from the theory of finite semigroups [17]. Rewriting a rational expression in SNF may result in an exponential blow-up, both in the number of summands and the size of each summand.

Example 2. For the expression $E = (a + b)^n$ for some $n > 0$, it can be shown that any equivalent expression in SNF will have at least 2^n summands. For $E' = \$(E\#)^* \subseteq \{\$, \#, a, b\}^*$, any equivalent SNF expression will have at least one summand of exponential size, and the expression $E \cdot E'$ in SNF will have exponentially many summands of exponential size.

The union operation of rational relations, unlike the product and Kleene closure, preserves conjugacy, i.e., if R_1 and R_2 are conjugate, then $R_1 \cup R_2$ is also conjugate. Therefore for proving Theorem 2 it suffices to decide the conjugacy of a rational relation given by a sumfree expression.

The second crucial definition is the notion of a common witness of a relation, inherited from Lyndon-Schützenberger's theorem. A *witness* of a conjugate pair (u, v) is a word z such that either $uz = zv$ (*inner witness*) or $zu = vz$ (*outer witness*). A word z is a *common inner (resp. outer) witness* of a relation, if for every pair (u, v) in the relation, z is an inner witness (*resp.* outer witness) of (u, v). By Theorem 1, if a relation has a common witness then it is conjugate. However, the converse is easily shown to be false.

We show that a sumfree rational relation is conjugate if and only if it has a common witness, i.e., either a common inner witness or a common outer witness, (but not necessarily both). This characterisation of conjugacy is a main contribution of our paper. It is in fact a generalisation of the Lyndon-Schützenberger theorem characterising conjugacy of two words.

There are two interesting questions regarding common witnesses:

I. *Is there a common witness for the relation R?*
II. *Given a word z, is it a common witness of R?*

Question II proves to be comparatively more tractable, as it can be reduced to verifying whether the rational relation $R' = \{(uz, zv) \mid (u, v) \in R\}$ (or, $R' = \{(zu, vz) \mid (u, v) \in R\}$) consists of only identical pairs [16]. In fact, the decidability of the twinning property of a transducer is connected to Question II.

Question I, on the other hand, is more difficult *a priori* as we do not have a bound on the size of a possible common witness. We provide a decision procedure for Question I. This is another main contribution of the article. Our characterisation of conjugacy via common witness, together with this procedure, yields an algorithm for deciding conjugacy.

1.1   Related Work

A problem much related is the *Conjugate Post Correspondence problem*: given a finite set of pairs G, does there exist of a pair $(u, v) \in G^*$ such that u and v are conjugate? This problem is shown to be undecidable by reduction to the word problem of a special type of semi-Thue systems [10]. In Sect. 3, we show that the universal version of this problem—checking if all the pairs in G^* are conjugate—is decidable.

A generalisation of Lyndon-Schützenberger's theorem to infinite sets, though with no comparison to ours, is considered in [4,12], where solutions to the language equation $XZ = ZY$, where X, Y, Z are sets of words, are given for special cases. The general solution is still open.

2   Conjugacy of Sumfree Expressions

We now proceed to solve the conjugacy problem for sumfree expressions. We use pairs of lowercase Greek letters (α, β) with suitable modifications to denote pairs of words over $A^* \times A^*$. Clearly $\emptyset$ and (ϵ, ϵ) are conjugates. For an expression of the form (α, β), it is straightforward to check conjugacy. Thus, the conjugacy problem is decidable for the class of expressions $\mathcal{F}_0$.

To show the decidability of the conjugacy problem for the whole family $\mathcal{F}$, it suffices to show that if the problem is decidable for $\mathcal{F}_i$, $i \geq 0$, then it is also decidable for $\mathcal{K}\mathcal{F}_i$ and $\mathcal{F}_{i+1} = \mathcal{M}\mathcal{K}\mathcal{F}_i$. Then by induction on i, the decidability extends to the whole family $\mathcal{F}$.

Assume that conjugacy is decidable for $\mathcal{F}_i$. Assume that $E \in \mathcal{F}_i$. Since $L(E) \subseteq L(E^*)$, if the expression E^* is conjugate then necessarily E is conjugate. Because conjugacy is decidable for $\mathcal{F}_i$, we can check this necessary condition. Therefore, to show the decidability of conjugacy for $\mathcal{K}\mathcal{F}_i$, it suffices to show the decidability of the following question.

Question 1 (Conjugacy of Kleene Closures). Given a conjugate sumfree expression E, is E^* conjugate?

Next, assume that the conjugacy is decidable for $\mathcal{K}\mathcal{F}_i$. Let

$$E = (\alpha_0, \beta_0)E_1^*(\alpha_1, \beta_1) \cdots E_k^*(\alpha_k, \beta_k)$$

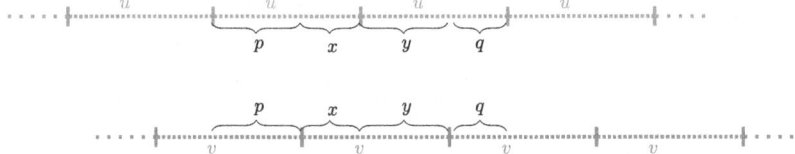

Fig. 1. v as infix of uu.

be an expression in $\mathcal{MKF}_i$ where $E_1^*, \ldots, E_k^*$ are from $\mathcal{KF}_i$. Analogous to the case of Kleene closures, E is conjugate only if $E_1^*, \ldots, E_k^*$ are conjugate, as the next lemma shows.

Lemma 1. *If the expression $E = (\alpha_0, \beta_0)F^*(\alpha_1, \beta_1)$ is conjugate, then F^* is conjugate.*

Proof. If F^* contains only the empty pair, then it is conjugate. Otherwise, assume that (u, v) is a nonempty pair in $L(F^*)$. Therefore, (u^ℓ, v^ℓ) for each $\ell \geq 0$ is also in $L(F^*)$. We can safely assume that $|u| = |v|$, otherwise each iteration will increase the difference in length between u^ℓ and v^ℓ, leading to nonconjugacy of E.

Let $k = |\alpha_0| + |\beta_0| + |\alpha_1| + |\beta_1|$. Consider the pair $(\alpha_0, \beta_0)(u^\ell, v^\ell)(\alpha_1, \beta_1)$ where ℓ is some value much larger than k, say 2^k. Since ℓ is much larger than k and $(\alpha_0 u^\ell \alpha_1, \beta_0 v^\ell \beta_1)$ is conjugate, there exist large factors of u^ℓ and v^ℓ that match as shown in Fig. 1. Since $|u| = |v|$, we can infer that u is a factor of vv, and v is a factor of uu.

Since v is an infix of uu, the following holds as shown in Fig. 1. There exist words x, y, p, and q such that $v = xy$ and $u = px = yq$. Since $|u| = |v|$, the length of p and the length of y are the same, which implies $p = y$ (since $u = px = yq$). Therefore, $u = yx$. Hence u and v are conjugate words. Since the pair (u, v) was arbitrary, F^* is conjugate.

We can generalise the above lemma to the general form of sumfree expressions.

Corollary 1. *If the expression $E = (\alpha_0, \beta_0)E_1^*(\alpha_1, \beta_1) \cdots E_k^*(\alpha_k, \beta_k)$ is conjugate, then each of $E_1^*, E_2^*, \ldots, E_k^*$ is conjugate.*

Since the conjugacy of $\mathcal{KF}_i$ is decidable, we can check whether $E_1^*, \ldots, E_k^*$ are conjugate expressions. Thus, to show the decidability of $\mathcal{MKF}_i$, it suffices to show the decidability of the following question.

Question 2 (Conjugacy of Monoid Closures). Given conjugate sumfree expressions $E_1^*, \ldots, E_k^*$, is the expression $E = (\alpha_0, \beta_0)E_1^*(\alpha_1, \beta_1) \cdots E_k^*(\alpha_k, \beta_k)$ conjugate?

We show that Question 1 and Question 2 can be effectively answered. The idea is to use the notion of common witness that we mentioned in the beginning. We present two common witness theorems that address the above questions:

1. Let G be an arbitrary set of conjugate pairs. The set G^* is conjugate if and only if G has a common witness (Theorem 3).
2. Let $G_1^*, \ldots, G_k^*$, $k > 0$, be arbitrary sets of conjugate pairs. The set

$$(\alpha_0, \beta_0) G_1^*(\alpha_1, \beta_1) \cdots G_k^*(\alpha_k, \beta_k),$$

called a *sumfree set*, is conjugate if and only if it has a common witness (Theorem 4).

Remark 1. Note that the assumption of conjugacy of the sets $G, G_1^*, \ldots, G_k^*$ is not necessary. However, if they are not conjugate then the corresponding sets will neither have a common witness nor be conjugate, and the statements will be vacuously true.

Item 2 is a generalisation of Item 1, and its proof relies on Item 1. Both theorems are generalisations of the Lyndon-Schützenberger theorem.

In both the theorems above, the common witness for the bigger expression can be computed in polynomial time from those of the subexpressions. When $G, G_1^*, \ldots, G_k^*$ are rational sumfree expressions of pairs, the above theorems are *effective*, that is a common witness, if it exists, is computable in polynomial time in the length of the expression (Sect. 4). Hence, we have the decidability result—Theorem 2.

3   Common Witness Theorems

In this section, it is shown that an infinite set of pairs generated by a sumfree set is conjugate if and only if a word is witnessing its conjugacy.

3.1   Common Witness and Its Characterisations

A word u is *primitive* if it cannot be expressed as a power of any strictly smaller word. For example, aba is primitive, but $abab$ is not. A word ρ is called a *primitive root* of a word u if $u = \rho^n$ for $n \geq 1$ and ρ is a primitive word. Every word u has a unique primitive root, denoted by ρ_u ([14], Proposition 1.3.1). We lift the notion of primitive root to a pair and a relation as follows: $R(u, v) = (\rho_u, \rho_v)$, and $R(G) = \{R(u, v) \mid (u, v) \in G\}$. For instance, if $G = \{(abab, baba), (bb, abb)\}$, then $R(G) = \{(ab, ba), (b, abb)\}$.

Recall from Theorem 1 that a pair of words (u, v) is conjugate, then there exists a word z such that $uz = zv$ where $u = xy$, $v = yx$ and $z \in (xy)^*x$. By symmetry of conjugacy, there also exists a word z' such that $z'u = vz'$ where $z' \in (yx)^*y$. We call z (resp. z') in the above characterisation as an *inner witness* (resp. *outer witness*) of the pair (u, v) (since z is appended to the inner ends). Given a conjugate pair (u, v), the set of all inner witnesses of (u, v) is $\{z \mid uz = zv\} = \cup_{\{(x,y)\mid u=xy,v=yx\}}(xy)^*x$. Similarly, the set of all outer witnesses of (u, v) is $\{z \mid zu = vz\} = \cup_{\{(x,y)\mid u=xy,v=yx\}}(yx)^*y$. For example, the pair (aba, baa) has inner witnesses $(aba)^*a$ and outer witnesses $(baa)^*ba$.

There is a connection between a conjugate pair and its primitive root. It is known that if a pair (u, v) is conjugate, then their primitive root (ρ_u, ρ_v) is also conjugate. Moreover, $(u, v) = (\rho_u, \rho_v)^n$ for some $n \geq 1$ (Lemma 1 of [6]). In fact, their witnesses are the same.

Proposition 1. *A word z is an inner (resp. outer) witness of a conjugate pair (u, v) iff z is an inner (resp. outer) witness of the primitive root (ρ_u, ρ_v).*

We generalise the notion of a witness of a pair to a set of pairs.

Definition 1 (Common Witness). *A word is a* common inner witness *of a set of pairs P if it is an inner witness of each pair in P. Similarly, a word is a* common outer witness *of P if it is an outer witness of each pair in P.*

A set of pairs has a common witness *if it has either a common inner witness or a common outer witness.*

The structure of a common witness of a set of pairs is obtained from Theorem 1.

Proposition 2. *Let P be a set of pairs of words. The following are equivalent.*

1. *z is a common inner witness of P.*
2. *$z \in \bigcap_{(u,v) \in P} \bigcup_{\{(x,y) \mid u=xy, v=yx\}} (xy)^* x$.*

The statement for common outer witness is analogous.

Example 3. Consider the set $P = \{(ab, ba), (abab, baba)\}$. The pair (ab, ba) has a unique cut (a, b), and the pair $(abab, baba)$ has two cuts: (a, bab) and (aba, b). The word a is a common inner witness of P since a belongs to both $(ab)^* a$ and $(abab)^* a$ (using the first cut). Similarly, aba is also a common inner witness of P since aba belongs to both $(ab)^* a$ and $(abab)^* aba$ (using the second cut). Notice that aba is not in the intersection of $(ab)^* a$ and $(abab)^* a$.

Proposition 1 connecting witness of a conjugate pair and its root can be lifted to a set of conjugate pairs and its root as follows.

Proposition 3. *The common witnesses of a set of conjugate pairs G and its root $R(G)$ are the same, i.e., a word z is a common inner (resp. outer) witness of G iff z is a common inner (resp. outer) witness of $R(G)$.*

When a set is not conjugate, clearly it has no common witness. However, even when a set is conjugate, it may have both common inner and outer witnesses, or only common inner witness, or only common outer witness, or neither of them as shown below.

Example 4. Consider the set $P = \{(ab, ba), (ac, ca)\}$. The pair (ab, ba) has inner witnesses $(ab)^* a$ and outer witnesses $(ba)^* b$. Similarly, the pair (ac, ca) has inner witnesses $(ac)^* a$ and outer witnesses $(ca)^* c$. According to Proposition 2, the set P has a unique common inner witness $a = (ab)^* a \cap (ac)^* a$, but it does not have any common outer witness since $(ba)^* b \cap (ca)^* c = \emptyset$.

The set $\{(ab, ba), (abab, baba)\}$ has both common inner witnesses $(ab)^*a = (ab)^*a \cap ((abab)^*aba \cup (abab)^*a)$ as well as common outer witnesses $(ba)^*b = (ba)^*b \cap ((baba)^*b \cup (baba)^*bab)$.

However, the set $\{(ab, ba), (ba, ab)\}$ has no common witnesses since $(ab)^*a \cap (ba)^*b = \emptyset$.

Proposition 4. *Let G be a set of pairs of words. The following are equivalent.*

1. *G has more than one common witness.*
2. *G has infinitely many common witnesses.*
3. *G has infinitely many common inner witnesses.*
4. *G has infinitely many common outer witnesses.*
5. *All the pairs in G have the same primitive root.*

Therefore, a set of pairs can have no common witness, a unique common witness, or infinitely many common witnesses.

3.2   Common Witness Theorem for Kleene Closure

We generalise the notion in Theorem 1 to an infinite set of pairs closed under concatenation. The question we ask is: "Given an arbitrary set of pairs G, is G^* conjugate?"

If G^* has a common witness, then each pair in G^* has a witness and thus, G^* is conjugate. We prove the converse, namely, if G^* is conjugate, then it has a common witness. The below theorem characterises the conjugacy of a freely generated set of pairs of words.

Theorem 3 (Common Witness Theorem for Kleene Closure). *Let G be an arbitrary set of conjugate pairs of words. The following are equivalent.*

1. *G^* is conjugate.*
2. *G^* has a common witness z.*
3. *G has a common witness z.*
4. *$R(G)$ has a common witness z.*

Proof (sketch). We prove (4) $\Rightarrow$ (3) $\Rightarrow$ (2) $\Rightarrow$ (1) $\Rightarrow$ (4).

(4) $\Rightarrow$ (3) Follows from Proposition 3.

(3) $\Rightarrow$ (2) WLOG assume that z is a common inner witness of the set G. Hence $\forall(u, v) \in G$, $uz = zv$. Let (u', v') be any arbitrary element from G^*, i.e., $(u', v') = (u_1 \cdots u_n, v_1 \cdots v_n)$ for some $n \geq 1$ and $(u_i, v_i) \in G$ for $1 \leq i \leq n$. By induction on n, we equate $u'z = zv'$ as follows. Thus, z is a common inner witness of G^*.

$$
\begin{aligned}
u'z &= u_1 \cdots u_{n-1}u_n z \\
&= u_1 \cdots u_{n-1}z v_n && \text{(Since } u_n z = z v_n) \\
&= z v_1 \cdots v_{n-1} v_n && \text{(Inductive Hypothesis)} \\
&= z v'
\end{aligned}
$$

(2) ⇒ (1) Follows from Theorem 1.

(1) ⇒ (4) The proof idea is to first prove when G is finite by case analysis and then extend it for a countably infinite set of pairs using a compactness argument—If every finite subset of an infinite set of pairs G has a common witness, then G has a common witness.

As a corollary, we get that E^* is conjugate iff E is conjugate for any rational expression of pairs E. Below is an instance of the common witness theorem for a set of pairs that is not rational.

Example 5. Let $G = \{(ab^p, b^p a) \mid p$ is a prime number$\}$. The set G has a common inner witness $a \in \bigcap_p \in \mathbb{N}, p$ is a prime$(ab^p)^* a$. It is also easy to verify that G^* is conjugate and a is a common inner witness of G^*.

3.3   Common Witness Theorem for Monoid Closure

Next, we give the common witness theorem for monoid closures, i.e., sumfree sets of the form $(\alpha_0, \beta_0)G_1{}^*(\alpha_1, \beta_1)G_2{}^* \cdots (\alpha_{k-1}, \beta_{k-1})G_k{}^*(\alpha_k, \beta_k), k > 0$ where $G_1^*, G_2^*, \ldots, G_k^*$ are arbitrary sets of conjugate pairs. It is shown that such a set is conjugate if and only if it has a common witness. Note that this does not generalise to arbitrary sets of pairs, in particular, rational sets using sum.

Example 6. $(ab, ba)^* + (ba, ab)^*$ is an infinite conjugate set with *no* common witness.

Definition 2 (Redux, Singleton Redux). *Let M be the sumfree set*

$$(\alpha_0, \beta_0)G_1{}^*(\alpha_1, \beta_1)G_2{}^* \cdots (\alpha_{k-1}, \beta_{k-1})G_k{}^*(\alpha_k, \beta_k) \ .$$

The redux *of M is the pair $(\alpha_0 \alpha_1 \cdots \alpha_k, \beta_0 \beta_1 \cdots \beta_k)$ obtained by substituting each G_i^* by the empty pair (ϵ, ϵ). A* singleton redux *of M is a set obtained by substituting all but one of the G_i^*'s by the empty pair (ϵ, ϵ). They are of the form $(\alpha_0 \cdots \alpha_{i-1}, \beta_0 \cdots \beta_{i-1})G_i{}^*(\alpha_i \cdots \alpha_k, \beta_i \cdots \beta_k)$ where $1 \leq i \leq k$.*

Example 7. Consider the set $M = (a, a)(baa, aba)^*(b, a)(aab, baa)^*(a, b)$. The redux of M is (aba, aab), and its singleton reduxes are $(a, a)(baa, aba)^*(ba, ab)$ and $(ab, aa)(aab, baa)^*(a, b)$.

If a sumfree set has a common witness, it is conjugate. We prove the converse, i.e., if a sumfree set is conjugate, then it has a common witness which is in the intersection of the common witnesses of the singleton reduxes of the set. Towards this, we need the following definition.

Definition 3 (Prefix Delay and Suffix Delay). *If u and v are words such that one of them is a prefix (resp. suffix) of another, we define the* prefix delay *(resp. suffix delay), denoted as $[u, v]_L$ (resp. $[u, v]_R$) between u and v as*

$$[u, v]_L = \begin{cases} u^{-1}v & \text{if } u \text{ is a prefix of } v \\ v^{-1}u & \text{if } v \text{ is a prefix of } u \end{cases} \qquad [u, v]_R = \begin{cases} vu^{-1} & \text{if } u \text{ is a suffix of } v \\ uv^{-1} & \text{if } v \text{ is a suffix of } u \end{cases}$$

Following is the common witness theorem for a sumfree set with only one Kleene star, i.e., $M = (\alpha_0, \beta_0)G^*(\alpha_1, \beta_1)$. In short, it states that such a set is conjugate if and only if it has a common witness that is determined by the common witnesses of $G \cup \{(\alpha_1\alpha_0, \beta_1\beta_0)\}$.

Proposition 5. *Let $M = (\alpha_0, \beta_0)G^*(\alpha_1, \beta_1)$ be a sumfree set with nonempty redux. The following are equivalent.*

1. *M is conjugate.*
2. *There exists a common witness of $G \cup \{(\alpha_1\alpha_0, \beta_1\beta_0)\}$.*
3. *M has a common witness. Furthermore,*
   (a) *If the set $G \cup \{(\alpha_1\alpha_0, \beta_1\beta_0)\}$ has a unique common inner witness, say z', then M has a unique common witness $z = [\alpha_0 z', \beta_0]_R = [\alpha_1, z'\beta_1]_L$. Moreover, if $|\alpha_0 z'| \geq |\beta_0|$ or equivalently $|\alpha_1| \leq |z'\beta_1|$, then z is a common inner witness, otherwise it is a common outer witness.*
   (b) *If the set $G \cup \{(\alpha_1\alpha_0, \beta_1\beta_0)\}$ has a unique common outer witness, say z', then M has a unique common witness $z = [\alpha_0, \beta_0 z']_R = [z'\alpha_1, \beta_1]_L$. Moreover, if $|z'\alpha_1| \geq |\beta_1|$ or equivalently $|\alpha_0| \leq |\beta_0 z'|$, then z is a common outer witness, otherwise it is a common inner witness.*
   (c) *If $G \cup \{(\alpha_1\alpha_0, \beta_1\beta_0)\}$ has infinitely many common witnesses, then M is a subset of powers of the primitive root of its redux. Thus, M has infinitely many common witnesses.*

Example 8. Let $M = (\alpha_0, \beta_0)G^*(\alpha_1, \beta_1)$ be a sumfree set with one Kleene star where

$$(\alpha_0, \beta_0) = (ab, b), G = \{(bab, abb)\}, (\alpha_1, \beta_1) = (b, ab).$$

The redux of M is $(\alpha_0\alpha_1, \beta_0\beta_1) = (abb, bab)$. The set $G \cup \{(\alpha_1\alpha_0, \beta_1\beta_0)\} = \{(bab, abb)\} \cup \{(bab, abb)\} = \{(bab, abb)\}$ and, hence it has infinitely many common witnesses. By Proposition 5 (c), M is a subset of powers of the primitive root of the redux, i.e., $M = (abb, bab)^+$. Therefore, M has infinitely many witnesses, the same as those of (abb, bab).

A singleton redux of a sumfree set is nothing but a sumfree set with only one Kleene star. Given any sumfree set M, if M is conjugate, each of its singleton reduxes is conjugate. From Proposition 5, a singleton redux of M has a common witness. Further, we prove that M has a common witness that is the common witness of each of its singleton reduxes. The below theorem characterises the conjugacy of a general sumfree set.

Theorem 4 (Common Witness Theorem for Monoid Closure). *Let M be a sumfree set. The following are equivalent.*

1. *M is conjugate.*
2. *Each of the singleton reduxes of M has a common witness z.*
3. *M has a common witness z.*

Example 9. Let $M = (\alpha_0, \beta_0)G_1^*(\alpha_1, \beta_1)G_2^*(\alpha_2, \beta_2)$ be a sumfree set with two Kleene star where $(\alpha_0, \beta_0) = (b, a), G_1 = \{(ac, ca)\}, (\alpha_1, \beta_1) = (ab, b), G_2 = \{(bab, bab)\}, (\alpha_2, \beta_2) = (\epsilon, b)$. The redux of M is $(\alpha_0\alpha_1\alpha_2, \beta_0\beta_1\beta_2) = (bab, abb)$. The set M has two singleton reduxes,

$$M_1 = (\alpha_0, \beta_0)G_1^*(\alpha_1\alpha_2, \beta_1\beta_2) = (b, a)(ac, ca)^*(ab, bb), \text{ and}$$

$$M_2 = (\alpha_0\alpha_1, \beta_0\beta_1)G_2^*(\alpha_2, \beta_2) = (bab, ab)(bab, bab)^*(\epsilon, b).$$

The set $G_1 \cup \{(\alpha_1\alpha_2\alpha_0, \beta_1\beta_2\beta_0)\} = \{(ac, ca), (abb, bba)\}$ has a unique common inner witness, say $z_1 = a = (ac)^*a \cap (abb)^*a$ and no common outer witness since $(ca)^*c \cap (bba)^*bb = \emptyset$. By Proposition 5 3a, the unique common inner witness of the singleton redux M_1 is $[\alpha_0 z_1, \beta_0]_R = [ba, a]_R = b$.

The set $G_2 \cup \{(\alpha_2\alpha_0\alpha_1, \beta_2\beta_0\beta_1)\} = \{(bab, bab)\}$ has infinitely many common witnesses. Thus, the singleton redux M_2 is a subset of powers of the primitive root of the redux using Proposition 5 3c, i.e., $M_2 = (bab, abb)^+$. Thus M_2 has infinitely many common inner witnesses $(bab)^*b$ and common outer witnesses $(abb)^*ab$.

By Theorem 4, M has a unique common inner witness $b \cap (bab)^*b = b$, that equals to the intersection of the common inner witness of its singleton reduxes M_1 and M_2.

4   Computing Witness of a Sumfree Expression

In this section, we give a decision procedure to compute the common witness of a sumfree expression, if it exists. The set of common witnesses (abbreviated as the *witness set*) of a sumfree expression is either empty, singleton, or infinite. Whenever there are infinitely many common witnesses for an expression, the witnesses are the same as those of its primitive root (Proposition 4). In that case, we compute the primitive root as their finite representation.

The following proposition shows that there is a bound to the size of the unique common witness of two conjugate pairs if it exists, which aids in computing the common witness of two pairs in polynomial time.

Proposition 6. *If two conjugate pairs (u_1, v_1) and (u_2, v_2) have a unique common witness z, then $|z| \leq 2 \cdot \max(|u_1|, |u_2|)$.*

The witness set of a sumfree expression is equal to the intersection of witness sets of each of its singleton reduxes. So first, we compute the witness set of a singleton redux.

Lemma 2. *Let $M = (\alpha_0, \beta_0)E^*(\alpha_1, \beta_1)$ be a sumfree expression. Given the witness set of E, we can compute the witness set of M in time $\mathcal{O}((m + n)^2)$ where m is the size of the expression M, and n is the size of the witness of E.*

Proof (sketch). If the redux of M is the empty word, then the witness set of M is equal to the witness set of E by Theorem 3. Now if M has a nonempty redux, M has a common witness iff $E \cup \{(\alpha_1\alpha_0, \beta_1\beta_0)\}$ has a common witness by Proposition 5. If it exists, the common witness of M can be computed from the common witness of $E \cup \{(\alpha_1\alpha_0, \beta_1\beta_0)\}$ using Proposition 5 3a, 3b, 3c.

Using the above algorithm, we compute the common witness of a general sumfree expression.

Lemma 3. *Let M be a sumfree expression. Given the witness set of each Kleene star in M, we can compute the witness set of M in time $\mathcal{O}(m \cdot (m+n)^2)$ where m is the size of the expression and n is the maximum size among the given witnesses.*

Proof (sketch). From Theorem 4, the witness set of M is the intersection of the witness sets of its singleton reduxes. The idea is that we compute the witness set of each singleton reduxes, if it exists, using Lemma 2. Assume M has a nonempty redux. If all the singleton reduxes have infinitely many witnesses, then M is a subset of powers of the primitive root of the redux of M by Proposition 5 3c and thus, M has infinitely many common witnesses. If there exists a singleton redux with a unique common witness, say z, then for all other singleton reduxes of M with a unique witness z', check if $z = z'$ (for all other singleton reduxes z is already a witness by virtue of being a witness of the redux of M). If so, z is the unique common witness of M; otherwise, M has no common witness. The case where M has an empty redux is similar.

Computation of the Witness Set: Given a sumfree expression M, we compute its witness set bottom-up. We start from the innermost Kleene star. It is a pair of words (u, v). First, we check if (u, v) is conjugate. If yes, then there are infinitely many common witnesses for $(u, v)^*$, namely the witnesses of its primitive root, otherwise M has no witness. This step can be done in a time polynomial in the length of (u, v). Now, we recursively use Lemma 3 to compute the common witness of the expression under the Kleene star in each level. If there is no common witness for any level of Kleene star expression, then M is not conjugate. To find out the complexity of the decision procedure, it suffices to estimate the maximum length of a witness involved in the computation.

Length of the Witness of a Sumfree Expression: We claim that if a sumfree expression M is conjugate, then there exists a witness of length linear in size of M. If M has infinitely many witnesses, M is a set of powers of a primitive root by Proposition 4. Thus, there exists a witness of length less than that of the length of the primitive root. Next, suppose M has a unique common witness. In that case, there exists a subexpression E_i^* such that E_i^* has a unique common witness, and all Kleene stars appearing in E_i have infinitely many witnesses. Thus, all of them have a common witness of length at most $|E_i|$. Therefore, there is a singleton redux M_i of E_i^* that has a unique witness z_i. The size of z_i is linear in M_i and the size of the witnesses of subexpressions of E_i. Both are upper bounded by the size of M. Furthermore, the common witnesses for all subsequent levels are unique (if they exist), and their length is bounded by $|M|$.

Complexity of the Algorithm: Since the size of the common witness of M is linear in $|M|$, by Lemma 3, the overall complexity of computing a common witness of a sumfree expression is $\mathcal{O}(h \cdot m^3)$ where h is the *star height* of M and m is the length of the expression.

5   Conclusion

The current decision procedure proceeds through the analysis of rational expressions. In its essence, it is analogous to the boundedness checking of distance automata using factorisation trees [8], though explicit use of factorisation trees are avoided using sumfree rational expressions instead. An obvious question is the existence of an automata-theoretic proof. Factorisation forests remain the primary tool to settle boundedness questions on automata and by that standard the proof approach taken in this paper is natural and quite possibly the most intuitive.

Computing a witness of a given sumfree expression, if one exists, can be done in polynomial time. However, converting a rational expression into a sum of sumfree expressions may result in an exponential blow-up. Thus, the algorithm presented in the paper is of exponential time. It remains to find the precise complexity of this problem.

It is natural to look at the conjugacy problem of more general classes, for instance functions definable by a deterministic two-way transducers (regular functions [9]), or by two-way pebble automata (polyregular functions [3]). The corresponding problem over free groups is another interesting problem.

References

1. Aiswarya, C., Manuel, A., Sunny, S.: Deciding conjugacy of a rational relation. CoRR abs/2307.06777 (2023). https://doi.org/10.48550/arXiv.2307.06777
2. Aiswarya, C., Manuel, A., Sunny, S.: Edit distance of finite state transducers. In: Bringmann, K., Grohe, M., Puppis, G., Svensson, O. (eds.) 51st International Colloquium on Automata, Languages, and Programming (ICALP 2024). Leibniz International Proceedings in Informatics (LIPIcs), vol. 297, pp. 125:1–125:20. Schloss Dagstuhl – Leibniz-Zentrum f'ur Informatik, Dagstuhl, Germany (2024). https://doi.org/10.4230/LIPIcs.ICALP.2024.125. https://drops.dagstuhl.de/entities/document/10.4230/LIPIcs.ICALP.2024.125. ISBN 978-3-95977-322-5. ISSN 1868-8969
3. Bojańczyk, M.: Transducers of polynomial growth. In: LICS 2022, pp. 1–27. ACM (2022). https://doi.org/10.1145/3531130.3533326
4. Cassaigne, J., Karhumäki, J., Manuch, J.: On conjugacy of languages. RAIRO Theor. Inform. Appl. **35**(6), 535–550 (2001). https://doi.org/10.1051/ita:2001130
5. Choffrut, C.: Une caractérisation des fonctions séquentielles et des fonctions sousséquentielles en tant que relations rationnelles. Theor. Comput. Sci. **5**(3), 325–337 (1977). https://doi.org/10.1016/0304-3975(77)90049-4
6. Choffrut, C., Karhumäki, J.: Combinatorics of words. In: Rozenberg, G., Salomaa, A. (eds.) Handbook of Formal Languages, pp. 329–438. Springer, Heidelberg (1997). https://doi.org/10.1007/978-3-642-59136-5_6
7. Choffrut, C., Schutzenberger, M.P.: Decomposition de fonctions rationnelles. In: Monien, B., Vidal-Naquet, G. (eds.) STACS 1986. LNCS, vol. 210, pp. 213–226. Springer, Heidelberg (1986). https://doi.org/10.1007/3-540-16078-7_78
8. Colcombet, T.: The factorisation forest theorem. In: Handbook of Automata Theory, pp. 653–693. European Mathematical Society Publishing House (2021). https://doi.org/10.4171/AUTOMATA-1/18

9. Engelfriet, J., Hoogeboom, H.J.: MSO definable string transductions and two-way finite-state transducers. ACM Trans. Comput. Log. **2**(2), 216–254 (2001). https://doi.org/10.1145/371316.371512
10. Finkel, O., Halava, V., Harju, T., Sahla, E.: On bi-infinite and conjugate post correspondence problems. RAIRO Theor. Inform. Appl. **57**, 7 (2023). https://doi.org/10.1051/ITA/2023008
11. Jecker, I., Filiot, E.: Multi-sequential word relations. Int. J. Found. Comput. Sci. **29**(2), 271–296 (2018). https://doi.org/10.1142/S0129054118400075
12. Karhumäki, J.: Combinatorial and computational problems on finite sets of words. In: Margenstern, M., Rogozhin, Y. (eds.) MCU 2001. LNCS, vol. 2055, pp. 69–81. Springer, Heidelberg (2001). https://doi.org/10.1007/3-540-45132-3_4
13. Lombardy, S., Sakarovitch, J.: Sequential? Theor. Comput. Sci. **356**(1–2), 224–244 (2006). https://doi.org/10.1016/J.TCS.2006.01.028
14. Lothaire, M.: Combinatorics on Words. Cambridge Mathematical Library, Cambridge University Press (1997). https://doi.org/10.1017/CBO9780511566097
15. Peifer, D.: Max Dehn and the origins of topology and infinite group theory. Am. Math. Mon. **122**(3), 217–233 (2015). https://doi.org/10.4169/amer.math.monthly.122.03.217
16. Sakarovitch, J.: Elements of Automata Theory. Cambridge University Press (2009). https://doi.org/10.1017/CBO9781139195218
17. Simon, I.: Factorization forests of finite height. Theor. Comput. Sci. **72**(1), 65–94 (1990). https://doi.org/10.1016/0304-3975(90)90047-L

Logic and Languages
of Higher-Dimensional Automata

Amazigh Amrane[1], Hugo Bazille[1(✉)], Uli Fahrenberg[1], and Marie Fortin[2]

[1] EPITA Research Laboratory (LRE), Paris, France
hugo@lrde.epita.fr
[2] Université Paris Cité, CNRS, IRIF, Paris, France

Abstract. In this paper we study finite higher-dimensional automata (HDAs) from the logical point of view. Languages of HDAs are sets of finite bounded-width interval pomsets with interfaces ($\mathsf{iiPoms}_{\leq k}$) closed under order extension. We prove that languages of HDAs are MSO-definable. For the converse, we show that the order extensions of MSO-definable sets of $\mathsf{iiPoms}_{\leq k}$ are languages of HDAs. Furthermore, both constructions are effective. As a consequence, unlike the case of all pomsets, the order extension of any MSO-definable set of $\mathsf{iiPoms}_{\leq k}$ is MSO-definable.

1 Introduction

Connections between logic and automata play a key role in several areas of theoretical computer science – logic being used to specify the behaviours of automata models in formal verification, and automata being used to prove the decidability of various logics. The first and most well-known result of this kind is the equivalence in expressive power of finite automata and monadic second-order logic (MSO) over finite words, proved independently by Büchi [5], Elgot [10] and Trakhtenbrot [37] in the 60's. This was soon extended to infinite words [6] as well as finite and infinite trees [8,32,34].

Finite automata over words are a simple model of sequential systems with a finite memory, each word accepted by the automaton corresponding to an execution of the system. For concurrent systems, executions may be represented as *pomsets* (partially ordered multisets or, equivalently, labelled partially ordered sets). Several classes of pomsets and matching automata models have been defined in the literature, corresponding to different communication models or different views of concurrency. In that setting, logical characterisations of classes of automata in the spirit of the Büchi-Elgot-Trakhtenbrot theorem have been obtained for several cases, such as asynchronous automata and Mazurkiewicz traces [35,40], branching automata and series-parallel pomsets [3,29], step transition systems and local trace languages [21,30], or communicating finite-state machines and message sequence charts [23].

Higher-dimensional automata (HDAs) [31,38] are another automaton-based model of concurrent systems that matches more closely an interval-based view

J. D. Day and F. Manea (Eds.): DLT 2024, LNCS 14791, pp. 51–67, 2024.
https://doi.org/10.1007/978-3-031-66159-4_5

of events. Initially studied from a geometrical or categorical point of view, the language theory of HDAs has become another focus for research in the past few years [13]. The language of an HDA is defined as a set of *interval pomsets with interfaces (interval ipomsets)* [15]. The idea is that each event in the execution of an HDA corresponds to an interval of time where some process is active.

Examples with three activity intervals labelled a, b, and c are shown in the top of Fig. 1 below. These events are then partially ordered as follows: two events are ordered if the first one ends before the second one starts, and they are concurrent if they overlap. This gives rise to a pomset as shown in the middle of Fig. 1. We allow some events to be started before the beginning (this is the case for the a-labelled events in Fig. 1), and some events might never be terminated. Such events are called interfaces, or respectively sources and targets.

In addition, if we shorten some intervals in one possible behaviour of the HDA, we obtain another valid behaviour for the HDA. In terms of pomsets, this means that the language of an HDA is closed under *subsumption* (called *order extension* in [21]). In addition, it also has bounded width, meaning that each set of pairwise concurrent events has size at most k for some k.

The interest of HDAs as a model for concurrency stems from their convenient automata-theoretic and geometric properties. They provide a natural extension of standard automata to higher dimensions and lend themselves to automata-theoretic reasoning. Several theorems of classical automata theory have already been extended to HDAs, including a Kleene theorem [14] and a Myhill-Nerode theorem [17], and [1,12] provide an extension to higher-dimensional *timed* automata. On the other hand, the precubical sets on which HDAs are based have been studied in geometry and algebraic topology for a long time [4,25,33] and have led to interesting results for example about state-space reduction [16,18,20,41] or about behavioural equivalences [9,11,27,28], see [19] for a recent overview.

The automata-theoretic closure properties of HDAs were studied in [2]. In particular, their languages are not closed under complement, but they are closed under *bounded width complement*: the subsumption closure of the complement of the language restricted to interval ipomsets of bounded width.

In this paper, we explore the relationship between HDAs and MSO. We prove that a set of interval ipomsets is regular if and only if it is simultaneously MSO-definable, of bounded width, and downward-closed for subsumption. The latter two assumptions are necessary as it is possible to define in MSO sets with unbounded width or sets that are not downward-closed.

The HDA-to-MSO direction is proved similarly to the original Büchi-Elgot-Trakhtenbrot theorem. We use one second-order variable for each *upstep* (starting events) or *downstep* (terminating events) of the HDA. The main difference with words is that each upstep or downstep involves several events. We rely on the existence of a canonical *sparse step decomposition* for any interval ipomset. We prove that this decomposition can be effectively "defined" in MSO.

On the other hand, the usual approach for the MSO-to-automata direction, which works by induction and relies on the closure properties of regular languages, does not work for HDAs, as they are not closed under complement.

One could try to use the bounded-width complement instead, but the downward closures present some difficulties. Instead, we rely on a known connection [2] between regular languages of interval ipomsets and regular languages of *step decompositions*. A step decomposition of an ipomset P is a sequence of discrete ipomsets (that is, pomsets where all events are concurrent) such that their composition is equal to P. We prove that for every MSO-definable language L of width at most k, the language of all step decompositions of ipomsets in L, viewed as words over a finite alphabet of discrete ipomsets, is regular. To do so, we give an effective translation from MSO formulas over ipomsets to MSO formulas over words with this new alphabet. It was shown in [14] that the downward closure of L is then effectively regular.

The paper is organised as follows. Interval pomsets with interfaces and step decompositions are defined in Sect. 2, and higher-dimensional automata in Sect. 3. In Sect. 4, we introduce monadic second-order logic and state our main result. Section 5 gives the proof for the MSO-to-HDA direction, and Sect. 6 for the HDA-to-MSO one.

2   Pomsets with Interfaces

We fix a finite alphabet Σ throughout this paper. A *pomset with interfaces*, or *ipomset*, is a structure $(P, <, \dashrightarrow, S, T, \lambda)$ comprising a finite set P, a (strict) partial order[1] $< \subseteq P \times P$ called the *precedence order*, an irreflexive and asymmetric relation $\dashrightarrow \subseteq P \times P$ called the *event order*, subsets $S, T \subseteq P$ called *source* and *target* sets, and a *labelling* $\lambda \colon P \to \Sigma$. We require the following properties:

– for all $e \neq e' \in P$, exactly one of $e < e'$, $e' < e$, $e \dashrightarrow e'$, or $e' \dashrightarrow e$ holds;
– for all $e_1 \in S$, $e_2 \in P$, and $e_3 \in T$, $e_2 \not< e_1$ and $e_3 \not< e_2$.

That is, all points in P are related by precisely one of the orders, sources are $<$-minimal, and targets are $<$-maximal. We may add subscripts "$_P$" to the elements above if necessary. The source and target *interfaces* of P are the conclists $(S, \dashrightarrow_{\restriction S \times S}, \lambda_{\restriction S})$ and $(T, \dashrightarrow_{\restriction T \times T}, \lambda_{\restriction T})$, where "$_{\restriction}$" denotes restriction.

Ipomsets are a generalisation of standard pomsets (see for example [24]) obtained by adding interfaces and event order. Both are needed in order to properly connect them with HDAs. In particular, event order is necessary in order to define gluing composition, see below. Further, a central tool in the language theory of HDAs (which we, however, do not need here) are so-called track objects, which provide an embedding of ipomsets into HDAs which essentially needs the event order; see [13] for details. In [13] and other works, a transitively closed event order is used instead of the relation we use here; we find it more convenient to use the non-transitive version which otherwise is equivalent.

An ipomset P is a *word* (with interfaces) if $<$ is total and *discrete* if $< = \emptyset$. P is a *pomset* if $S = T = \emptyset$, a *conclist* (short for "concurrency list") if it is a discrete pomset, a *starter* if it is discrete and $T = P$, a *terminator* if it is discrete and $S = P$, and an *identity* if it is both a starter and a terminator.

---

[1] A strict partial order is a relation which is irreflexive, asymmetric and transitive. We will omit the qualifier "strict".

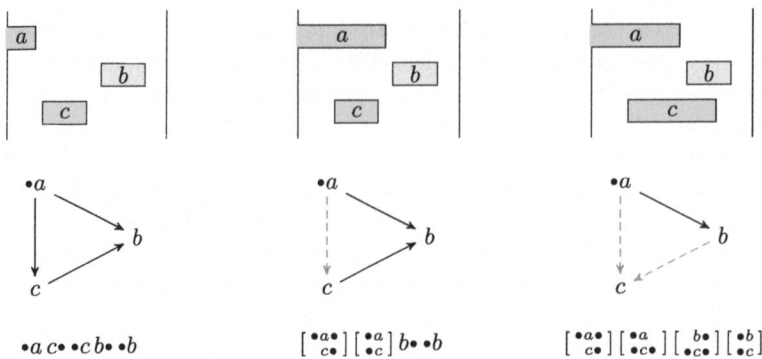

Fig. 1. Activity intervals of events (top), corresponding ipomsets (middle), and notation as a word over starters and terminators (bottom), *cf.* Example 1. Full arrows indicate precedence order; dashed arrows indicate event order; bullets indicate interfaces.

Figure 1 shows some simple examples. Source and target events are marked by "•" at the left or right side. Precedence $<$ and event order $\dashrightarrow$ are intended to order sequential and concurrent events, respectively. Another representation that we will use is as words of starters and terminators. In this representation, the ipomset is decomposed into a sequence of starters and terminators representing which events are in parallel and in which order. In the starter/terminator representation, event order is omitted and always assumed to go downward.

Ipomsets P and Q are *isomorphic* if there exists a bijection $f\colon P \to Q$ for which

1. $f(S_P) = S_Q$, $f(T_P) = T_Q$, $\lambda_Q \circ f = \lambda_P$, and
2. $f(e_1) <_Q f(e_2) \iff e_1 <_P e_2$ and $e_1 \dashrightarrow_P e_2 \iff f(e_1) \dashrightarrow_Q f(e_2)$.

That is, f respects interfaces and labels and the two relations. Because of the requirement that all elements are related by $<$ or $\dashrightarrow$, there is at most one isomorphism between any two ipomsets [13]. That means that we may switch between ipomsets and their isomorphism classes, and we will do so often in the sequel.

An ipomset P is *interval* if $<_P$ is an interval order [22]; that is, if it admits an interval representation given by functions $f, g\colon (P, <_P) \to (\mathbb{R}, <_{\mathbb{R}})$ such that $f(e) \leq_{\mathbb{R}} g(e)$ for all $e \in P$ and $e_1 <_P e_2$ iff $g(e_1) <_{\mathbb{R}} f(e_2)$ for all $e_1, e_2 \in P$. Given that our ipomsets represent activity intervals of events, any of the ipomsets we will encounter will be interval, and we omit the qualification "interval". We emphasise that this is *not* a restriction, but rather induced by the semantics, [39]. The *width* wid(P) of an ipomset P is the cardinality of a maximal $<$-antichain.

We let iiPoms denote the set of (interval) ipomsets and iiPoms$_{\leq k} = \{P \in$ iiPoms $\mid$ wid(P) $\leq k\}$. We write St, Te, Id $\subseteq$ iiPoms for the sets of starters, terminators, and identities and let $\Omega = $ St $\cup$ Te. Further, for $S \in \{$St, Te, Id, $\Omega\}$, $S_{\leq k} = S \cap$ iiPoms$_{\leq k}$. Note that Id $=$ St $\cap$ Te and Id$_{\leq k} = $ St$_{\leq k} \cap$ Te$_{\leq k}$.

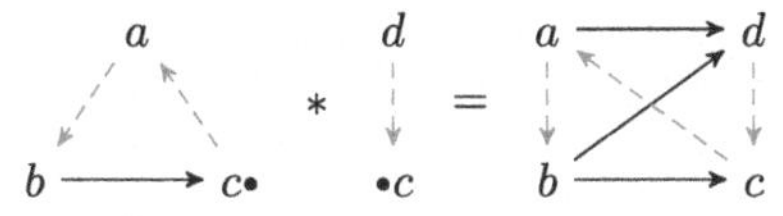

Fig. 2. Gluing composition of ipomsets.

We introduce special notation for starters and terminators and write $_A{\uparrow}U = (U, \emptyset, {\dashrightarrow}, U \setminus A, U, \lambda)$ and $U{\downarrow}_B = (U, \emptyset, {\dashrightarrow}, U, U \setminus B, \lambda)$ (with ${\dashrightarrow}$ and λ induced from U). The intuition is that the starter $_A{\uparrow}U$ does nothing but start the events in $A = U \setminus S_U$ and the terminator $U{\downarrow}_B$ terminates the events in $B = U \setminus T_U$.

Ipomsets may be *refined* by shortening activity intervals, potentially removing concurrency and expanding precedence. The inverse to refinement is called *subsumption*. Formally, for ipomsets P and Q we say that P refines Q, or that Q subsumes P, and write $P \sqsubseteq Q$ if there is a bijection $f \colon P \to Q$ for which

(1) $f(S_P) = S_Q$, $f(T_P) = T_Q$, and $\lambda_Q \circ f = \lambda_P$,
(2) $f(e_1) <_Q f(e_2) \implies e_1 <_P e_2$, and $e_1 {\dashrightarrow}_P e_2 \implies f(e_1) {\dashrightarrow}_Q f(e_2)$.

This definition adapts the one of [24] to event orders and interfaces. Intuitively, P has more order and less concurrency than Q. Note that isomorphisms are precisely those subsumptions whose inverses are also subsumptions.

For a subset $A \subseteq \mathsf{iiPoms}$ we let

$$A{\downarrow} = \{P \in \mathsf{iiPoms} \mid \exists Q \in A : P \sqsubseteq Q\}$$

denote its closure under subsumptions. A *language* is a subset $L \subseteq \mathsf{iiPoms}$ for which $L{\downarrow} = L$.

Example 1. In Fig. 1 there is a sequence of subsumptions from left to right. An event e_1 is smaller than e_2 in the precedence order if e_1 is terminated before e_2 is started; e_1 is smaller than e_2 in the event order if they are concurrent and e_1 is above e_2 in the respective conclist.

The *gluing* $P * Q$ of ipomsets P and Q is defined if $T_P - S_Q = P \cap Q$ *as conclists* (i.e., ${\dashrightarrow}_{P{\restriction}T_P \times T_P} = {\dashrightarrow}_{Q{\restriction}S_Q \times S_Q}$ and $\lambda_{P{\restriction}T_P} = \lambda_{Q{\restriction}S_Q}$), and then $P * Q = (P \cup Q, <, {\dashrightarrow}, S_P, T_Q, \lambda)$, where $< \, = \, <_P \cup <_Q \cup (P \setminus T_P) \times (Q \setminus S_Q)$, ${\dashrightarrow} = ({\dashrightarrow}_P \cup {\dashrightarrow}_Q)^+$, and $\lambda = \lambda_P \cup \lambda_Q$. (Here $^+$ denotes transitive closure.) Gluing is associative, and ipomsets in Id are identities for $*$. Figure 2 shows an example.

Any ipomset P can be decomposed as a gluing of starters and terminators $P = P_1 * \cdots * P_n$ [15,26]. Such a presentation we call a *step decomposition*. If starters and terminators are alternating, the step decomposition is called *sparse*.

Lemma 2 ([17]). *Every ipomset P has a unique sparse step decomposition.*

We will also use the following notion, introduced in [2]. A word $P_1 \ldots P_n \in \Omega^*$ is *coherent* if the gluing $P_1 * \cdots * P_n$ is defined. We denote by $\mathsf{Coh} \subseteq \Omega^*$ the set of coherent words and $\mathsf{Coh}_{\le k} = \mathsf{Coh} \cap \Omega^*_{\le k}$.

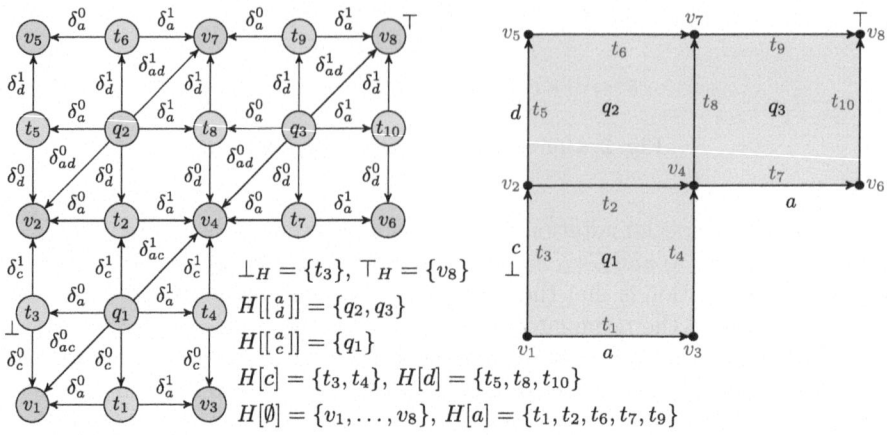

Fig. 3. A two-dimensional HDA $\mathcal{H}$ on $\Sigma = \{a, c, d\}$, see Example 3.

3 Higher-Dimensional Automata

Let $\square$ denote the set of conclists. A *precubical set*

$$\mathcal{H} = (H, \mathsf{ev}, \{\delta^0_{A,U}, \delta^1_{A,U} \mid U \in \square, A \subseteq U\})$$

consists of a set of *cells* H together with a function $\mathsf{ev} \colon H \to \square$ which to every cell assigns a conclist of concurrent events which are active in it. We write $H[U] = \{q \in H \mid \mathsf{ev}(q) = U\}$ for the cells of type U. For every $U \in \square$ and subset $A \subseteq U$ there are *face maps* $\delta^0_{A,U}, \delta^1_{A,U} \colon H[U] \to H[U \setminus A]$ (we omit the subscript U from now) which are required to satisfy $\delta^\nu_A \delta^\mu_B = \delta^\mu_B \delta^\nu_A$ for $A \cap B = \emptyset$ and $\nu, \mu \in \{0, 1\}$. The *upper* face maps δ^1_A terminate events in A and the *lower* face maps δ^0_A transform a cell q into one in which the events in A have not yet started. A *higher-dimensional automaton* (*HDA*) $\mathcal{H} = (\mathcal{H}, \bot_H, \top_H)$ is a finite precubical set together with subsets $\bot_H, \top_H \subseteq H$ of *start* and *accept* cells. The *dimension* of an HDA $\mathcal{H}$ is $\dim(\mathcal{H}) = \max\{|\mathsf{ev}(q)| \mid q \in H\} \in \mathbb{N}$.

A standard automaton is the same as a one-dimensional HDA $\mathcal{H}$ with the property that for all $q \in \bot_H \cup \top_H$, $\mathsf{ev}(q) = \emptyset$: cells in $H[\emptyset]$ are states, cells in $H[\{a\}]$ for $a \in \Sigma$ are a-labelled transitions, and face maps $\delta^0_{\{a\}}$ and $\delta^1_{\{a\}}$ attach source and target states to transitions. In contrast to ordinary automata we allow start and accept *transitions* instead of merely states, so languages of one-dimensional HDAs may contain words with interfaces.

Example 3. Figure 3 shows a two-dimensional HDA as a combinatorial object (left) and in a geometric realisation (right). It consists of 21 cells: states $H_0 = \{v_1, \ldots, v_8\}$ in which no event is active ($\mathsf{ev}(v_i) = \emptyset$), transitions $H_1 = \{t_1, \ldots, t_{10}\}$ in which one event is active (*e.g.*, $\mathsf{ev}(t_3) = \mathsf{ev}(t_4) = c$), squares $H_2 = \{q_1, q_2, q_3\}$ where $\mathsf{ev}(q_1) = [\begin{smallmatrix} a \\ c \end{smallmatrix}]$ and $\mathsf{ev}(q_2) = \mathsf{ev}(q_3) = [\begin{smallmatrix} a \\ d \end{smallmatrix}]$. The arrows between cells in the left representation correspond to the face maps connecting them. For example, the upper face map δ^1_{ac} maps q_1 to v_4 because the latter is the cell in which the active events a and c of q_1 have been terminated. On the right, face maps are used to glue cells, so that for example $\delta^1_{ac}(q_1)$ is glued to the top right of q_1. In this and other geometric realisations, when we have two concurrent events a and c with $a \dashrightarrow c$, we will draw a horizontally and c vertically.

Computations of HDAs are *paths*, *i.e.*, sequences $\alpha = (q_0, \varphi_1, q_1, \ldots, q_{n-1}, \varphi_n, q_n)$ consisting of cells $q_i \in H$ and symbols φ_i which indicate face map types: for every $i \in \{1, \ldots, n\}$, $(q_{i-1}, \varphi_i, q_i)$ is either

- $(\delta^0_A(q_i), \nearrow^A, q_i)$ for $A \subseteq \mathsf{ev}(q_i)$ (an *upstep*)
- or $(q_{i-1}, \searrow_A, \delta^1_A(q_{i-1}))$ for $A \subseteq \mathsf{ev}(q_{i-1})$ (a *downstep*).

Downsteps terminate events, following upper face maps, whereas upsteps start events by following inverses of lower face maps. We denote by $\mathsf{ups}(\mathcal{H})$ and $\mathsf{downs}(\mathcal{H})$ the finite set of upsteps and downsteps of $\mathcal{H}$.

The *source* and *target* of α as above are $\mathsf{src}(\alpha) = q_0$ and $\mathsf{tgt}(\alpha) = q_n$. A path α is *accepting* if $\mathsf{src}(\alpha) \in \bot_H$ and $\mathsf{tgt}(\alpha) \in \top_H$. Paths α and β may be concatenated if $\mathsf{tgt}(\alpha) = \mathsf{src}(\beta)$; their concatenation is written $\alpha * \beta$.

Path equivalence is the congruence $\simeq$ generated by $(q \nearrow^A r \nearrow^B p) \simeq (q \nearrow^{A \cup B} p)$, $(p \searrow_A r \searrow_B q) \simeq (p \searrow_{A \cup B} q)$, and $\gamma \alpha \delta \simeq \gamma \beta \delta$ whenever $\alpha \simeq \beta$. This relation allows to assemble subsequent upsteps or downsteps into one bigger step.

The *event ipomset* $\mathsf{ev}(\alpha)$ of a path α is defined recursively as follows:

- if $\alpha = (q)$, then $\mathsf{ev}(\alpha) = \mathsf{id}_{\mathsf{ev}(q)}$;
- if $\alpha = (q \nearrow^A p)$, then $\mathsf{ev}(\alpha) = {}_A\!\!\uparrow\!\mathsf{ev}(p)$;
- if $\alpha = (p \searrow_B q)$, then $\mathsf{ev}(\alpha) = \mathsf{ev}(p)\!\downarrow_B$;
- if $\alpha = \alpha_1 * \cdots * \alpha_n$ is a concatenation, then $\mathsf{ev}(\alpha) = \mathsf{ev}(\alpha_1) * \cdots * \mathsf{ev}(\alpha_n)$.

Note that upsteps in α correspond to starters in $\mathsf{ev}(\alpha)$ and downsteps correspond to terminators. Path equivalence $\alpha \simeq \beta$ implies $\mathsf{ev}(\alpha) = \mathsf{ev}(\beta)$ [14].

Example 4. The HDA $\mathcal{H}$ of Example 3 (Fig. 3) admits several accepting paths, for example $t_3 \nearrow^a q_1 \searrow_c t_2 \nearrow^d q_2 \searrow_a t_8 \nearrow^a q_3 \searrow_{ad} v_8$. Its event ipomset is

$$
{}_a\!\uparrow\!\begin{bmatrix} a \\ c \end{bmatrix} * \begin{bmatrix} a \\ c \end{bmatrix}\!\downarrow_c * {}_d\!\uparrow\!\begin{bmatrix} a \\ d \end{bmatrix} * \begin{bmatrix} a \\ d \end{bmatrix}\!\downarrow_a * {}_a\!\uparrow\!\begin{bmatrix} a \\ d \end{bmatrix} * \begin{bmatrix} a \\ d \end{bmatrix}\!\downarrow_{ad} = \begin{bmatrix} \begin{array}{c} a \longrightarrow a \\ \downarrow \nearrow \downarrow \\ \bullet c \longrightarrow d \end{array} \end{bmatrix}
$$

which is a sparse step decomposition. This path is equivalent to $t_3 \nearrow^a q_1 \searrow_c t_2 \nearrow^d q_2 \searrow_a t_8 \nearrow^a q_3 \searrow_a t_{10} \searrow_d v_8$ which induces the coherent word w_1 of Fig. 4 below.

The *language* of an HDA $\mathcal{H}$ is $L(\mathcal{H}) = \{\mathsf{ev}(\alpha) \mid \alpha \text{ accepting path in } \mathcal{H}\}$. A language is *regular* if it is the language of a finite HDA. Languages of HDAs are closed under subsumption, that is, if L is regular, then $L\!\downarrow = L$ [13,14].

A language is *rational* if it is constructed from $\emptyset$, $\{\mathsf{id}_\emptyset\}$ and discrete ipomsets using $\cup$, $*$ and $^+$ (Kleene plus) [14]. These operations have to take subsumption closure into account; in particular,

$$
L_1 * L_2 = \{P * Q \mid P \in L_1, Q \in L_2\}\!\downarrow.
$$

Theorem 5 ([14]). *A language is regular if and only if it is rational.*

The *width* of a language L is $\mathsf{wid}(L) = \sup\{\mathsf{wid}(P) \mid P \in L\}$. For $k \geq 0$ and $L \subseteq$ iiPoms, denote $L_{\leq k} = \{P \in L \mid \mathsf{wid}(P) \leq k\}$.

Lemma 6 ([14]). *Any regular language has finite width.*

It immediately follows that the universal language iiPoms is *not* rational.

4   MSO

Monadic second-order (MSO) logic is an extension of first-order logic allowing to quantify existentially and universally over elements as well as subsets of the domain of the structure. It uses second-order variables $X, Y, \ldots$ interpreted as subsets of the domain in addition to the first-order variables $x, y, \ldots$ interpreted as elements of the domain of the structure, and a new binary predicate $x \in X$ interpreted commonly. We refer the reader to [36] for more details about MSO.

We interpret MSO over iiPoms. Thus we consider the signature $\mathcal{S} = \{<, \dashrightarrow, (a)_{a \in \Sigma}, s, t\}$ where $<$ and $\dashrightarrow$ are binary relation symbols and the a's, s and t are unary predicates (over first-order variables). We associate to every ipomset $(P, <, \dashrightarrow, S, T, \lambda)$ the relational structure $S = (P; <; \dashrightarrow; (a)_{a \in \Sigma}; \mathsf{s}; \mathsf{t})$ where $<$ and $\dashrightarrow$ are interpreted as the orderings $<$ and $\dashrightarrow$ over P, and $a(x)$, $\mathsf{s}(x)$ and $\mathsf{t}(x)$ hold respectively if and only if $\lambda(x) = a$, $x \in S$ and $x \in T$. We say that a relation $R \subseteq P^n \times (2^P)^m$ is *MSO-definable* in S if and only if there exists an MSO-formula $\psi(x_1, \ldots, x_n, X_1, \ldots, X_m)$ over $\mathcal{S}$ which is satisfied if and only if the interpretation of the free variables $(x_1, \ldots, x_n, X_1, \ldots, X_m)$ in S is a tuple of R. The well-formed MSO formulas are built using the following grammar:

$$\psi ::= a(x) \mid \mathsf{s}(x) \mid \mathsf{t}(x) \mid x < y \mid x \dashrightarrow y \mid x \in X$$
$$\exists x.\, \psi \mid \forall x.\, \psi \mid \exists X.\, \psi \mid \forall X.\, \psi \mid \psi_1 \wedge \psi_2 \mid \psi_1 \vee \psi_2 \mid \neg \psi$$

In order to shorten formulas we use several notations and shortcuts such as $\psi_1 \implies \psi_2$. We define $x \prec y := x < y \wedge \neg(\exists z. x < z < y)$.

Let $\psi(x_1, \ldots, x_n, X_1, \ldots, X_m)$ be an MSO formula whose free variables are $x_1, \ldots, x_n, X_1, \ldots, X_m$ and let $P \in$ iiPoms. The pair of functions $\nu = (\nu_1, \nu_2)$ where $\nu_1 \colon \{x_1, \ldots, x_n\} \to P$ and $\nu_2 \colon \{X_1, \ldots, X_m\} \to 2^P$ is called a *valuation* or an *interpretation*. We write $P \models_\nu \psi$, or, by a slight abuse of notation, $P \models \psi(\nu(x_1), \ldots, \nu(x_n), \nu(X_1), \ldots, \nu(X_m))$, if ψ holds when x_i and X_j are interpreted as $\nu(x_i)$ and $\nu(X_j)$. A *sentence* is a formula without free variables. In this case no valuation is needed. Given an MSO sentence ψ, we define $L(\psi) = \{P \in$ iiPoms $\mid P \models \psi\}$. Note that this may not be closed under subsumption, hence not a language in our sense. A set $L \subseteq$ iiPoms is MSO-definable if and only if there exists an MSO sentence ψ over $\mathcal{S}$ such that $L = L(\psi)$.

Example 7. Let $\varphi = \exists x \, \exists y.\, a(x) \wedge b(y) \wedge \neg(x < y) \wedge \neg(y < x)$. That is, there are at least two concurrent events, one labelled a and the other b. $L(\varphi)$ is not width-bounded, as φ is satisfied, among others, by any conclist which contains at least one a and one b, nor closed under subsumption, given that $\left[\begin{smallmatrix} a \\ b \end{smallmatrix}\right] \models \varphi$ but $ab, ba \not\models \varphi$. Note, however, that $L(\varphi)_{\leq k}\!\downarrow$ is a regular language for any k.

We will also use MSO over words of $\Omega_{\leq k}^*$. (Note that, up to isomorphisms, $\Omega_{\leq k}$ is a finite set.) The definitions above can be easily adapted to this case by considering a word of $\Omega_{\leq k}^*$ as a structure of the form $(W, <, \lambda\colon W \to \Omega_{\leq k})$: a totally ordered set W labelled by $\Omega_{\leq k}$, and the signature $\{<, (D)_{D \in \Omega_{\leq k}}\}$: the atomic predicates are $D(x)$ for $D \in \Omega_{\leq k}$, $x < y$ and $x \in X$, with first-order variables ranging over positions in the word and second-order variables over sets of positions. We denote by MSO_Ω^k the set of MSO formulas over $\Omega_{\leq k}^*$. For example the following MSO_Ω^2 formula where $P_i \in \Omega_{\leq 2}$ stands for the ith discrete ipomset of w_1 in Fig. 4 is satisfied only by w_1.

$$\varphi' := \exists y_1, \ldots, y_7. \bigwedge_{1 \leq i \leq 7} P_i(y_i) \wedge y_1 \prec \cdots \prec y_7 \wedge \forall y. \bigvee_{1 \leq i \leq 7} y = y_i$$

The main result of this paper is the following:

Theorem 8. *For all $L \subseteq \mathsf{iiPoms}$,*

1. *if L is MSO-definable, then $L_{\leq k}\downarrow$ is regular for all $k \in \mathbb{N}$.*
2. *if L is regular, then it is MSO-definable.*

Moreover, the constructions are effective in both directions.

Corollary 9. *For all $k \in \mathbb{N}$, a language $L \subseteq \mathsf{iiPoms}_{\leq k}$ is regular if and only if it is MSO-definable.*

The next two sections are devoted to the proof of Theorem 8. For the first assertion we effectively build an HDA $\mathcal{H}$ from a sentence φ and an integer k such that $L(\mathcal{H}) = L(\varphi)_{\leq k}\downarrow$. Since emptiness of HDAs is decidable [2], asking, given a formula φ such that $L(\varphi) = L(\varphi)_{\leq k}\downarrow$, if there exists $P \in \mathsf{iiPoms}$ such that $P \models \varphi$ and if $L(\mathcal{H}) \subseteq L(\varphi)$ for some HDA $\mathcal{H}$ are decidable.

Corollary 10. *For MSO sentences φ such that $L(\varphi) = L(\varphi)_{\leq k}\downarrow$, the satisfiability problem and the model-checking problem for HDAs are both decidable.*

Actually, looking more closely at our construction which goes through finite automata accepting step sequences, we get the same result for MSO formulas even without the assumption that $L(\varphi)$ is downward-closed (but still over $\mathsf{iiPoms}_{\leq k}$, and not iiPoms). This could also be shown alternatively by observing that $\mathsf{iiPoms}_{\leq k}$ has bounded treewidth (in fact, even bounded pathwidth), and applying Courcelle's theorem [7]. In fact our implied proof of decidability is relatively similar, using step sequences instead of path decompositions.

For the second assertion of the theorem, we show that regular languages of HDAs are MSO-definable, again using an effective construction. Thus, using both directions of Theorem 8 and the closure properties of HDAs, we also get the following:

Corollary 11. *For all $k \in \mathbb{N}$ and MSO-definable $L \subseteq \mathsf{iiPoms}_{\leq k}$, $L\downarrow$ is MSO-definable.*

Note that this property does *not* hold for the class of *all* pomsets [21].

5   From MSO to HDAs

Given an MSO sentence φ over iiPoms we effectively build an HDA $\mathcal{H}$ such that $L(\mathcal{H}) = L(\varphi)_{\leq k}\!\downarrow$. The first step is to define an MSO-interpretation of interval ipomsets of width at most k into words of $\Omega^+_{\leq k}$, so that:

Lemma 12. *For every MSO sentence φ over iiPoms and every k there exists $\widehat{\varphi} \in \mathrm{MSO}^k_\Omega$ such that for all $P_1 \ldots P_n \in (\Omega_{\leq k} \setminus \{\mathrm{id}_\emptyset\})^+$, we have $P_1 \ldots P_n \models \widehat{\varphi}$ if and only if $P = P_1 * \cdots * P_n$ is well-defined and $P \models \varphi$.*

We will treat the case of the empty ipomset $\mathrm{id}_\emptyset$ separately afterwards. Prior to proving the lemma, we introduce key concepts and provide an overview. Informally speaking, we want a word $P_1 \ldots P_n$ of $(\Omega_{\leq k} \setminus \{\mathrm{id}_\emptyset\})^+$ to satisfy $\widehat{\varphi}$ if and only if the gluing composition $P = P_1 * \cdots * P_n$ is a model for φ. Thus $\widehat{\varphi}$ must accept only coherent words. This is MSO^k_Ω-definable by:

$$\mathbf{Coh}_k := \forall x \, \forall y. \, x \prec y \implies \bigvee_{P_1 P_2 \in \mathsf{Coh}_{\leq k} \cap \Omega^2_{\leq k}} P_1(x) \wedge P_2(y).$$

That is, discrete ipomsets of $\Omega_{\leq k}$ at consecutive positions x and y may be glued.

Hence, $\widehat{\varphi}$ will be the conjunction of $\mathbf{Coh}_k$ and an MSO^k_Ω formula φ' which we will build by induction on φ. Therefore, we have to consider formulas φ that contain free variables. We will construct φ' so that its free variables will be all the free first-order variables of φ and second-order variables $X_1, \ldots, X_k$ for every free second-order variable X of φ. In addition, we make sure during the construction that when $P \models \varphi$ then for every $w = P_1 \ldots P_n$ such that $P_1 * \cdots * P_n = P$, when the free variable x is interpreted as some event e of P in φ it is interpreted in φ' as some position in w where e occurs, and X_i will contain positions of w having as ith event according to $\dashrightarrow$ the interpretation of some element of X (see Example 13).

More formally, let $w = P_1 \ldots P_n \in \mathsf{Coh}_{\leq k}$ and $P = P_1 * \cdots * P_n$. Let $E = \{1, \ldots, n\} \times \{1, \ldots, k\}$. Our construction is built on a partial function $evt \colon E \to P$ defined as follows: if P_ℓ consists of events $e_1 \dashrightarrow \cdots \dashrightarrow e_r$, then for every $i \leq r$, $evt(\ell, i) = e_i$. We sometimes abuse notation and write $evt(P_\ell, i)$. Since $e \in P$ may occur in consecutive P_ℓ within w, one must determine when $evt(\ell, i) = evt(\ell', j)$. This can be done in MSO^k_Ω when $\ell' = \ell + 1$ as follows. For all $i, j \leq k$, let $M_{i,j} = \{P_1 P_2 \in \Omega^2_{\leq k} \mid evt(1, i) = evt(2, j)\}$. Then

$$glue_{i,j}(x, y) := x \prec y \wedge \bigvee_{P_1 P_2 \in M_{i,j}} P_1(x) \wedge P_2(y).$$

More generally, let us define the equivalence relation $\sim$ on E generated by $(\ell, i) \sim (\ell', i')$ if and only if $glue_{i,i'}(\ell, \ell')$ holds. Then for all $(\ell_1, i), (\ell_2, j) \in E$, $(\ell_1, i) \sim (\ell_2, j)$ if and only if $evt(\ell_1, i) = evt(\ell_2, j)$. The relation $\sim$ is MSO^k_Ω-definable:

$$(x, i) \sim (y, j) := \forall X_1, \ldots, X_k. \left(x \in X_i \wedge \bigwedge_{i,j \leq k} \forall x, y. \right.$$

$$\left. x \in X_i \wedge (glue_{i,j}(x, y) \vee glue_{j,i}(y, x)) \implies y \in X_j \right) \implies y \in X_j.$$

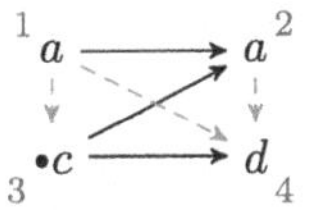

$$w_1 = \begin{bmatrix} a\bullet \\ \bullet c\bullet \end{bmatrix} \begin{bmatrix} \bullet a\bullet \\ \bullet c \end{bmatrix} \begin{bmatrix} \bullet a\bullet \\ d\bullet \end{bmatrix} \begin{bmatrix} \bullet a \\ \bullet d\bullet \end{bmatrix} \begin{bmatrix} a\bullet \\ \bullet d\bullet \end{bmatrix} \begin{bmatrix} \bullet a \\ \bullet d\bullet \end{bmatrix} \begin{bmatrix} \bullet d \end{bmatrix}$$

Fig. 4. Ipomset and corresponding coherent word (numbers indicate positions).

As mentioned before, we want a free first-order variable x to be interpreted in φ' as a position $P_\ell \in \Omega_{\leq k}$ of a coherent word $P_1 \ldots P_n$ where the interpretation $e \in P_1 * \cdots * P_n$ of x in φ occurs. That is there exists i such that $evt(l, i) = e$. Precisely, we construct a formula φ'_τ relative to a function τ which associates with every free first-order variable x of φ some $\tau(x) \in \{1, \ldots, k\}$. We sometimes leave τ implicit. Our aim is to have the following *invariant property* at each step of the induction: $P \models_\nu \varphi$ if and only if $w \models_{\nu'} \varphi'_\tau$ for any valuations ν, ν' satisfying the following:

1. $evt(\nu'(x), \tau(x)) = \nu(x)$ and
2. $\bigcup_{1 \leq i \leq k} \{evt(e, i) \mid e \in \nu'(X_i)\} = \nu(X)$.

Example 13. Figure 4 displays an ipomset P and the coherent word $w_1 = P_1 \ldots P_7$ such that $P_1 * \cdots * P_7 = P$. Let $e_1, \ldots, e_4$ be the events of P labelled respectively by the left a, the right a, c, and d and let $p_1, \ldots, p_7$ the positions on w_1 from left to right. Assume that $P \models_\nu \varphi(x, X)$ for some MSO-formula φ and the valuation $\nu(x) = e_1$ and $\nu(X) = \{e_2, e_3\}$. Then, $w_1 \models_{\nu'} \varphi'_{[x \mapsto 1]}(x, X_1, X_2)$ when, for example, $\nu'(x) = p_2$, $\nu'(X_1) = \{p_6\}$ and $\nu'(X_2) = \{p_3\}$ since this valuation satisfies the invariant property. For $\sim$ we have $(p_1, 1) \sim \cdots \sim (p_4, 1)$, $(p_1, 2) \sim (p_2, 2)$, $(p_3, 2) \sim \cdots \sim (p_6, 2) \sim (p_7, 1)$ and $(p_5, 1) \sim (p_6, 1)$. In particular $(p_1, 1) \not\sim (p_5, 1)$ since neither $glue_{1,1}(p_4, p_5)$ nor $glue_{2,1}(p_4, p_5)$ hold.

We are now ready to prove Lemma 12.

Proof (of Lemma 12). First, we let $\hat{\varphi} := \mathsf{Coh}_k \wedge \varphi'$ and we build φ' by induction on φ as follows. When φ is $\psi_1 \vee \psi_2$ or $\neg \psi$, then we let φ' be $\psi_1' \vee \psi_2'$ or $\neg \psi'$, respectively. For $\varphi = \exists X \psi$ we let $\varphi' := \exists X_1, \ldots, X_k.\psi'$. The function τ emerges in the case $\varphi = \exists x \psi$, where we let $\varphi'_\tau := \bigvee_{1 \leq i \leq k} \exists x \psi'_{[x \mapsto i]}$. When $\varphi = x \in X$, we let

$$\varphi'_{[x \mapsto i]} := \bigvee_{1 \leq j \leq k} \exists y \, (x, i) \sim (y, j) \wedge y \in X_j$$

For $\varphi = \mathsf{s}(x)$, we let $\varphi'_{[x \mapsto i]} := \bigwedge_{1 \leq j \leq k} \forall y \, (x, i) \sim (y, j) \implies \mathsf{s}(y, j)$, where $\mathsf{s}(y, j)$ is defined as the disjunction of all $D(y)$ where $evt(D, j) \in S_D$. We define $\varphi'_{[x \mapsto i]}$ similarly when $\varphi = \mathsf{t}(x)$. For $\varphi = x < y$ we let

$$\varphi'_{[x \mapsto i, y \mapsto j]} := \bigwedge_{1 \leq i', j' \leq k} \forall x', y'.\big((x', i') \sim (x, i) \wedge (y', j') \sim (y, j)\big) \implies x' < y'.$$

For $\varphi = x \dashrightarrow y$ we let

$$\varphi'_{[x \mapsto i, y \mapsto j]} := \bigvee_{1 \leq i' < j' \leq k} \exists z \, (z, i') \sim (x, i) \wedge (z, j') \sim (y, j).$$

Finally, when $\varphi = a(x)$, then we let $\varphi'_{[x \mapsto i]}$ be the disjunction of all $D(x)$ where $evt(D, i)$ is labelled by a. $\qquad \square$

As a consequence, we obtain:

Proposition 14. *Let φ be an MSO sentence over* iiPoms, $k \in \mathbb{N}$, *and* $L = \{P \in$ iiPoms$_{\leq k} \mid P \models \varphi\}\!\downarrow$. *Then L is effectively regular.*

Proof. Let $K = \{P \in$ iiPoms$_{\leq k} \mid P \models \varphi\}$. By Lemma 12, $L' = \{P_1 \ldots P_n \in (\Omega_{\leq k} \setminus \{id_\emptyset\})^+ \mid P_1 * \cdots * P_n \in K\}$ is effectively MSO$^k_\Omega$-definable, and thus so is $\bar{L}'' = \{P_1 \ldots P_n \in \Omega^+_{\leq k} \mid P_1 * \cdots * P_n \in K\}$. Indeed, whether $id_\emptyset$ satisfies φ is decidable and so is whether $id_\emptyset$ is in L''. By the standard Büchi and Kleene theorems, L'' is obtained from $\emptyset$ and $\Omega_{\leq k}$ using $\cup$, $\cdot$ and $^+$. By replacing concatenation of words by gluing composition, L is rational and thus effectively regular by Theorem 5. □

6   From HDAs to MSO

In this section we prove the second assertion of Theorem 8. The proof adapts the classical construction, encoding accepting paths of an automaton, to the case of HDAs. Our construction relies on the uniqueness of the sparse step decomposition (Lemma 2) and the MSO-definability of the relation: "an event is started/terminated before another event is started/terminated" in a sparse step decomposition (Lemma 17 below).

More formally, let $P \in$ iiPoms, then P admits a unique sparse step decomposition $P = P_1 * \cdots * P_n$. Given $e \in P \setminus S_P$, we denote by $\mathsf{St}(e)$ the step where e is started in the decomposition, *i.e.*, the minimal i such that $e \in P_i$. For $e \in P \setminus T_P$, we similarly denote by $\mathsf{Te}(e)$ the step where e is terminated. For $x \in S_P$ we let $\mathsf{St}(x) = -\infty$ and for $x \in T_P$, $\mathsf{Te}(x) = +\infty$. Then P_i contains precisely all $e \in P$ such that $\mathsf{St}(e) \leq i \leq \mathsf{Te}(e)$, that is all events which are started before or at P_i (or never) and are terminated after or at P_i (or never). In particular, if P_i is a starter, then it starts all e such that $\mathsf{St}(e) = i$, and if it is a terminator, it terminates all e such that $\mathsf{Te}(e) = i$. Note that $\mathsf{St}(e) < \mathsf{Te}(e)$ for all $e \in P$.

Example 15.   Proceeding with Example 13, let $w_2 = P_1 \ldots P_6 = \left[\begin{smallmatrix} a \bullet \\ \bullet c \bullet \end{smallmatrix}\right]\left[\begin{smallmatrix} \bullet a \bullet \\ \bullet c \end{smallmatrix}\right]\left[\begin{smallmatrix} \bullet a \bullet \\ d \bullet \end{smallmatrix}\right]\left[\begin{smallmatrix} \bullet a \\ \bullet d \bullet \end{smallmatrix}\right] \left[\begin{smallmatrix} a \bullet \\ \bullet d \bullet \end{smallmatrix}\right]\left[\begin{smallmatrix} \bullet a \\ \bullet d \end{smallmatrix}\right]$ be the sparse step decomposition of P (see also Example 4). We have $\mathsf{St}(e_3) = -\infty, \mathsf{St}(e_1) = 1, \mathsf{St}(e_4) = 3$ and $\mathsf{St}(e_2) = 5$. Also, $\mathsf{Te}(e_3) = 2, \mathsf{Te}(e_1) = 4$ and $\mathsf{Te}(e_2) = \mathsf{Te}(e_4) = 6$. Further, P_1 contains e_1 since $\mathsf{St}(e_1) = 1$ and e_3 because $\mathsf{St}(e_3) \leq 1 \leq \mathsf{Te}(e_3)$; P_4 contains e_1 since $\mathsf{Te}(e_1) = 4$ and e_4 because $\mathsf{St}(e_4) \leq 4 \leq \mathsf{Te}(e_4)$.

The example above will be continued in Example 19 at the end of this section. It will be pertinent in particular for the next lemma describing the existence of an accepting path inducing a sparse step decomposition as the existence of labellings $\rho_\nearrow$ and $\rho_\searrow$ mapping each started or terminated event of P to the upstep or downstep of the HDA performing it.

Lemma 16. *Let $\mathcal{H}$ be an HDA and $P \in$ iiPoms $\setminus$ Id whose sparse step decomposition is $P_1 * \cdots * P_n$. We have $P \in L(\mathcal{H})$ if and only if there exist $\rho_\nearrow : P \setminus S_P \to$ ups$(\mathcal{H})$ and $\rho_\searrow : P \setminus T_P \to$ downs$(\mathcal{H})$ such that, for all $e_1, e_2 \in P$:*

1. *if* $\mathsf{St}(e_1) = \mathsf{St}(e_2)$ *then* $\rho_{\nearrow}(e_1) = \rho_{\nearrow}(e_2)$;
2. *if* $\mathsf{Te}(e_1) = \mathsf{Te}(e_2)$ *then* $\rho_{\nwarrow}(e_1) = \rho_{\nwarrow}(e_2)$;
3. *if* $\mathsf{St}(e_2) = \mathsf{Te}(e_1) + 1$ *then* $\mathsf{src}(\rho_{\nearrow}(e_2)) = \mathsf{tgt}(\rho_{\nwarrow}(e_1))$;
4. *if* $\mathsf{Te}(e_2) = \mathsf{St}(e_1) + 1$ *then* $\mathsf{src}(\rho_{\nwarrow}(e_2)) = \mathsf{tgt}(\rho_{\nearrow}(e_1))$;
5. *if* $\rho_{\nearrow}(e_1) = (p, \nearrow^A, q)$ *then*

$$A = (U = \{e \mid \mathsf{St}(e) = \mathsf{St}(e_1)\}, \dashrightarrow_{P_{\uparrow U}}, \lambda_{P_{\uparrow U}}),$$
$$\mathsf{ev}(q) = (V = \{e \mid \mathsf{St}(e) \leq \mathsf{St}(e_1) < \mathsf{Te}(e)\}, \dashrightarrow_{P_{\uparrow V}}, \lambda_{P_{\uparrow V}});$$

6. *if* $\rho_{\nwarrow}(e_1) = (p, \nwarrow_A, q)$ *then*

$$A = (U = \{e \mid \mathsf{Te}(e) = \mathsf{Te}(e_1)\}, \dashrightarrow_{P_{\uparrow U}}, \lambda_{P_{\uparrow U}}),$$
$$\mathsf{ev}(p) = (V = \{e \mid \mathsf{St}(e) < \mathsf{Te}(e_1) \leq \mathsf{Te}(e)\}, \dashrightarrow_{P_{\uparrow V}}, \lambda_{P_{\uparrow V}});$$

7. *if* $\mathsf{St}(e_1) = 1$ *then* $\mathsf{src}(\rho_{\nearrow}(e_1)) \in \bot_H$;
8. *if* $\mathsf{Te}(e_1) = 1$ *then* $\mathsf{src}(\rho_{\nwarrow}(e_1)) \in \bot_H$;
9. *if* $\mathsf{St}(e_1) = n$ *then* $\mathsf{tgt}(\rho_{\nearrow}(e_1)) \in \top_H$;
10. *if* $\mathsf{Te}(e_1) = n$ *then* $\mathsf{tgt}(\rho_{\nwarrow}(e_1)) \in \top_H$.

As $P \notin \mathsf{Id}$, $\rho_{\nearrow}$ or $\rho_{\nwarrow}$ must be defined for at least one element of P above.

Our goal is to show that the conditions given by Lemma 16 can be expressed in MSO. We want to define a formula $\exists X_1 \ldots \exists X_m . \exists Y_1 \ldots \exists Y_n . \varphi$ with one X_i (resp. Y_j) for each upstep (resp. downstep) of the HDA. Intuitively, each X_i (Y_j) will contain all the events started (terminated) by performing the corresponding upstep (downstep). The sentence φ expresses that each event belongs to exactly one X_i (unless it is a source, in which case it belongs to none) and one Y_i (unless it is a target), and that the resulting labellings $\rho_{\nearrow}$ and $\rho_{\nwarrow}$ satisfy the conditions of the lemma. Hence, identity events do not belong to any X_i or Y_j. Nevertheless, conditions 5 and 6 ensure that they are consistent with the encoded path.

Let us first prove that the relations used in Lemma 16 are MSO-definable.

Lemma 17. *For $f, g \in \{\mathsf{St}, \mathsf{Te}\}$ and $\bowtie \in \{=, <, >\}$, the relations $f(x) \bowtie g(y)$, $\min(f)$ and $\max(f)$ are MSO-definable.*

Proof. We first define $\mathsf{Te}(x) < \mathsf{St}(y)$ as the formula $x < y$, together with $\mathsf{St}(x) < \mathsf{Te}(y) := \neg(\mathsf{Te}(y) < \mathsf{St}(x))$. Because starters and terminators alternate in the sparse step decomposition, we can then let

$$\mathsf{St}(x) < \mathsf{St}(y) := \exists z . \mathsf{St}(x) < \mathsf{Te}(z) \wedge \mathsf{Te}(z) < \mathsf{St}(y),$$
$$\mathsf{St}(x) = \mathsf{St}(y) := \neg(\mathsf{St}(x) < \mathsf{St}(y)) \wedge \neg(\mathsf{St}(y) < \mathsf{St}(x)) \wedge \neg \mathsf{s}(x) \wedge \neg \mathsf{s}(y)$$
$$\min(\mathsf{St}(x)) := \neg \mathsf{s}(x) \wedge \neg \exists y . \mathsf{Te}(y) < \mathsf{St}(x)$$
$$\max(\mathsf{Te}(x)) := \neg \mathsf{t}(x) \wedge \neg \exists y . \mathsf{St}(y) > \mathsf{Te}(x) .$$

The other formulas are defined similarly. □

We can also define $\mathsf{St}(y) = \mathsf{Te}(x) + 1$ and $\mathsf{Te}(y) = \mathsf{St}(x) + 1$ using standard techniques. Observe that $\mathsf{Te}(x) < \mathsf{St}(y)$ implies $\neg \mathsf{t}(x) \wedge \neg \mathsf{s}(y)$, given that the end of the x-event precedes the beginning of the y-event. As a consequence $\mathsf{St}(x) < \mathsf{St}(y)$ implies $\neg \mathsf{s}(y)$. On the other hand $\mathsf{St}(x) < \mathsf{Te}(y)$ holds in particular when x or y are interpreted as identities.

Proposition 18. *Given an HDA $\mathcal{H}$, one can effectively construct an MSO sentence φ such that $L(\mathcal{H}) = \{P \in \mathsf{iiPoms} \mid P \models \varphi\}$.*

Proof. We define

$$\varphi := (\exists x.\ \neg s(x) \vee \neg t(x)) \implies \exists X_1, \dots, X_m. \exists Y_1, \dots, Y_n. \bigwedge\nolimits_{i=0,\dots,10} \varphi_i$$

$$\wedge\ (\forall y.\ s(y) \wedge t(y)) \implies \bigvee_{\substack{p \in \perp_H \cap \top_H \\ \mathsf{ev}(p) \neq \emptyset}} \exists y_1, \dots, y_{|\mathsf{ev}(p)|}.\mathsf{ev}(p)(y_1, \dots, y_{|\mathsf{ev}(p)|}).$$

where φ_0 checks that the X_i's and Y_i's define labellings $\rho_\nearrow$ and $\rho_\searrow$ as in Lemma 16, that is, each event belongs to at most one X_i (is associated with at most one upstep) and one Y_i, and to no X_i iff it is a source and to no Y_i iff it is a target. The other formulas φ_i check condition i of Lemma 16. The second line of φ is satisfied by all non-empty identities accepted by $\mathcal{H}$. Thus $L(\varphi) = L(\mathcal{H}) \setminus \{\mathsf{id}_\emptyset\}$. If $\mathsf{id}_\emptyset \in L(\mathcal{H})$ then $L(\mathcal{H}) = L(\varphi \vee \neg \exists x.\ \mathsf{true})$. □

Example 19. Let $\mathcal{H}$ be the HDA of Fig. 3 and P the ipomset of Fig. 4. Recall that P is accepted by $\mathcal{H}$ by (among others) the path

$$t_3 \nearrow^a q_1 \searrow_c t_2 \nearrow^d q_2 \searrow_a t_8 \nearrow^a q_3 \searrow_{ad} v_8$$

which induces the sparse step decomposition

$$\left[\begin{smallmatrix} a\bullet \\ \bullet c\bullet \end{smallmatrix}\right] * \left[\begin{smallmatrix} \bullet a\bullet \\ \bullet c \end{smallmatrix}\right] * \left[\begin{smallmatrix} \bullet a\bullet \\ d\bullet \end{smallmatrix}\right] * \left[\begin{smallmatrix} \bullet a \\ \bullet d\bullet \end{smallmatrix}\right] * \left[\begin{smallmatrix} a\bullet \\ \bullet d\bullet \end{smallmatrix}\right] * \left[\begin{smallmatrix} \bullet a \\ \bullet d \end{smallmatrix}\right].$$

Let

$$\rho_\nearrow(e_1) = t_3 \nearrow^a q_1, \quad \rho_\nearrow(e_2) = t_8 \nearrow^a q_3, \quad \rho_\nearrow(e_4) = t_2 \nearrow^d q_2,$$

$$\rho_\searrow(e_1) = q_2 \searrow_a t_8, \quad \rho_\searrow(e_2) = \rho_\searrow(e_4) = q_3 \searrow_{ad} v_8, \quad \rho_\searrow(e_3) = q_1 \searrow_c t_2.$$

These definitions of $\rho_\nearrow$ and $\rho_\searrow$ satisfy the conditions of Lemma 16. Likewise, let $\varphi_\mathcal{H}$ be the MSO sentence built from $\mathcal{H}$ as in Proposition 18. Then $P \models \varphi_\mathcal{H}$, since the following interpretation satisfies $\bigwedge_{i=0,\dots,10} \varphi_i$:

$$X_{t_3 \nearrow^a q_1} = \{e_1\}, \quad X_{t_2 \nearrow^d q_2} = \{e_4\}, \quad X_{t_8 \nearrow^a q_3} = \{e_2\},$$

$$Y_{q_1 \searrow_c t_2} = \{e_3\}, \quad Y_{q_2 \searrow_a t_8} = \{e_1\}, \quad Y_{q_3 \searrow_{ad} v_8} = \{e_2, e_4\},$$

and the other X_u, Y_d for $u \in \mathsf{ups}(\mathcal{H})$ and $d \in \mathsf{downs}(\mathcal{H})$ are empty. Note that $\rho_\nearrow$ is not defined for e_3 since it is a source for P. For the same reason e_3 does not belong to any interpretation of X_u for any $u \in \mathsf{ups}(\mathcal{H})$.

7   Conclusion

This paper enriches the language theory of higher-dimensional automata with a Büchi-Elgot-Trakhtenbrot-like theorem. We have shown that the subsumption closures of MSO-definable subsets of $\mathrm{iiPoms}_{\leq k}$ are regular and that regular languages of HDAs are MSO-definable, both with effective constructions. Also, the MSO theory of $\mathrm{iiPoms}_{\leq k}$ and the MSO model-checking for HDAs are decidable.

Theorem 8 induces also a construction, for an MSO sentence φ over $\mathrm{iiPoms}_{\leq k}$, of $\varphi{\downarrow}$ such that $L(\varphi{\downarrow}) = L(\varphi){\downarrow}$. This property fails when we consider non-interval pomsets. However, the construction of $\varphi{\downarrow}$ is not efficient, as the current workflow is to transform φ to an HDA and then get $\varphi{\downarrow}$. We are wondering whether a more direct construction is possible.

Our work could be continued by considering logics weaker than MSO. For example, the study of the expressive power of first order logic over $\mathrm{iiPoms}_{\leq k}$ would be useful for model-checking purposes. In this regard, another operational model that would naturally arise is a class of ω-HDAs: HDAs over infinite ipomsets.

References

1. Amrane, A., Bazille, H., Clement, E., Fahrenberg, U.: Languages of higher-dimensional timed automata. In: PETRI NETS, 2024 (2024). Accepted. https://arxiv.org/abs/2401.17444
2. Amrane, A., Bazille, H., Fahrenberg, U., Ziemiański, K.: Closure and decision properties for higher-dimensional automata. In: Ábrahám, E., Dubslaff, C., Tarifa, S.L.T. (eds.) Theoretical Aspects of Computing – ICTAC 2023. ICTAC 2023. LNCS, vol. 14446, pp. 295–312. Springer, Cham (2023). https://doi.org/10.1007/978-3-031-47963-2_18
3. Bedon, N.: Logic and branching automata. Log. Methods Comput. Sci. **11**(4) (2015)
4. Brown, R., Higgins, P.J.: On the algebra of cubes. J. Pure Appl. Alg. **21**, 233–260 (1981)
5. Richard Büchi, J.: Weak second order arithmetic and finite automata. Zeitschrift für Mathematische Logik und Grundlagen der Mathematik **6**, 66–92 (1960)
6. Richard Büchi, J.: On a decision method in restricted second order arithmetic. In: Nagel, E., Suppes, P., Tarski, A. (eds.), LMPS'60, pp. 1–11. Stanford University Press (1962)
7. Courcelle, B.: The monadic second-order logic of graphs. I. Recognizable sets of finite graphs. Inf. Comput. **85**(1), 12–75 (1990)
8. Doner, J.: Tree acceptors and some of their applications. J. Comput. Syst. Sci. **4**(5), 406–451 (1970)
9. Dubut, J., Goubault, É., Goubault-Larrecq, J.: Natural homology. In: Halldórsson, M.M., Iwama, K., Kobayashi, N., Speckmann, B. (eds.) ICALP 2015. LNCS, vol. 9135, pp. 171–183. Springer, Heidelberg (2015). https://doi.org/10.1007/978-3-662-47666-6_14
10. Elgot, C.C.: Decision problems of finite automata design and related arithmetics. Trans. Am. Math. Soc. **98**, 21–52 (1961)
11. Fahrenberg, U.: A category of higher-dimensional automata. In: Sassone, V. (ed.) FoSSaCS 2005. LNCS, vol. 3441, pp. 187–201. Springer, Heidelberg (2005). https://doi.org/10.1007/978-3-540-31982-5_12
12. Fahrenberg, U.: Higher-dimensional timed and hybrid automata. Leibniz Trans. Embed. Syst. **8**(2), 03:1–03:16 (2022)

13. Fahrenberg, U., Johansen, C., Struth, G., Ziemiański, K.: Languages of higher-dimensional automata. Math. Struct. Comput. Sci. **31**(5), 575–613 (2021)
14. Fahrenberg, U., Johansen, C., Struth, G., Ziemiański, K.: A Kleene theorem for higher-dimensional automata. In: Klin, B., Lasota, S., Muscholl, A. (eds.), CONCUR, volume 243 of Leibniz International Proceedings in Informatics, pp. 29:1–29:18. Schloss Dagstuhl - Leibniz-Zentrum für Informatik (2022)
15. Fahrenberg, U., Johansen, C., Struth, G., Ziemiański, K.: Posets with interfaces as a model for concurrency. Inf. Comput. **285**(B), 104914 (2022)
16. Fahrenberg, U., Raussen, M.: Reparametrizations of continuous paths. J. Homotopy Relat. Struct. **2**(2), 93–117 (2007)
17. Fahrenberg, U., Ziemiański, K.: A myhill-nerode theorem for higher-dimensional automata. In: Gomes, L., Lorenz, R. (eds.) Application and Theory of Petri Nets and Concurrency. PETRI NETS 2023. LNCS, vol. 13929, pp. 167–188. Springer, Cham (2023). https://doi.org/10.1007/978-3-031-33620-1_9
18. Fajstrup, L., Goubault, É., Haucourt, E., Mimram, S., Raussen, M.: Trace spaces: an efficient new technique for state-space reduction. In: Seidl, H. (ed.) ESOP 2012. LNCS, vol. 7211, pp. 274–294. Springer, Heidelberg (2012). https://doi.org/10.1007/978-3-642-28869-2_14
19. Fajstrup, L., Goubault, E., Haucourt, E., Mimram, S., Raussen, M.: Directed Algebraic Topology and Concurrency. Springer, Cham (2016). https://doi.org/10.1007/978-3-319-15398-8
20. Fajstrup, L., Raussen, M., Goubault, E., Haucourt, E.: Components of the fundamental category. Appl. Categ. Struct. **12**, 81–108 (2004)
21. Fanchon, J., Morin, R.: Pomset languages of finite step transition systems. In: Franceschinis, G., Wolf, K. (eds.) PETRI NETS 2009. LNCS, vol. 5606, pp. 83–102. Springer, Heidelberg (2009). https://doi.org/10.1007/978-3-642-02424-5_7
22. Fishburn, P.C.: Interval Orders and Interval Graphs: A Study of Partially Ordered Sets. Wiley, Hoboken (1985)
23. Genest, B., Kuske, D., Muscholl, A.: A Kleene theorem and model checking algorithms for existentially bounded communicating automata. Inf. Comput. **204**(6), 920–956 (2006)
24. Grabowski, J.: On partial languages. Fundam. Inform. **4**(2), 427 (1981)
25. Grandis, M., Mauri, L.: Cubical sets and their site. Theory Appl. Categ. **11**(8), 185–211 (2003)
26. Janicki, R., Koutny, M.: Operational semantics, interval orders and sequences of antichains. Fundam. Inform. **169**(1–2), 31–55 (2019)
27. Kahl, T.: Topological abstraction of higher-dimensional automata. Theor. Comput. Sci. **631**, 97–117 (2016)
28. Kahl, T.: Weak equivalence of higher-dimensional automata. Discret. Math. Theor. Comput. Sci. **23**(1) (2021)
29. Kuske, D.: Infinite series-parallel posets: logic and languages. In: Montanari, U., Rolim, J.D.P., Welzl, E. (eds.) ICALP 2000. LNCS, vol. 1853, pp. 648–662. Springer, Heidelberg (2000). https://doi.org/10.1007/3-540-45022-X_55
30. Kuske, D., Morin, R.: Pomsets for local trace languages. J. Autom. Lang. Comb. **7**(2), 187–224 (2002)
31. Pratt, V.R.: Modeling concurrency with geometry. In: POPL, pp. 311–322. ACM Press, New York City (1991)
32. Rabin, M.O.: Decidability of second-order theories and automata on infinite trees. Trans. Am. Math. Soc. **141**, 1–35 (1969)
33. Serre, J.-P.: Homologie singulière des espaces fibrés. PhD thesis, Ecole Normale Supérieure, Paris, France (1951)

34. Thatcher, J.W., Wright, J.B.: Generalized finite automata theory with an application to a decision problem of second-order logic. Math. Syst. Theory **2**(1), 57–81 (1968)

35. Thomas, W.: On logical definability of trace languages. In: Algebraic and Syntactic Methods in Computer Science (ASMICS), Report TUM-I9002, Technical University of Munich, pp. 172–182 (1990)

36. Thomas, W.: Languages, automata, and logic. In: Rozenberg, G., Salomaa, A. (eds.) Handbook of Formal Languages, pp. 389–455. Springer, Heidelberg (1997). https://doi.org/10.1007/978-3-642-59126-6_7

37. Trakhtenbrot, B.A.: Finite automata and monadic second order logic. Sib. Math. J. **3**, 103–131 (1962). In Russian; English translation in Amer. Math. Soc. Transl. **59**(1966), 23–55

38. van Glabbeek, R.J.: Bisimulations for higher dimensional automata. Email message, June 1991. http://theory.stanford.edu/~rvg/hda

39. Wiener, N.: A contribution to the theory of relative position. Proc. Camb. Philos. Soc. **17**, 441–449 (1914)

40. Zielonka, W.: Notes on finite asynchronous automata. RAIRO - Informatique Théorique et Applications **21**(2), 99–135 (1987)

41. Ziemiański, K.: Stable components of directed spaces. Appl. Categ. Struct. **27**(3), 217–244 (2019)

Universal Rewriting Rules for the Parikh Matrix Injectivity Problem

Ingyu Baek[1], Joonghyuk Hahn[1], Yo-Sub Han[1(✉)], and Kai Salomaa[2]

[1] Yonsei University, Seoul, Republic of Korea
{ingyubaek,greghahn,emmous}@yonsei.ac.kr
[2] Queen's University, Kingston, Canada
salomaa@queensu.ca

Abstract. The injectivity problem of the Parikh matrix is closely related to the characterization of the M-equivalence. Current studies provide partial results such as a necessary condition of the M-equivalence or characterization of the M-equivalence over a binary or ternary alphabet. While these studies give rise to rewriting rules that construct M-equivalent strings of a given string over a binary or ternary alphabet, it has been open to designing general rewriting rules for M-equivalence independent of the alphabet size. We propose rewriting rules using exponent-strings, which are an extension of strings, as an intermediate representation. We introduce a special normal form for an exponent-string and prove that any string can be rewritten into an M-equivalent exponent-string in this special form. Then, we show that our rewriting rules characterize M-equivalence over an ordered alphabet of an arbitrary size.

Keywords: Exponent-string · Parikh matrix · M-equivalence · Strong M-equivalence

1 Introduction

The Parikh matrix is an extension of the Parikh vector, which maps strings to vectors in the form of a matrix. While the Parikh vector computes occurrences of characters of a given alphabet, the Parikh matrix considers subsequences that capture a broader range of numerical properties. Nevertheless, the Parikh matrix mapping is still not injective. Thus, we say that two strings u, v are M-*equivalent* [14] when u and v have the same Parikh matrix. One of the most fundamental problems is to characterize M-equivalence—to identify when the two strings are M-equivalent with simple conditions. While the sharpness of the Parikh vector mapping is fairly self-explanatory, it is a difficult task to present a natural characterization of the strings with the same Parikh matrix. Mateescu et al. [14] already mentioned this as a major open problem when they introduced the Parikh matrix.

Several researchers proposed various rewriting rules for characterizing M-equivalence [1–3,10–12] over binary or ternary alphabets. Atanasiu et al. [4] introduced the palindromicly amicable relation for the M-equivalence over a

J. D. Day and F. Manea (Eds.): DLT 2024, LNCS 14791, pp. 68–81, 2024.
https://doi.org/10.1007/978-3-031-66159-4_6

binary alphabet. Recently, Hahn et al. [12] characterized M-equivalence over a ternary alphabet. They proposed a $\cong$-relation based on rewriting rules that extended the rules from [4]. There are some partial characterizations of M-equivalence over an alphabet of size three or more including more numerical attributes of Parikh matrices such as weak M-relation and strong M-equivalence [5,7,9,16,18]. Hutchinson et al. [13] developed a toolkit program that implements various operations about Parikh matrices to support research on such properties. However, as far as we are aware, there is no known prior work providing a full characterization of M-equivalence.

We use *exponent-strings*—an extension of strings—and present rewriting rules for exponent-strings that characterize M-equivalence precisely. Roughly speaking, an exponent-string is a finite sequence of characters in which each character has a repetition value of positive rational as its exponent. It is not unusual to have an exponent value for a character of a string. For example, a string abb is often written as ab^2 to shorten the presentation; in this case, we can think of 2 as an exponent of b. In practice, it is common to use natural numbers for such repetitions for strings. We extend this notion and allow positive rational numbers to be exponent values. This helps us to have an exponent-string $ab^{1.7}$— one unit of a and 1.7 units of b. Our rewriting rules for exponent-strings are based on a modification of rewriting rules for M-equivalent strings. Note that the exponent-string is different from the fractional power of words that appears in Brandenburg [8].

Since exponent-strings have notations that are different from the traditional strings, we clarify the difference by presenting formal notations and definitions regarding both strings and exponent-strings in Sect. 2. We also recall general notations and definitions in the field of Parikh matrices [14,17]. In Sect. 3, we present our main result of characterizing M-equivalence using new rewriting rules defined over exponent-strings. Then, we extend our approach for the strong M-equivalence problem and establish partial results in Sect. 4, and conclude the paper in Sect. 5.

2   Preliminaries

2.1   Basic Notations and Definitions

An *ordered alphabet* $\Sigma = \{a_1 < a_2 < \cdots < a_k\}$ is a finite set of k characters with the order $a_1 < a_2 < \cdots < a_k$. For example, given $\Sigma = \{a < b < c\}$, $|\Sigma| = 3$. Note that $\{a < b < c\} \neq \{a < c < b\}$ as the characters are ordered differently. We use the same basic notations of an alphabet also for an ordered alphabet.

Parikh [15] introduced the Parikh vector to represent the number of occurrences of characters in a string as a vector.

Definition 1. *Let* $\Sigma = \{a_1 < a_2 < \cdots < a_k\}$ *be an ordered alphabet. The Parikh mapping is a monoid morphism* $\Psi : \Sigma^* \to \mathbb{N}^k$ *defined as* $\Psi(w) = (|w|_{a_1}, |w|_{a_2}, \ldots, |w|_{a_k})$. *Then,* $\Psi(w)$ *is the Parikh vector of* $w \in \Sigma^*$.

The Parikh vector extends to the Parikh matrix [14], which is an upper *unitriangular matrix* of nonnegative integers. A unitriangular matrix of dimension k is a square matrix $m = (m_{i,j})_{1 \leq i,j \leq k}$ such that (1) $m_{i,j} \in \mathbb{N}$, (2) $m_{i,j} = 0$ for all $1 \leq j < i \leq k$, and (3) $m_{i,i} = 1$ for all $1 \leq i \leq k$. The set of all unitriangular matrices of dimension $k \geq 1$ is denoted by $\mathcal{M}_k$. Definition 2 presents the formal definition of a Parikh matrix and its mapping. Proposition 1 follows right after Definition 2 that illustrates how to compute each element of the Parikh matrix.

Definition 2. *Let $\Sigma = \{a_1 < a_2 < \cdots < a_k\}$ be an ordered alphabet. The Parikh matrix mapping is a monoid morphism $\Psi_\Sigma : \Sigma^* \to \mathcal{M}_{k+1}$ defined as follows. For $a_t \in \Sigma$, if $\Psi_\Sigma(a_t) = (m_{i,j})_{1 \leq i,j \leq k+1}$, then $m_{i,i} = 1$ for $1 \leq i \leq k+1$, $m_{t,t+1} = 1$, and all the other entries are zero. Then, $\Psi_\Sigma(w)$ is the Parikh matrix of $w \in \Sigma^*$.*

Proposition 1 (Mateescu et al. [14]). *Let $\Sigma = \{a_1 < a_2 < \cdots < a_k\}$ be an ordered alphabet. We denote by $a_{i,j}$ the string $a_i a_{i+1} \cdots a_j$ for $1 \leq i \leq j \leq k$. For $w \in \Sigma^*$, its Parikh matrix $\Psi_\Sigma(w)$ has the following properties:*

1. *$m_{i,j} = 0$, for all $1 \leq j < i \leq k+1$,*
2. *$m_{i,i} = 1$, for all $1 \leq i \leq k+1$,*
3. *$m_{i,j+1} = |w|_u$ where $u = a_{i,j}$ for all $1 \leq i \leq j \leq k$.*

We illustrate the examples of a Parikh vector and a Parikh matrix in Example 1.

Example 1. Consider a string $w = abaaca$ over $\Sigma = \{a < b < c\}$. Following is the Parikh vector $\Psi(w)$ and the Parikh matrix $\Psi_\Sigma(w)$ of a string w:

$$\Psi(w) = (|w|_a, |w|_b, |w|_c) = (4, 1, 1),$$

$$\Psi_\Sigma(w) = \begin{pmatrix} 1 & |w|_a & |w|_{ab} & |w|_{abc} \\ 0 & 1 & |w|_b & |w|_{bc} \\ 0 & 0 & 1 & |w|_c \\ 0 & 0 & 0 & 1 \end{pmatrix} = \begin{pmatrix} 1 & 4 & 1 & 1 \\ 0 & 1 & 1 & 1 \\ 0 & 0 & 1 & 1 \\ 0 & 0 & 0 & 1 \end{pmatrix}.$$

M-equivalence is the relation between strings with the same Parikh matrix defined as follows.

Definition 3. *Given two strings u and v over Σ, u and v are M-equivalent if $\Psi_\Sigma(u) = \Psi_\Sigma(v)$, denoted as $u \equiv_M v$.*

Proposition 2 (Atanasiu et al. [4]). *For an ordered alphabet Σ of size k and $a, b \in \Sigma$, the rewriting rules I, II always generate M-equivalent strings:*

I If a and b are not consecutive in Σ, then $ab \leftrightarrow ba$.
II If a and b are consecutive characters in Σ, then $abba \leftrightarrow baab$.

For a positive integer k and a square matrix M, M^k is the k-th power of M. We allow k to be a non-integer value using the binomial theorem as follows [19]:

$$M^k = (I + X)^k = I + kX + \frac{k(k-1)}{2!}X^2 + \frac{k(k-1)(k-2)}{3!}X^3 + \cdots,$$

where $M = I + X$ and I is the identity matrix. Based on observations in Waugh and Abel [19], the series defining M^k always converges when M is a Parikh matrix, which justifies the extended Parikh matrix in Sect. 3.

2.2   Exponent-Strings

String is usually denoted using exponents; $aabbbccaa = a^2b^3c^2a^2$. Informally, an exponent-string is an extended string, where an exponent can be an arbitrary value. We briefly review the terminologies of the exponent-string introduced by some of the authors in [6].

Definition 4 (S-exponent-string). *For an alphabet Σ and a semigroup $(S,+)$, an S-exponent-string p over Σ is a finite sequence of pairs $p = (\sigma_1, s_1), (\sigma_2, s_2), \ldots, (\sigma_n, s_n)$, where each pair (σ_i, s_i) consists of a base character $\sigma_i \in \Sigma$ and its exponent $s_i \in S$, and two consecutive pairs must have different base characters—namely, $\sigma_i \neq \sigma_{i+1}$ for $1 \leq i \leq n - 1$. We use λ to denote the empty sequence.*

The rest of this section introduces several technical concepts and notations associated with exponent-strings.

Representation. For an S-exponent-string $p = (\sigma_1, s_1), (\sigma_2, s_2), \ldots, (\sigma_n, s_n)$, we can represent p in a string form $p = a_1^{q_1} a_2^{q_2} \cdots a_k^{q_k}$ if and only if there exist integers $1 = k_1 < k_2 < \cdots < k_{n+1} = k+1$ such that $s_i = q_{k_i} + q_{k_i+1} + \cdots + q_{k_{i+1}-1}$ and $\sigma_i = a_{k_i} = a_{k_i+1} = \cdots = a_{k_{i+1}-1}$ for $1 \leq i \leq n$. In this way, p can have more than one string form representation. For example, let S be the semigroup of positive real numbers equipped with addition. Then, an S-exponent-string $p = (c, \frac{7}{3}), (a, \pi), (b, \sqrt{2})$ can be represented as $p = c^1 c^{\frac{4}{3}} a^\pi b^{\sqrt{2}} = c^{1.5} c^{0.5} c^{\frac{1}{3}} a^{\pi-2} a^2 b^{0.5} b^{\sqrt{2}-1} b^{0.5} = \cdots$. Note that an equality like $p = c^1 c^{\frac{4}{3}} a^\pi b^{\sqrt{2}}$, strictly speaking, abuses notation because p is an S-exponent-string and the right side is a particular representation of the exponent-string. Among multiple string representations of S-exponent-string $p = (\sigma_1, s_1), (\sigma_2, s_2), \ldots, (\sigma_n, s_n)$, $p = \sigma_1^{s_1} \sigma_2^{s_2} \cdots \sigma_n^{s_n}$ (string representation with no term-breaking in p) is a *contraction form* of p. In the later parts of this section, we only consider $\mathbb{Q}^+$-exponent-strings, where the binary operation of the semigroup $\mathbb{Q}^+$ is an addition. We denote the set of every S-exponent-string over Σ as Σ_S^*. For instance, the set of $\mathbb{Q}^+$-exponent-string is denoted $\Sigma_{\mathbb{Q}^+}^*$.

Factor. For the string form representation of a nonempty $\mathbb{Q}^+$-exponent-string $p = a_1^{s_1} a_2^{s_2} \cdots a_n^{s_n}$, each $a_i^{s_i}$ for $1 \leq i \leq n$ is a *factor* of p. If the base characters $a_{i_1}, a_{i_2}, \cdots, a_{i_k}$ satisfy $\sigma = a_{i_1} = a_{i_2} = \cdots = a_{i_k}$, each $a_{i_j}^{s_{i_j}}$ for $1 \leq j \leq k$ is a σ-*factor* of p. For instance, given an $\mathbb{Q}^+$-exponent-string representation $p = a^{2.3} a^1 b^2$, there are two types of factors—a-factor and b-factor—and both $a^{2.3}$ and a^1 are a-factors.

Concatenation. For two nonempty $\mathbb{Q}^+$-exponent-strings $p = a_1^{s_1} a_2^{s_2} \cdots a_n^{s_n}$ and $q = b_1^{t_1} b_2^{t_2} \cdots b_m^{t_m}$, a concatenation of p and q, denoted as $p \cdot q$ is $p \cdot q = a_1^{s_1} a_2^{s_2} \cdots a_n^{s_n} b_1^{t_1} b_2^{t_2} \cdots b_m^{t_m}$. For the empty $\mathbb{Q}^+$-exponent-string λ and an arbitrary $\mathbb{Q}^+$-exponent-string p, we define $p \cdot \lambda = \lambda \cdot p = p$. We often omit the concatenation symbol $\cdot$ when its role is clear in the context and also omit exponents of value 1 for simplicity. Notice that $\mathbb{N}$-exponent-strings only have integer exponents and thus, we can consider such representation of $\mathbb{N}$-exponent-strings as traditional strings.

Length, Factor-Length of Contraction Form. For $p = \sigma_1^{s_1}\sigma_2^{s_2}\cdots\sigma_n^{s_n}$, length of p is defined as $len(p) := \sum_{i=1}^{n} s_i$. The factor length of p, $Flen(p)$ is the number of factors in the contraction form of p. In other words, if $p = \tau_1^{t_1}\tau_2^{t_2}\cdots\tau_k^{t_k}$ is a $\mathbb{Q}^+$-exponent-string in the contraction form, $Flen(p) = k$.

Bracket Notation. For $p = \sigma_1^{s_1}\sigma_2^{s_2}\cdots\sigma_n^{s_n}$, $p[x]$ is a base character of position x.

$$p[x] = \begin{cases} \sigma_1, \text{ if } 0 \leq x < s_1, \\ \sigma_2, \text{ if } s_1 \leq x < s_1 + s_2, \\ \vdots \\ \sigma_n, \text{ if } \sum_{i=1}^{n-1} s_i \leq x < len(p) = \sum_{i=1}^{n} s_i. \end{cases}$$

Prefix, Infix and Suffix. For $p, q \in \Sigma_{\mathbb{Q}^+}^*$, q is an infix (prefix, suffix) of p if there exist $u, v \in \Sigma_{\mathbb{Q}^+}^*$ such that $p = uqv$ ($p = qv$, $p = uq$, respectively). For instance, a $\mathbb{Q}^+$-exponent-string $\mathsf{s}^{1.5}\mathsf{ci}^{\frac{7}{3}}\mathsf{ence}^{1.2}$, has a prefix $\mathsf{s}^{1.5}\mathsf{ci}^{\frac{1}{3}}$, infix $\mathsf{i}^2\mathsf{e}^{0.2}$ and suffix $\mathsf{ce}^{1.2}$.

Palindrome. For $\mathbb{Q}^+$-exponent-string p, if $p = \tau_1^{t_1}\tau_2^{t_2}\cdots\tau_k^{t_k} = \tau_k^{t_k}\cdots\tau_2^{t_2}\tau_1^{t_1}$ in contraction form, p is a palindrome. Figure 1 illustrates palindrome and non-palindrome $\mathbb{Q}^+$-exponent-strings.

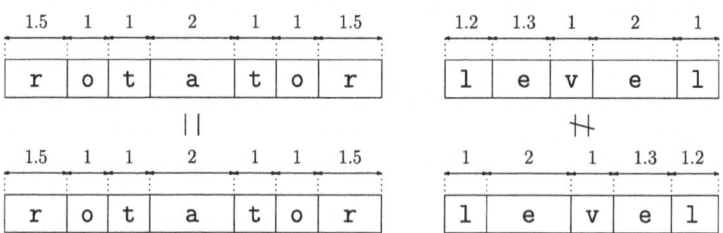

Fig. 1. A comparison of a palindrome $\mathbb{Q}^+$-exponent-string $\mathsf{r}^{1.5}\mathsf{ota}^2\mathsf{tor}^{1.5}$ and a non-palindrome $\mathbb{Q}^+$-exponent-string $\mathsf{1}^{1.2}\mathsf{e}^2\mathsf{ve}^{1.3}\mathsf{1}^{1.2}$

3   Characterization of M-Equivalence

The characterization of M-equivalence and M-ambiguity is a much-investigated problem that has been open for decades. Recently, Hahn et al. [12] proposed a characterization of M-equivalence over a ternary alphabet. We present a characterization of M-equivalence over an arbitrary alphabet using exponent-strings.

3.1   Parikh Matrix and Exponent-Strings

We extend the Parikh matrix mapping domain from Σ^* to $\Sigma^*_{\mathbb{Q}^+}$. Note that Σ is an ordered alphabet.

Definition 5. *An extended Parikh matrix mapping $\Psi_\Sigma^{\mathbb{Q}^+}$ of a $\mathbb{Q}^+$-exponent-string is a mapping from $\Sigma^*_{\mathbb{Q}^+}$ to a set of $(|\Sigma| + 1) \times (|\Sigma| + 1)$ matrices. The mapping preserves exponent-string exponentiation to matrix exponentiation, and exponent-string concatenation aligns with matrix multiplication as follows:*

$$\Psi_\Sigma^{\mathbb{Q}^+}(\lambda) = I$$
$$\Psi_\Sigma^{\mathbb{Q}^+}(a_1^{s_1} a_2^{s_2} \cdots a_n^{s_n}) = \Psi_\Sigma(a_1)^{s_1} \Psi_\Sigma(a_2)^{s_2} \cdots \Psi_\Sigma(a_n)^{s_n}$$

where $a_1, a_2, \cdots, a_n \in \Sigma, s_1, s_2, \cdots, s_n \in \mathbb{Q}^+$. We call the superdiagonal entries of an extended Parikh matrix an extended Parikh vector.

It is straightforward that the value of an extended Parikh matrix mapping is not specific to different representations of a $\mathbb{Q}^+$-exponent-string, as the concept of fractional power of matrix extends the notion of power in a well-defined manner. Example 2 illustrates an extended Parikh matrix for a $\mathbb{Q}^+$-exponent-string.

Example 2. Given an ordered alphabet $\Sigma = \{a < b < c\}$, let $p = a^3 c^{2.5} b^{\frac{1}{3}} \in \Sigma^*_{\mathbb{Q}^+}$. Then,

$$\Psi_\Sigma^{\mathbb{Q}^+}(p) = \Psi_\Sigma^{\mathbb{Q}^+}(a^3 c^{2.5} b^{\frac{1}{3}}) = \begin{pmatrix} 1 & 1 & 0 & 0 \\ 0 & 1 & 0 & 0 \\ 0 & 0 & 1 & 0 \\ 0 & 0 & 0 & 1 \end{pmatrix}^3 \begin{pmatrix} 1 & 0 & 0 & 0 \\ 0 & 1 & 0 & 0 \\ 0 & 0 & 1 & 1 \\ 0 & 0 & 0 & 1 \end{pmatrix}^{2.5} \begin{pmatrix} 1 & 0 & 0 & 0 \\ 0 & 1 & 1 & 0 \\ 0 & 0 & 1 & 0 \\ 0 & 0 & 0 & 1 \end{pmatrix}^{\frac{1}{3}}$$

$$= \begin{pmatrix} 1 & 1 & 0 & 0 \\ 0 & 1 & 0 & 0 \\ 0 & 0 & 1 & 0 \\ 0 & 0 & 0 & 1 \end{pmatrix}^{2+1} \begin{pmatrix} 1 & 0 & 0 & 0 \\ 0 & 1 & 0 & 0 \\ 0 & 0 & 1 & 1 \\ 0 & 0 & 0 & 1 \end{pmatrix}^{1.5+1} \begin{pmatrix} 1 & 0 & 0 & 0 \\ 0 & 1 & 1 & 0 \\ 0 & 0 & 1 & 0 \\ 0 & 0 & 0 & 1 \end{pmatrix}^{\frac{1}{6}+\frac{1}{6}} = \Psi_\Sigma^{\mathbb{Q}^+}(a^2 a^1 c^{1.5} c^1 b^{\frac{1}{6}} b^{\frac{1}{6}})$$

$$= \begin{pmatrix} 1 & 1 & 0 & 0 \\ 0 & 1 & 0 & 0 \\ 0 & 0 & 1 & 0 \\ 0 & 0 & 0 & 1 \end{pmatrix}^2 \begin{pmatrix} 1 & 1 & 0 & 0 \\ 0 & 1 & 0 & 0 \\ 0 & 0 & 1 & 0 \\ 0 & 0 & 0 & 1 \end{pmatrix}^1 \begin{pmatrix} 1 & 0 & 0 & 0 \\ 0 & 1 & 0 & 0 \\ 0 & 0 & 1 & 1 \\ 0 & 0 & 0 & 1 \end{pmatrix}^{1.5} \begin{pmatrix} 1 & 0 & 0 & 0 \\ 0 & 1 & 0 & 0 \\ 0 & 0 & 1 & 1 \\ 0 & 0 & 0 & 1 \end{pmatrix}^1 \begin{pmatrix} 1 & 0 & 0 & 0 \\ 0 & 1 & 1 & 0 \\ 0 & 0 & 1 & 0 \\ 0 & 0 & 0 & 1 \end{pmatrix}^{\frac{1}{6}} \begin{pmatrix} 1 & 0 & 0 & 0 \\ 0 & 1 & 1 & 0 \\ 0 & 0 & 1 & 0 \\ 0 & 0 & 0 & 1 \end{pmatrix}^{\frac{1}{6}}$$

$$= \begin{pmatrix} 1 & 3 & 1 & 0 \\ 0 & 1 & \frac{1}{3} & 0 \\ 0 & 0 & 1 & 2.5 \\ 0 & 0 & 0 & 1 \end{pmatrix}.$$

   When exponents are integers, it is clear that the extended definition is identical to the original definition of the Parikh matrix mapping defined over strings. This leads us to extend the notion of M-equivalence defined over strings to $\mathbb{Q}^+$-exponent-strings.

Definition 6. *[Extended M-equivalence]* $\mathbb{Q}^+$*-exponent-strings* p_1 *and* p_2 *are EM-equivalent if and only if*

$$\Psi_{\Sigma}^{\mathbb{Q}^+}(p_1) = \Psi_{\Sigma}^{\mathbb{Q}^+}(p_2).$$

We denote by $p_1 \equiv_{EM} p_2$ *when* p_1 *and* p_2 *are EM-equivalent.*

Proposition 3. *For two strings* p_1 *and* p_2, *we have that* $p_1 \equiv_{EM} p_2$ *if and only if* $p_1 \equiv_M p_2$.

Proof. Since $\Psi_{\Sigma}^{\mathbb{Q}^+}(p_1) = \Psi_{\Sigma}(p_1)$ and $\Psi_{\Sigma}^{\mathbb{Q}^+}(p_2) = \Psi_{\Sigma}(p_2)$, it is straightforward to verify that the statement holds.   □

From Proposition 3, it is immediate that the characterization of *EM*-equivalence will give one for *M*-equivalence of traditional strings. Specifically, rewriting rules in Theorem 1 characterizes *EM*-equivalence as well as *M*-equivalence, by employing a $\mathbb{Q}^+$-exponent-string as an intermediate form between two strings.

Proposition 4 introduces an extended Parikh matrix mapping formulation that aligns equivalently with Proposition 1, which addresses the Parikh matrix mapping.

Proposition 4. *Given an* $\mathbb{Q}^+$*-exponent-string* $p = r_1^{q_1} r_2^{q_2} \cdots r_n^{q_n}$ *over an ordered alphabet* $\Sigma = \{a_1 < a_2 < \cdots < a_k\}$, *its extended Parikh matrix mapping* $\Psi_{\Sigma}^{\mathbb{Q}^+}(p) = (m_{i,j})_{1 \leq i,j \leq (k+1)}$ *has the following properties:*

1. $m_{i,j} = 0$ *for* $1 \leq j < i \leq k+1$.
2. $m_{i,i} = 1$ *for* $1 \leq i \leq k+1$.
3. *Let* $S_{i,j}$ *be a set of index sequences* $(d_1, d_2, \ldots, d_{j-i+1})$ *such that the string* $a_i a_{i+1} \cdots a_j$ *occurs in* $r_1 r_2 \cdots r_n$ *as a subsequence,*

$$S_{i,j} = \{(d_1, d_2, \cdots, d_{j-i+1}) \mid a_i = r_{d_1}, a_{i+1} = r_{d_2}, \cdots, a_j = r_{d_{j-i+1}}$$
$$\text{and } 1 \leq d_1 < d_2 < \cdots < d_{j-i+1} \leq n\}.$$

Then,

$$m_{i,j+1} = \sum_{(d_1, d_2, \cdots, d_{j-i+1}) \in S_{i,j}} \prod_{l=1}^{j-i+1} q_{d_l} \quad \text{for } 1 \leq i \leq j \leq k.$$

4. *Let* c *be a common denominator of* $q_1, q_2, \ldots, q_n$ *and* $w = r_1^{cq_1} r_2^{cq_2} \cdots r_n^{cq_n} \in \Sigma^*$. *Then,*

$$m_{i,j+1} = \frac{1}{c^{j-i+1}} [\Psi_{\Sigma}(w)]_{i,j+1} = \frac{1}{c^{j-i+1}} |w|_{a_i a_{i+1} \cdots a_j} \quad \text{for } 1 \leq i \leq j \leq k.$$

In the spirit of extending the occurrence notation of a string used in Proposition 1, we denote $|p|_{a_i a_{i+1} \cdots a_j} = m_{i,j+1}$ for $p \in \Sigma_{\mathbb{Q}^+}^*$.

3.2   Characterization by Universal Rewriting Rule

We extend the Parikh matrix mapping for $\mathbb{Q}^+$-exponent-strings from the previous section. By modifying the rewriting rules in Proposition 2, we propose rewriting rules for $\mathbb{Q}^+$-exponent-strings that characterize EM-equivalence over an alphabet of any size.

Theorem 1. *For two $\mathbb{Q}^+$-exponent-strings p and q over an ordered alphabet $\Sigma = \{a_1 < a_2 < \cdots < a_k\}$ of an arbitrary size k, there exists a finite rewriting sequence consisting of*

R1. $a_i^{x_1} a_j^{x_2} \leftrightarrow a_j^{x_2} a_i^{x_1}$ *for* $1 \leq i, j \leq k, |i - j| \geq 2$ *and* $x_1, x_2 \in \mathbb{Q}^+$,
R2. $a_i^{x_1} a_{i+1}^{2x_2} a_i^{x_1} \leftrightarrow a_{i+1}^{x_2} a_i^{2x_1} a_{i+1}^{x_2}$ *for* $1 \leq i \leq k - 1$ *and* $x_1, x_2 \in \mathbb{Q}^+$,

that transforms p to q if and only if p and q are EM-equivalent.

We first prove the [only-if] direction of Theorem 1, which is rather straightforward, in Lemma 1. For the [if] direction, for the sake of readability, we first present the idea of the proof by considering the binary case in Lemma 2. Then, we prove the [if] direction for an alphabet of any size by showing the merging factors of p (Lemma 3), converting p into an RM form (Lemma 4) and transforming p in RM form to q in RM form by using R1 and R2 (Lemma 5).

Before giving formal proof, we first demonstrate how the two rules work. For example, two $\mathbb{Q}^+$-exponent-strings $p = a^3 b^{7.2} c^2 a^{2.5}$ and $q = a^2 b^{3.6} a^2 b^{3.6} a^{0.5} c^2 a$ over $\Sigma = \{a < b < c\}$ can be rewritten to each other by the following finite rewriting sequence from p to q (at each step, the underlined infix is being rewritten),

$$a^3 b^{7.2} \underline{c^2 a} a^{1.5} \xrightarrow{R1} a^2 \underline{a b^{7.2} a c^2} a^{1.5} \xrightarrow{R2} a^2 b^{3.6} a^2 b^{3.6} \underline{c^2 a^{0.5}} a \xrightarrow{R1} a^2 b^{3.6} a^2 b^{3.6} a^{0.5} c^2 a.$$

We use $p \overset{*}{\leftrightarrow} q$ to denote that p and q can be obtained from each other by a rewriting sequence consisting of R1 and R2.

Lemma 1. *Given two $\mathbb{Q}^+$-exponent-strings p and q, if there exists a finite rewriting sequence consisting of R1 and R2 that transforms p to q, then p and q are EM-equivalent.*

We first examine the palindromic amicability relation in Proposition 5 and then prove the [if] direction of Theorem 1 for the binary case in Lemma 2.

Proposition 5. [Extended Palindromic Amicability] *For an ordered binary alphabet $\Sigma = \{a < b\}$, any palindrome $\mathbb{Q}^+$-exponent-string p can be rewritten into one of the forms $a^x, b^y, a^x b^y a^x$.*

Corollary 1. *For an ordered binary alphabet $\Sigma = \{a < b\}$, let A, B be two EM-equivalent $\mathbb{Q}^+$-exponent-strings that are palindromes. Then, $p_1 A p_2 \overset{*}{\leftrightarrow} p_1 B p_2$ for arbitrary $\mathbb{Q}^+$-exponent-strings p_1, p_2.*

Lemma 2. *For an ordered binary alphabet $\Sigma = \{a < b\}$ and a pair of $\mathbb{Q}^+$-exponent-strings p, q, $p \equiv_{EM} q$ implies $p \overset{*}{\leftrightarrow} q$.*

Proof (Sketch). M-equivalence over binary strings is characterized by using palindromicly amicable relation in Atanasiu et al. [4]. Instead of directly rewriting from p to q, we transform the two EM-equivalent $\mathbb{Q}^+$-exponent-strings into M-equivalent strings. Then, there exists a rewriting sequence between strings. If we transform the strings back to the $\mathbb{Q}^+$-exponent-strings, we get the rewriting sequence from p to q. □

We introduce a method for merging factors in p and q, which allows us to broaden the approach in Lemma 2 to an arbitrary size alphabet.

Lemma 3 (Merging Lemma). *Every $\mathbb{Q}^+$-exponent-string p over $\Sigma = \{a_1 < a_2 < \cdots < a_n\}$ containing an a_n-factor can be rewritten in the form of $v_1 a_n{}^q v_2$ such that v_1 and v_2 are $\mathbb{Q}^+$-exponent-strings without any a_n-factors and q is a rational number. In other words, we can merge all a_n-factors in p into a single a_n-factor by applying R1 and R2 finitely many times.*

Proof. We assume that p is in contraction form. If p has a single a_n-factor, then p itself is already in the desired form. Now consider the case when p has two or more a_n-factors. We say that two a_n-factors in p without any other a_n-factors between them are *near*. For example, given $p = u a_n^x v a_n^y w$, where u, v and w are $\mathbb{Q}^+$-exponent-strings and v does not contain a_n-factors, we say that a_n^x and a_n^y are near.

If we can merge these two near a_n-factors, we can repeat the merging of two near a_n-factors until we have only one a_n-factor. Thus, it is enough to show that we can convert an infix $a_n^x v a_n^y$ into $v_1 a_n^{x+y} v_2$, where v, v_1 and v_2 are $\mathbb{Q}^+$-exponent-strings without a_n-factors.

For the proof, we use induction on n—the alphabet size. For the base case, if $n = 2$, the $\mathbb{Q}^+$-exponent-string v is just a single a_1-factor. Therefore, by applying Lemma 2, we can rewrite $a_n^x v a_n^y$:

$$a_2^x \, a_1^z \, a_2^y \overset{*}{\leftrightarrow} a_1^{\frac{y}{x+y}z} \, a_2^{x+y} \, a_1^{\frac{x}{x+y}z}$$

One can easily verify that $a_2^x a_1^z a_2^y$ and $a_1^{\frac{y}{x+y}z} a_2^{x+y} a_1^{\frac{x}{x+y}z}$ are EM-equivalent. For the induction step, we assume the claim holds for $\mathbb{Q}^+$-exponent-strings over $|\Sigma| = n < k$. Applying the inductive hypothesis to v gives:

$$a_k^x \, v \, a_k^y \overset{*}{\leftrightarrow} a_k^x \, p_1' \, a_{k-1}^z \, p_2' \, a_k^y,$$

where p_1', p_2' are $\mathbb{Q}^+$-exponent-strings without any a_{k-1}-factors and a_k-factors. By applying $R1$,

$$a_k^x \, p_1' \, a_{k-1}^z \, p_2' \, a_k^y \overset{*}{\leftrightarrow} p_1' \, a_k^x \, a_{k-1}^z \, a_k^y \, p_2'.$$

Then, by Lemma 2, we have

$$p_1' \, a_k^x \, a_{k-1}^z \, a_k^y \, p_2' \overset{*}{\leftrightarrow} p_1' \, a_{k-1}^{\frac{y}{x+y}z} \, a_k^{x+y} \, a_{k-1}^{\frac{x}{x+y}z} \, p_2'.$$

□

Next, we consider a special form of $\mathbb{Q}^+$-exponent-string obtained by repeatedly applying the merging procedure in Lemma 3. For instance, $\mathbb{Q}^+$-exponent-string $babc^2bcab$ over $\Sigma = \{a < b < c\}$ can be rewritten as follows:

$$\underline{babc^2bcab} \xleftarrow{\;\text{merge } c\;} babb^{\frac{1}{3}}c^3 b^{\frac{2}{3}} \underline{ab} \xleftarrow{\;\text{merge } b\;} a^{\frac{4}{7}}b^{\frac{3}{3}}a^{\frac{3}{7}}c^3 a^{\frac{3}{5}}b^{\frac{5}{3}}a^{\frac{2}{5}}.$$

The resulting $\mathbb{Q}^+$-exponent-string is in the 3-depth recursively merged form according to Definition 7.

Definition 7 (n-depth recursively merged form). *We say that a $\mathbb{Q}^+$-exponent-string p over $\Sigma = \{a_1 < a_2 < \cdots < a_N\}$ is in the n-depth recursively merged form (in short, n-RM form) if*

$$p = \begin{cases} a_1^{\square} \text{ or } \lambda, & \text{when } n = 1; \\ q\, a_n^{\square}\, r \text{ or } q, & \text{otherwise,} \end{cases}$$

where $\square$ denotes a positive rational number, q and r are $\mathbb{Q}^+$-exponent-strings in $(n-1)$-RM form, and $n \leq N$ are positive integers (See Fig. 2 for an example).

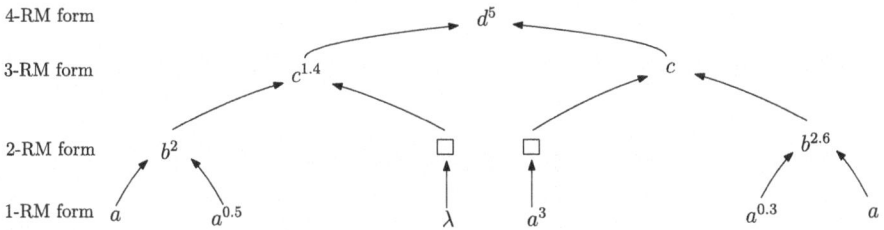

Fig. 2. A pictorial illustration of $\mathbb{Q}^+$-exponent-string $ab^2 a^{0.5} c^{1.4} d^5 a^3 ca^{0.3} b^{2.6} a$ over $\Sigma = \{a < b < c < d\}$; that is in the 4-RM form.

Lemma 4. *An arbitrary $\mathbb{Q}^+$-exponent-string over $\Sigma = \{a_1 < a_2 < \cdots < a_n\}$ can be rewritten in the n-RM form.*

Proof. We prove the statement by induction on n. For $n = 1$, it is immediate.

For the inductive step, let $n = k$ for an integer $k > 1$. Our inductive hypothesis is that a $\mathbb{Q}^+$-exponent-string on $\Sigma = \{a_1 < a_2 < \cdots < a_{k-1}\}$ can be rewritten in the $(k-1)$-RM form. Each $\mathbb{Q}^+$-exponent-string on $\Sigma = \{a_1 < a_2 < \cdots < a_k\}$ that does not contain a_k-factor can be rewritten in the $(k-1)$-RM form by the inductive hypothesis. By Definition 7, every $\mathbb{Q}^+$-exponent-string in the $(k-1)$-RM form is in the k-RM form.

If a $\mathbb{Q}^+$-exponent-string on $\Sigma = \{a_1 < a_2 < \cdots < a_k\}$ contains an a_k-factor, by Lemma 3, we know that it can be rewritten as $v_1 a_k^x v_2$ such that v_1, v_2 are $\mathbb{Q}^+$-exponent-strings without a_k-factors. Then, by the inductive hypothesis, v_1, v_2 can be rewritten as $\mathbb{Q}^+$-exponent-strings v_1', v_2', that are in the $(k-1)$-RM form. By Definition 7, $v_1' a_k^x v_2'$ is in the k-RM form. □

Notice that two EM-equivalent $\mathbb{Q}^+$-exponent-strings can be written differently in k-RM form. For instance, $abac^2aba$ and $aba^{0.5}c^2a^{1.5}ba$ in 3-RM form are EM-equivalent.

Lemma 5. *Two EM-equivalent $\mathbb{Q}^+$-exponent-strings p and q in the n-RM form over $\Sigma = \{a_1 < a_2 < \cdots < a_n\}$ can be rewritten to each other by R1 and R2.*

Proof (Sketch). We prove the statement by induction on n. Notice that $\mathbb{Q}^+$-exponent-strings in n-RM form have at most one a_n-factor, and thus we relocate a_n-factor to prefix, and apply the inductive hypothesis to the remaining suffix. For instance, consider two EM-equivalent $\mathbb{Q}^+$-exponent-strings over $\Sigma = \{a < b < c < d\}$, $p = a^2b^3ac^{\frac{1}{2}}abadabac^{\frac{1}{2}}aba$ and $q = a^2b^3ac^{\frac{1}{2}}a^2db^2a^2c^{\frac{1}{2}}aba$. Both p and q are in 4-RM form. By applying R1,

$$p \overset{*}{\leftrightarrow} (a^2b^3c^{\frac{1}{2}}d)aabaabac^{\frac{1}{2}}aba$$

and

$$q \overset{*}{\leftrightarrow} (a^2b^3c^{\frac{1}{2}}d)aa^2b^2a^2c^{\frac{1}{2}}aba.$$

Note that the rewriting sequences of p and q results in $\mathbb{Q}^+$-exponent-strings with the common prefix $a^2b^3c^{\frac{1}{2}}d$.

In general, since p and q are in n-RM form, we can rewrite

$$p \overset{*}{\leftrightarrow} (a_i^{\square}a_{i+1}^{\square}\cdots a_n^{\square})x$$

and

$$q \overset{*}{\leftrightarrow} (a_i^{\square}a_{i+1}^{\square}\cdots a_n^{\square})y,$$

where x and y are the remaining suffixes of the resulting $\mathbb{Q}^+$-exponent-strings from p and q that do not contain an a_n-factor. Then, from the Parikh matrix of p and q, we can compute the exact exponent of each $a_j^{\square}$ for $i \leq j \leq n$ by Proposition 4. This guarantees that these two $\mathbb{Q}^+$-exponent-strings have the common prefix. Now we can convert x and y in $(n-1)$-RM form by Lemma 4. Because the remaining suffixes x and y are EM-equivalent, there is a rewriting sequence between x and y by the inductive hypothesis.   □

Lemmas 4 and 5 prove the [if]-part of Theorem 1.

4   Strong M-Equivalence and Center of Mass

We say that two strings p and q are strongly M-equivalent ($p \overset{s}{\equiv} q$), if they are M-equivalent for any possible order of an alphabet [18]. In the spirit of extending the notation of strings to exponent-strings, we say that two $\mathbb{Q}^+$-exponent-strings p, q are strongly EM-equivalent ($p \overset{s}{\equiv}_{EM} q$) if they are EM-equivalent for any possible order of an alphabet. Proposition 6 shows that the characterization of strong EM-equivalence characterizes strong M-equivalence. The following statement is immediate from applying Proposition 3 for any alphabet order.

Proposition 6. *For two strings p and q, we have that $p \overset{s}{\equiv} q$ if and only if $p \overset{s}{\equiv}_{EM} q$.*

We present a new rewriting rule $SR1$, which is a modification of rewriting rules in Theorem 1, for the injectivity problem of strong EM-equivalence.

Theorem 2 (Rewriting Rules). *For an alphabet Σ and $q_1, q_2 \in \mathbb{Q}^+$, we can apply the following rewriting rule on an infix of a $\mathbb{Q}^+$-exponent-string p and obtain a strongly EM-equivalent $\mathbb{Q}^+$-exponent-string q.*

$SR1.$ $a^{q_1} b^{2q_2} a^{q_1} \leftrightarrow b^{q_2} a^{2q_1} b^{q_2}$ *for $a, b \in \Sigma$ and $q_1, q_2 \in \mathbb{Q}^+$.*

Proof. Let $a, b \in \Sigma$ and let us fix an order of the alphabet that is arbitrarily chosen. If a and b are not consecutive, then we have $a^{q_1} b^{2q_2} a^{q_1} \leftrightarrow b^{q_2} a^{q_1} b^{q_2} a^{q_1} \leftrightarrow b^{q_2} a^{2q_1} b^{q_2}$ by applying R1. On the other hand, if a and b are consecutive, then we have $a^{q_1} b^{2q_2} a^{q_1} \leftrightarrow b^{q_2} a^{2q_1} b^{q_2}$ by applying R2. Thus, these two guarantee that $SR1$ is a valid rewriting rule for strong EM-equivalence.   □

We next define the center of mass—an important property of $\mathbb{Q}^+$-exponent-strings—in Definition 8.

Definition 8. *For a $\mathbb{Q}^+$-exponent-string p on the alphabet $\Sigma = \{a_1, a_2, \cdots, a_n\}$ and $1 \leq i \leq n$, center of mass $\mathbb{CM}_{a_i}(p)$ of a_i in p is defined as follows:*

$$\mathbb{CM}_{a_i}(p) = \frac{1}{|p|_{a_i}} \int_0^{len(p)} x\delta(a_i, p[x]) dx,$$

where $\delta : \Sigma^2 \to \{0, 1\}$ is

$$\delta(\sigma_1, \sigma_2) = \begin{cases} 1, & \text{if } \sigma_1 = \sigma_2; \\ 0, & \text{otherwise.} \end{cases}$$

Informally, $\mathbb{CM}_{a_i}(p)$ represents the center of mass of physics by treating p as a 1-dimensional bar, where only the segments corresponding to a_i have mass. For example, $\mathbb{CM}_\mathbf{a}(\mathbf{b}^{0.5} \mathbf{c}^{0.5} \mathbf{a}^1 \mathbf{b}^{1.75} \mathbf{a}^{1.25} \mathbf{c}^1)$ is 3.09722 as illustrated with corresponding 1-dimensional bar in Fig. 3. Parts of $\mathbf{a}$ in the bar are colored darker than the other parts, that do not have mass.

Theorems 3 and 4 show the importance of the center of mass in characterizing strong EM-equivalence.

Theorem 3. *For any symbol in Σ, an application of $SR1$ to a $\mathbb{Q}^+$-exponent-string does not change its center of mass.*

Theorem 4. *Given two $\mathbb{Q}^+$-exponent-strings p and q over an alphabet Σ, if $p \overset{s}{\equiv}_{EM} q$, then $\mathbb{CM}_a(p) = \mathbb{CM}_a(q)$ for any symbol $a \in \Sigma$.*

Theorem 4 says that any rewriting rules for strongly EM-equivalent $\mathbb{Q}^+$-exponent-strings should preserve the center of mass. However, it remains open to verify whether or not $SR1$ fully characterizes strong EM-equivalence (and strong M-equivalence).

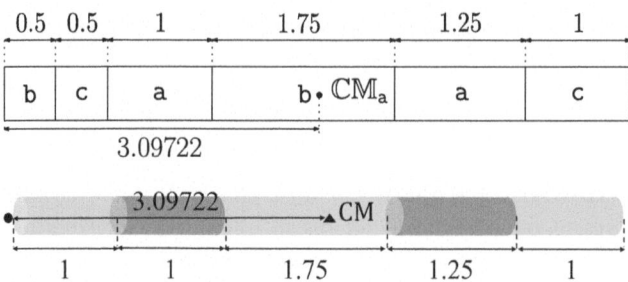

Fig. 3. Illustration of $\mathbb{CM}_a$ for $b^{0.5}c^{0.5}a^1b^{1.75}a^{1.25}c^1$. The point CM is the physical, center of mass of the 1-dimensional bar.

5   Conclusions

The Parikh matrix injectivity problem—characterization of M-equivalence of strings—has been open for decades. We have introduced universal rewriting rules applicable to $\mathbb{Q}^+$-exponent-string, an extension of strings developed by Baek et al. [6], and resolved the problem. Furthermore, we have studied strong M-equivalence and presented a rewriting rule along with the center of mass, which helps to characterize strong M-equivalence.

It remains open to develop an algorithm that constructs a rewriting sequence between two M-equivalent strings using R1 and R2. Another open problem is to design a rewriting rule that characterizes M-equivalence of any alphabet using traditional strings only.

Acknowledgements. Baek, Hahn and Han were supported by the NRF grant (RS-2023-00208094) and the AI Graduate School Program (No. RS-2020-II201361) funded by the Korean government (MSIT), and Salomaa was supported by the Natural Sciences and Engineering Research Council of Canada (NSERC). The first two authors contributed equally to this work.

References

1. Atanasiu, A.: Binary amiable words. Int. J. Found. Comput. Sci. **18**(2), 387–400 (2007)
2. Atanasiu, A.: Parikh matrix mapping and amiability over a ternary alphabet. In: Discrete Mathematics and Computer Science. In Memoriam Alexandru Mateescu (1952–2005), pp. 1–12 (2014)
3. Atanasiu, A., Atanasiu, R., Petre, I.: Parikh matrices and amiable words. Theoret. Comput. Sci. **390**(1), 102–109 (2008)
4. Atanasiu, A., Martín-Vide, C., Mateescu, A.: On the injectivity of the Parikh matrix mapping. Fund. Inf. **49**(4), 289–299 (2002)
5. Atanasiu, A., Poovanandran, G., Teh, W.C.: Parikh matrices for powers of words. Acta Informatica **56**(6), 521–535 (2019)
6. Baek, I., Hahn, J., Han, Y.: Exponent-strings and their edit distance (2024)

7. Bera, S., Ceterchi, R., Mahalingam, K., Subramanian, K.G.: Parikh q-matrices and q-ambiguous words. Int. J. Found. Comput. Sci. **31**(1), 23–36 (2020)
8. Brandenburg, F.: Uniformly growing k-th power-free homomorphisms. Theoret. Comput. Sci. **23**, 69–82 (1983)
9. Chern, Z.J., Subramanian, K.G., Ahmad, A., Teh, W.C.: A new study of Parikh matrices restricted to terms. Int. J. Found. Comput. Sci. **31**(5), 621–638 (2020)
10. Şerbănuţă, V.N., Şerbănuţă, T.F.: Injectivity of the Parikh matrix mappings revisited. Fund. Inf. **73**(1–2), 265–283 (2006)
11. Dick, J., Hutchinson, L.K., Mercas, R., Reidenbach, D.: Reducing the ambiguity of Parikh matrices. Theoret. Comput. Sci. **860**, 23–40 (2021)
12. Hahn, J., Cheon, H., Han, Y.: M-equivalence of Parikh matrix over a ternary alphabet. In: Proceedings of the 27th Implementation and Application of Automata, pp. 141–152 (2023)
13. Hutchinson, L.K., Mercas, R., Reidenbach, D.: A toolkit for Parikh matrices. In: Proceedings of the 26th Implementation and Application of Automata, pp. 116–127 (2022)
14. Mateescu, A., Salomaa, A., Salomaa, K., Yu, S.: A sharpening of the Parikh mapping. RAIRO Informatique Theorique et Applications **35**(6), 551–564 (2001)
15. Parikh, R.: On context-free languages. J. ACM **13**(4), 570–581 (1966)
16. Poovanandran, G., Teh, W.C.: Strong $(2 \cdot t)$ and strong $(3 \cdot t)$ transformations for strong M-equivalence. Int. J. Found. Comput. Sci. **30**(5), 719–733 (2019)
17. Salomaa, A.: On the injectivity of Parikh matrix mappings. Fund. Inf. **64**(1–4), 391–404 (2005)
18. Teh, W.C.: Parikh matrices and strong M-equivalence. Int. J. Found. Comput. Sci. **27**(5), 545–556 (2016)
19. Waugh, F.V., Abel, M.E.: On fractional powers of a matrix. J. Am. Stat. Assoc. **62**(319), 1018–1021 (1967)

O_2 is a Multiple Context-Free Grammar: An Implementation-, Formalisation-Friendly Proof

Marco B. Caminati[✉][iD]

School of Computing and Communications, Lancaster University in Leipzig,
Nikolaistrasse 10, 04109 Leipzig, Germany
m.caminati@lancaster.ac.uk

Abstract. Classifying formal languages according to the expressiveness of grammars able to generate them is a fundamental problem in computational linguistics and, therefore, in the theory of computation. Furthermore, such kind of analysis can give insight into the classification of abstract algebraic structure such as groups, for example through the correspondence given by the word problem. While many such classification problems remain open, others have been settled. Recently, it was proved that n-balanced languages (i.e., whose strings contain the same number of occurrences of letters a_i and A_i with $1 \leq i \leq n$) can be generated by multiple context-free grammars (MCFGs), which are one of the several slight extensions of context free grammars added to the classical Chomsky hierarchy to make the mentioned classification more precise. This paper analyses the existing proofs from the computational and the proof-theoretical point of views, systematically studying whether each proof can lead to a verified (i.e., checked by a proof assistant) algorithm parsing balanced languages via MCFGs. We conclude that none of the existing proofs is realistically suitable against this practical goal, and proceed to provide a radically new, elementary, extremely short proof for the crucial case $n \leq 2$. A comparative analysis with respect to the existing proofs is finally performed to justify why the proposed proof is a substantial step towards concretely obtaining a verified parsing algorithm for O_2.

1 Introduction

The classical connection between formal grammars, formal languages and abstract machines has been extensively and fruitfully explored. This led to an enhanced understanding of the capabilities and limitations of given models of languages and computability. Firstly, by establishing links between the traditional models (e.g., regular grammars and finite state machines); subsequently, by introducing new models, usually for domain-specific reasons (e.g., computational linguists introducing new notions of formal grammars to better model natural languages) and studying how these new models translate. This already fertile interplay has been made even more relevant, in later years, by noting how staple problems in computational abstract algebra (for example, combinatorial group theory, which in turn is a useful tool in geometry and topology) can be rephrased and solved using the definitions, ideas and methods of formal languages theory and computability theory [7].

© The Author(s), under exclusive license to Springer Nature Switzerland AG 2024
J. D. Day and F. Manea (Eds.): DLT 2024, LNCS 14791, pp. 82–97, 2024.
https://doi.org/10.1007/978-3-031-66159-4_7

Therefore, the typical computational linguistics problem of establishing how power-ful a formal grammar must be in order to generate and parse a given family of languages becomes relevant also in computability theory and computational abstract algebra. One widely studied example is given by the family of n-balanced languages, containing exactly the words with equal pairwise occurrences of n pairs of letters:

$$O_n := \left\{ w \in \left(\bigcup_{i=1}^{n} \{a_i, \overline{a}_i\} \right)^* . \ \forall i \leq n. \ |w|_{a_i} = |w|_{\overline{a}_i} \right\}. \tag{1}$$

Above, $|v|_x$ denotes the number of occurrences of the letter x in a word v, and the alpha-bet Σ_n is made up of the mutually distinct letters a_i's and $\overline{a}_i$'s. For a given i, we will say that $\overline{a}_i$ is the *conjugate* letter of a_i and vice versa. The general notation $\overline{x}$ will indi-cate the conjugate letter of any given $x \in \Sigma_n$. Another way to characterise a balanced language is via the introduction of the ancillary concept of *balance*: given a string w of Σ_n, its balance $\mu(w)$ is the n-tuple of integers $(|w|_{a_1} - |w|_{\overline{a_1}}, \ldots, |w|_{a_n} - |w|_{\overline{a_n}})$, making a string balanced if and only if all the entries of its balance are 0.

From the point of view of computational linguistics, it is interesting to locate the O_n family of languages within the known hierarchy of grammars proposed to model, e.g., natural languages. From the point of view of computational group theory, information about formal languages can be used to explore corresponding groups [7]. This led to a series of recent papers by several authors culminating with a result establishing that O_n can be generated by an n-multiple context-free grammar (n-MCFG, see below for a definition) for any n. All the existing proofs (e.g., [6,9,12]) develop and improve on the original argument of Salvati [13], being all essentially geometrical in nature, in stark contrast with the one given here. While this geometric idea turned out to be powerful, it also has a couple of disadvantages: first, all its variations ultimately rely on key results which offer little indications as to which rules are used at every parsing step; secondly, as we will see, they would all likely turn out to be extremely hard to formalise in a proof assistant to check their validity. This paper focuses on the computational and proof-theoretical aspects of that result, proposing a novel proof which has the advantage of being constructive and based on elementary ideas which are, in comparison, much more viable to a formalisation, and is the first step in a project aiming at delivering formally verified code which can parse O_2 words. The price to pay for these upsides is that the new proof is limited to the bidimensional $n = 2$ case which, however, is a crucial case, having originated the series of papers mentioned above, and having significant interest on its own, both from a formal languages and from a computational group theory point of view given the relationship between O_2 and the open problems related to the word problem for $\mathbb{Z}^2$ [8]: in particular, the problem of O_2 being or not being an indexed language, which is still open. Section 2 introduces the problem and the needed definitions. Section 3 gives the basic idea behind the new proof and the remarkably short proof itself. In Sect. 4, we will systematically examine the existing proofs and their drawbacks in terms of implementability and formalisability. Section 5 concludes.

Note 1 (Notation). In what follows, we will use the terms "tuple", "list" interchange-ably; the terms "string" and "word" will refer to the same notion as "tuple" and "list", but will usually be adopted when working within the context of formal languages. ℕ will denote the set of natural numbers including 0. $|p|$ or length (p) is the length of the list p. Sometimes we will regard a non-empty list as a function over an initial segment of N (indices), thereby writing $p\,(i-1)$ for the i-th entry (or the entry having *index i*) of p, so that $p = p\,(0)\,p\,(1)\ldots p\,(|p|-1)$. The notation $\underline{p}$ will indicate the reverse of a given string p: i.e., $\underline{p} = p\,(|p|-1)\ldots p\,(0)$, and $\underline{p}\,(0)$ is the last letter of p when p is non-empty. Finally, ε will denote the empty list.

2   The Context

By the eighties, a general consensus formed around the idea that filling the gap between context-free grammars (CFGs) and context-sensitive grammars (CSGs) in the classical Chomsky hierarchy could have helped to more adequately model natural languages. Under the loose term of *mildly context-sensitive* grammars, several models were gradually conceived to add some expressiveness to context free grammars, to add some of the power of context-sensitive grammars in a somehow controlled way [10]. As elaborated in Sect. 1, this created new interesting problems which have been studied across disciplines, with one main problem among these being the generability of particular languages given a particular mildly context-sensitive grammar. In Sect. 1, we also anticipated which one of these (now solved) problems we are focusing on from a computability and proof-theoretical point of view: that of the generability of O_n languages using the mildly context-sensitive grammar n-MCFGs. With the easy definition of O_n already given, we now focus on introducing MCFGs.

One way of looking at non-terminals in CFGs is as boolean functions of one string of terminals. In a parse tree, each non-terminal is the root of a subtree whose leaves form a substring of the final phrase. This can happen only for well-formed substrings (e.g., a valid noun phrase), and not for others; therefore, we can view at a given non-terminal as a function returning true exactly when this happens.

We also note that we can adopt this view of non-terminals as functions because the corresponding truth value does not depend on what is outside the given substring (this is given by context-freeness). Moreover, each substring matching a given non-terminal as in the example just given must be a contiguous substring of the final sentence. This latter phenomenon corresponds to the fact that, in the view above of non-terminals as functions, these functions only take one argument [4], and imposes substantial limitations on what a CFG can do. To work around this limitedness, the main idea behind MCFGs is to allow such functions to take a finite number n of arguments; then, n is called the *fanout*[1] of the corresponding non-terminal. This strictly increases the expressiveness of the grammar, but without yielding the same expressiveness as CSGs [10]. In particular, MCFGs allow to keep track of phrases with "gaps" in them [4] when parsing sentences. Now, on top of this main idea, several details are to be sorted out: for example, how do non-terminals interact with each other, now that they all can have several arguments? And how does this issue relate to the notion of transitively applying rules to obtain derivations that we had in other grammars? The following formal definition tackles these details. As with, say, CFGs, it involves terminals, non-terminals, rules and

---

[1] Some sources adopt the term "dimension" in lieu of "fanout".

a starting symbol. Besides the discussed difference in the nature of non-terminals, the other main consequent difference is in the structure of the rules.

Definition 1 (MCFGs). *A* multiple context-free grammar *(MCFG) is given by four mutually disjoint, non-empty, finite sets* Σ *(the set of* letters *or* terminals*),* N*,* X *(the set of* variables*),* R*, by one distinct element* $S \in N$*, and by two maps* f *and* ρ *associating a natural number to, respectively, each* non-terminal *of* N *and to each* rule *of* R*. The round brackets "()" arrow "$\rightarrow$" colon ":" and comma "," symbols are not included in any of the sets above, as are neither* f *nor* ρ*. Each rule* r *of* R *has the form*

$$r : A\left(s_1, \ldots, s_{f(A)}\right) \rightarrow B_1\left(x_{1,1}, \ldots, x_{1,f(B_1)}\right), \ldots, B_{\rho(r)}\left(x_{\rho(r),1}, \ldots, x_{\rho(r),f\left(B_{\rho(r)}\right)}\right), \tag{2}$$

where A *and the* B_i*'s denote generic elements of* N*, and the following constraints hold: 1. all the* $x_{i,j} \in X$ *occurring in each rule are pairwise distinct; the set they constitute is denoted by* X_r*; 2.* $s_1 \ldots s_{f(A)} \in (X_r \cup \Sigma)^*$*; 3. each element of* X_r *occurs exactly once in* $s_1 \ldots s_{f(A)}$*; 4.* $f(S) = 1$*.* f *is called the* fanout *function and* ρ *the* rank *function.* $\max_N f$ *is the* fanout *of the MCFG, and for any* $n \geq \max_N f$*, the grammar is said to be an* n*-MCFG.* Σ *is called the* alphabet *of the MCFG,* X *its set of* variables*, and* S *its* start symbol*.*

Equation (2) is the key ingredient of Definition 1: it allows to interleave into the arguments $s_1, \ldots, s_{f(A)}$ the substrings held by the various x's in the right-hand side, obtained by previous generative steps by using other non-terminals $B_1, \ldots, B_{\rho(r)}$. This implies that, as usual, the language is generated by finitely iterating rule applications until non-terminals (except S) are gone. This is made precise by the following definition.

Definition 2 (Instance, sentential form, derivability). *Given a MCFG G as the tuple* $(\Sigma, N, X, R, S, f, \rho)$*, an* instance *of any rule* $r \in R$ *is the string obtained by replacing both occurrences of each variable in the rule with one string of* Σ^**. Recursively, a* sentential form *for G is either the empty string or the left-hand side of an instance of some* $r \in R$ *such that all the comma-separated elements on its right-hand side are sentential forms. A string* w *of* Σ^* *is* derivable *(or a* sentence*) in G if* $S(w)$ *is a sentential form. The subset of* Σ^* *constituted by all derivable strings in G is called the* language generated by G*, indicated with* $L(G)$*, and styled a* multiple context-free language *(MCFL).*

We note that, while for CFGs it is customary to generate strings by applying rules left to right, with MCFGs, as from Definition 2, string generation is customarily obtained by applying rules right to left. Consequently, to do the parsing of a string (as opposed to generating), one applies MCFG rules from left to right. Additionally, we note that the recursive definition of a sentential form provided by Definition 2 implies that the first step in generating strings must always begin from rules with an empty right hand side (base case).

We now introduce a specific MCFG, called G_2, for two reasons: first, it will act as a simple example to clarify the definitions above; secondly, we will use G_2 to generate the language O_2, thereby giving the proof central to this paper. This means that the alphabet Σ of G_2 coincides with the alphabet of the language O_2, which is obtained by substituting n with 2 in (1). For notational convenience, however, we will denote its letters with $a, \overline{a}, b, \overline{b}$ in lieu of $a_1, \overline{a_1}, a_2, \overline{a_2}$, so that O_2 becomes

$$O_2 := \left\{ w \in \{a, \overline{a}, b, \overline{b}\}^* . \ |w|_a = |w|_{\overline{a}}, \ |w|_b = |w|_{\overline{b}} \right\}. \tag{3}$$

Definition 3 (G_2). *Let G_2 be the MCFG with $\Sigma = \{a, \overline{a}, b, \overline{b}\}$, $N = \{S, I\}$, $X = \{v, w, x, y\}$, R containing the following rules*

$$
\begin{aligned}
&r_0 : I\,(\varepsilon, \varepsilon) \rightarrow && r_l : I\,(vxw, y) \rightarrow I\,(v, w)\,, I\,(x, y)\\
&r_a : I\,(a, \overline{a}) \rightarrow && r_r : I\,(v, xwy) \rightarrow I\,(v, w)\,, I\,(x, y)\\
&r_{\overline{a}} : I\,(\overline{a}, a) \rightarrow && r_n : I\,(vx, yw) \rightarrow I\,(v, w)\,, I\,(x, y)\\
&r_b : I\,(b, \overline{b}) \rightarrow && r_s : I\,(vx, wy) \rightarrow I\,(v, w)\,, I\,(x, y)\\
&r_{\overline{b}} : I\,(\overline{b}, b) \rightarrow && r_z : S\,(vw) \rightarrow I\,(v, w)
\end{aligned}
$$

with f and ρ as easily gatherable from the ruleset above.

The reader should note that for some rules of Definition 3 the right hand side is empty, reflecting the fact that the function ρ is 0 for those rules (e.g., r_a). We want to prove that *any* string in O_2 admits a derivation in G_2, or, equivalently, that $O_2 \subseteq L(G_2)$. This will prove the main result that $O_2 = L(G_2)$ since the other set inclusion is obvious. We start by noticing that, in general, there is not a unique derivation; in fact, we will prove a stronger result: for any pair of non-empty strings (v, w) such that $vw \in O_2$, $I(v, w)$ is a sentential form. This means that *however* we choose to split *any* non-empty string in O_2 into two proper factors, we can find a derivation where the second topmost node of the tree leads to the chosen split. While this result is stronger than what we need, it is liable to a straightforward structural recursion argument: it is easy to check that we will just need to prove that any $I(v, w)$ is the left hand of an instance of some rule of G_2, as soon as $vw \in O_2$ and $|v|, |w| > 0$. This is immediate for strings with length not exceeding two, so that we can restrict our attention to longer ones. We also note that the special rule S will only be used at the very final step of the string generation, with only I playing a non-trivial role. As a consequence, we will just drop the symbol I to declutter the notations in our proofs. With this simplified notation, the result we need to prove is the following main theorem, which is preceded by the definitions needed to state it.

Definition 4 (Factorisation, decomposition). *A* factorisation *of a string s is a finite list of strings $(s_0, \dots, s_{n-1})$ such that $s = s_0 \dots s_{n-1}$; each of its entries is called a* factor *of s; a* left factor *if it is the first entry; a* right factor *if it is the last; a* proper factor *if it is not empty ;* ultra-proper *if it is proper and distinct from s; the factorisation is (ultra-)proper if each of its entries is (ultra-)proper.*
Given a finite, non-empty list of non-empty strings $(s_0, \dots, s_{n-1})$, choose a proper factorisation for each s_i, and put all the resulting factors into a list ω, in some arbitrary order. If ω can be partitioned into two lists $(p_0, \dots, p_{l-1})$ and $(q_0, \dots, q_{m-1})$ with $0 < l, m \leq n$ and $|p_0 \dots p_{l-1}|, |q_0 \dots q_{m-1}| < |s_0 \dots s_{n-1}|$, we will say that

$((p_0, \ldots, p_{l-1}), (q_0, \ldots, q_{m-1}))$ *is a* decomposition *of* $(s_0, \ldots, s_{n-1})$. *In this case, we will call each list* $(p_0, \ldots, p_{l-1})$ *and* $(q_0, \ldots, q_{m-1})$ a component *of the decomposition.*

A factorisation of a balanced word will also be called balanced. Finally, a decomposition the components of which are all balanced will be called balanced or a bal-decomposition.

Theorem 1. *Any proper, binary factorisation of a string of O_2 of length bigger than 2 admits a balanced decomposition. In other words, the set $\{(z_0, z_1) . z_0 z_1 \in O_2, |z_0 z_1| > 2, \min\{|z_0|, |z_1|\} > 0, (z_0, z_1)$ admits no balanced decomposition$\}$ is empty.*

In Theorem 1, the notion of balanced decomposibility reproduces that of derivability, but without any reference to a particular rule, in accordance to our decluttered notation introduced above:

Note that Definition 4 imposes that, to be decomposed, a list of strings must be a proper factorisation of some string. Additionally, it imposes that any of its components is also a proper factorisation of some string, with the additional condition of its length being bigger than 0 and at most the length of the original list. Finally, observe that the operation of factorisation applies to a string and returns a list of strings, while that of decomposition applies to a list of strings and return a list of lists of strings. For O_2, we will be particularly interested to the case in which the n occurring in Definition 4 is 1.

Note 2 (Notation). Occasionally, it will be convenient to denote a factor of a given non-empty string q using the correspondent set of indices, by writing $\langle q \rangle_i^j$ (or, alternatively, $\langle q \rangle_{[i,j]}$) to mean the factor $\Pi_{k=i}^{\min\{j,|q|-1\}}(k)$ for $i, j \in \mathbb{N}$. Note that this yields the empty string for some values of i, j, for example when $j < i$. We will just write $\langle q \rangle^j$ in lieu of $\langle q \rangle_0^j$, and $\langle q \rangle_i$ instead of $\langle q \rangle_i^{|q|}$. Additionally, we will use the notation $p - X$, where $X \subseteq \mathbb{N}$, to indicate the string obtained by removing all letters having indices in X: for example, $ab\bar{b}b\bar{a} - \{2, 4\} = abb$. Finally, the notation $p \preceq q$ will mean that p is a left factor of q.

The following propositions and corollaries establish the sufficiency of Theorem 1 to obtain the main result that $L(G_2) = O_2$ making thus O_2 a MCFL. Therefore, the rest of the paper, starting with next section, will be devoted to proving Theorem 1.

Proposition 1. *Given a factorisation (p, q) of a string of O_2, $I(p, q)$ is a sentential form for the grammar G_2 of Definition 3.*

Proof. The case $|pq| \leq 2$ is straightforward. Let us then consider a pair (p, q) violating the thesis and such that $|pq| \geq 4$ is the minimal possible.

1. Case $\{p, q\} \cap O_2 = \emptyset$: By Theorem 1, let us consider a bal-decomposition of (p, q) of components s and t. The length of either s or t must be 2 since p and q are not balanced. By symmetry of G_2, we can assume $|s| = 2$, so that $s = (s_0, s_1)$ and $I(s_0, s_1)$ must be a sentential form by minimality. If $|t| = 1$, so that $t = (t_0)$, it must be $p = s_0 t_0$ and $q = s_1$, with $I(t_0, \varepsilon)$ a sentential form, so that we could obtain

that $I(s_0t_0, s_1\varepsilon)$ is also a sentential form by, e.g., rule r_s. Hence, we can assume $|s| = 2 = |t|$, and that $I(s_0, s_1)$, $I(t_0, t_1)$, $I(s_0s_1, \varepsilon)$, $I(t_0t_1, \varepsilon)$ are all sentential forms. There are the following sub-cases, all leading to $I(p, q)$ by applying the respective rule of G_2:

$p = s_0s_1t_0$: rule r_s to $I(s_0s_1, \varepsilon)$ and $I(t_0, t_1)$.
$p = s_0, q = s_1t_0t_1$: rule r_s to $I(s_0, s_1)$ and $I(\varepsilon, t_0t_1)$.
$p = s_0, q = t_0s_1t_1$: rule r_r to $I(s_0, s_1)$ and $I(t_0, t_1)$.
$p = s_0t_0s_1$: rule r_l to $I(s_0, s_1)$ and $I(t_0, t_1)$.
$p = s_0t_0, q = t_1s_1$: rule r_n to $I(s_0, s_1)$ and $I(t_0, t_1)$.
$p = s_0t_0, q = s_1t_1$: rule r_s to $I(s_0, s_1)$ and $I(t_0, t_1)$.

2. Case $\{p, q\} \subseteq O_2 \setminus \{\varepsilon\}$: then $I(p, \varepsilon)$ and $I(\varepsilon, q)$ are both sentential forms by minimality, so that we can just apply rule r_n to obtain that $I(p, q)$ is a sentential form.
3. Case $\varepsilon \in \{p, q\}$: if $q = \varepsilon$, then $I(p(0), \langle p\rangle_1)$ is a sentential form by case (1), as is $I(\varepsilon, \varepsilon)$ using r_0. Hence $I(p(0)\varepsilon\langle p\rangle_1, \varepsilon)$ is also a sentential form by rule r_l. Similarly if $p = \varepsilon$.

Corollary 1 (of Proposition 1). $L(G_2) = O_2$.

Proof. It suffices to show that $O_2 \subseteq L(G_2)$. Any non-empty string s of O_2 is in $L(G_2)$ since that $I(s_0, \langle s\rangle_1)$ is a sentential form by Proposition 1 and by applying rule r_z to $I(s_0, \langle s\rangle_1)$.

3   The New Proof

Definition 5. *Consider the equivalence relation $\equiv_n$ (or just $\equiv$ when clear) induced over Σ_n^* by the balance map μ introduced in Sect. 1: that is, Σ_n^* is partitioned into classes of strings whereby two strings are in the same class if and only if they have the same balance. Each class will have a subset of strings of minimal length, which we call short. In other words, a short string is one not having two letters from the same pair occurring: if an a_i occurs in it, then $\overline{a_i}$ does not occur, and vice versa. A list made up entirely of short strings will also be called short.*

Now, going back to the case $n = 2$, we start from the restriction of Theorem 1 to the short case (recall from Note 1 that the notation $\underline{x}$ denotes the reversion of the list x, and, in particular, $\underline{x}(0)$ is the last entry of x if the latter is non-empty):

Lemma 1. *Let (x, y) be a proper, short factorisation of a string of O_2 of length > 2. Then, either $x(0) = y(0)$, or $\underline{x}(0) = \underline{y}(0)$, or there are p and q ultra-proper left factors of x and y, respectively, such that $pq \in O_2$.*

Proof. Assume the lemma does not hold, and consider among the counterexamples a particular (x, y) such that $|xy|$ is minimal. By shortness and without loss of generality, we can then assume $x = a\alpha x'$ and $y = \overline{b}^{1+l}\overline{a}y'\overline{b}$ for some $l \in \mathbb{N}$, with $\alpha \in \{a, b\}$, $x' \in \{a, b\}^*$ and $y' \in \{\overline{a}, \overline{b}\}^*$. Now if α is a, the thesis holds for $\left(\alpha x', \overline{b}^{1+l}y'\overline{b}\right)$ by minimality, implying that there must be ultra-proper left factors p and q of ax' and

$\overline{b}^{1+l}y'\overline{b}$ respectively such that pq is balanced. But then it would be $|q| > 1 + l$ in order to compensate the first letter (a) of p, so that we can easily obtain a decomposition of (x, y), contradicting the assumption (made at the beginning of the proof) that (x, y) does not obey the thesis of the lemma.

Therefore, it must be $\alpha = b$ and $l = m + 1$ for some $m \in \mathbb{N}$. Then, there must be two ultra-proper left factors p and q of ax' and of $\overline{b}^{m+1}\overline{a}y'\overline{b}$ respectively such that pq is balanced. Again, then $|q| > m + 1$ and $|p| > 1$, so that we can easily extend p and q to ultra-proper left factors of x and y, respectively, again contradicting the very first assumption of this proof.

Corollary 2. *Any proper, short, binary factorisation of a string of O_2 of length bigger than 2 admits a balanced decomposition.*

Proof. Lemma 1 asserts that it is possible to find such a decomposition of a very special form (equivalent to using either r_s or a restricted form of r_n). Hence, the corollary follows immediately.

Now that we settled the problem for the short case with Corollary 2, we can use it as a base case for the, general, non-short case. The idea is to reduce the latter to the former by eliminating pairs of conjugate letters. The following definition gives us a way to do this in a suitable, gradual manner.

Definition 6. *A minimal non-short factor of a string $p \in \Sigma_n^*$ will be called a bump of p, and will be indicated through its indices. More formally, $[i, j]$ is an x-bump (or just bump) for p if $\langle p \rangle_i^j$ is not short, all its ultra-proper factors are short, and $p(i) = x$. Note that this implies that $p(j) = \overline{x}$, and that (since all the ultra-proper factors of a bump are short) $p(k) \notin \{x, \overline{x}\} \, \forall k \in \,]i, j[$. x is called the direction of the bump. $B^x(p)$ is the set of all x-bumps of p, and we set $B^X(p) := \bigcup_{x \in X} B^x(p)$, $B(p) := \bigcup_{x \in \Sigma_n} B^x(p)$. If $[i, j]$ is a bump for p, the string $\langle p \rangle_i^j$ will also be called a bump for p.*

Note 3. p is short if and only if $B(p) = \emptyset$. Also note that, in the case $n = 2$, any bump has the form $\alpha \beta^m \overline{\alpha}$ for some $m \in \mathbb{N}$, with α and β non-conjugated letters of Σ_2.

3.1 The Non-short Case: Intuition

We will now suppose there is a balanced but not bal-decomposable factorisation of a string of O_2 longer than 2 (that is, violating Theorem 1), so that we can pick one such factorisation having a concatenation of minimal length. We first note that, by minimality, nowhere in this factorisation a bump of length 2 (that is, of the form $\alpha \overline{\alpha}$) can occur: this is easy to see and will be later formally implied by Proposition 2. Secondly, at least one of the two component strings (let us say the first one) will have a bump by Corollary 2 and Note 3, so that we can pick one of minimal length and cancel its first and last letters, thus preserving balancedness of their concatenation. By minimality, the "canceled" factorisation is now bal-decomposable, so that we have the situation depicted below (where the parentheses enclose each component of the factorisation):

$$(p_0 \acute{d}b \ldots b \ldots b \grave{a} p_1 \, p_2)(p_3).$$

$\underbrace{}_{A} \quad \underbrace{}_{B}$

We have used under-brackets to mark one of the two components of a bal-decomposition of the "canceled" factorisation. This means that the portion marked by the brackets is balanced, as is the unmarked one. Note that one bracket (here indicated with A) must straddle the canceled bump (otherwise we could reinstate the two canceled letters and still have a bal-decomposition), while the other (B) must not (otherwise we could again reinstate the two canceled letters and still have a bal-decomposition). The other bracket (B) could also mark a left or right factor of p_3 depending on the particular bal-decomposition: this would not change the following reasoning. We also assumed, without loss of generality, that the direction of the bump is a (see Definition 6).

Now, the first letter of p_2 cannot be a, otherwise we could reinstate the canceled letters and obtain a bal-decomposition. Similarly, the last letter of p_1 cannot be $\bar{a}$. There are two cases: the last letter of p_1 is a or b (the case $\bar{b}$ is similar to the latter). In the first case, the first letter of p_2 must be in $\{b, \bar{b}\}$, let us say it is b (again, the case $\bar{b}$ is similar):

$$(p_0 \overset{\diagup}{d} b \ldots b \ldots b \overset{\diagup}{d} p_1' a bb \ldots p_2')(p_3)$$

$$\underbrace{\phantom{(p_0 \overset{\diagup}{d} b \ldots b \ldots b}}_{A} \underbrace{\phantom{\overset{\diagup}{d} p_1' a bb \ldots p_2'}}_{B}$$

In this configuration, we can begin "sliding" the upwards edge of A and the upwards edge of B rightwards one letter at a time while preserving balancedness. If the upwards edge of A reaches the canceled $\bar{a}$, we can, again, reinstate the canceled letters. Otherwise, after several rightwards sliding steps, we will eventually find ourselves (supposing B has not shortened to length 0) in a configuration like this:

$$(p_0 \overset{\diagup}{d} b \ldots bbb \ldots b \ldots b \overset{\diagup}{d} p_1' abb \ldots b \, \bar{a} p_2'')(p_3).$$

$$\underbrace{\phantom{(p_0 \overset{\diagup}{d} b \ldots bbb \ldots b \ldots b}}_{A} \underbrace{\phantom{\overset{\diagup}{d} p_1' abb \ldots b \, \bar{a} p_2''}}_{B}$$

Note that the leftmost letter under B cannot be b (otherwise we could have kept sliding), cannot be a (otherwise we could reinstate the canceled letters and obtain a bal-decomposition), and cannot be $\bar{b}$ because, as previously observed, there cannot be bumps of length 2: hence it must be $\bar{a}$, as depicted above. But this means that the bump on the right in the figure above is of length strictly smaller than the canceled bump, which is impossible because we chose one of minimal length.

The case where the last letter of p_1 is b entails a similar sliding reasoning, and is left to the reader. Please note that this informal explanation leaves a number of corner cases unexplored which will be formally addressed in the forthcoming lemmas and propositions. For instance, in our example, the reasoning becomes problematic when the canceled bump touches one or both extrema of the factor (that is, the round brackets in the figures above): this is addressed in the last proof of this paper. Or, during the "sliding", the bracket B could collapse to a length of 0, which is addressed in Proposition 2. The discussion above was chiefly added to help the reader understanding the remaining formal results below.

3.2   The Non-short Case: Formal Proofs

We start from a result valid not only in the case $n = 2$, but in general. It states that balanced-decomposability is preserved upon replacing factors of length smaller than

2 with equivalent ($\equiv_n$ as from Definition 5) ones. In particular, this implies that a bal-indecomposable factorisation $(p_0, \ldots, p_n)$ cannot contain bumps of length 2 or 3 unless there is another bal-indecomposable factorisation whose concatenation has length smaller than that of $p_0 \ldots p_n$; a fact that we have already used in our informal "sliding" reasoning above. It will be used repeatedly in the subsequent proofs.

Proposition 2. *Assume* $(s_0, \ldots, s_{n-1})$ *is balanced-decomposable in* Σ_n^*, *and that* $s \equiv \langle s_0 \rangle_i^{i-1+\delta}$ *with* $\delta \in \{0, 1\}$.
Then $\left(\langle s_0 \rangle^{i-\delta} s \langle s_0 \rangle_{i+1}, \ldots, s_{n-1} \right)$ *is also balanced-decomposable.*

Proof. We can assume that the entry of index i of s_0 is the entry of index j of p_0, where $(p_0, \ldots, p_{l-1})$ is a component of a balanced decomposition of $(s_0, \ldots, s_{n-1})$. Then $p_0 \equiv p_0' := \langle p_0 \rangle^{j-\delta} s \langle p_0 \rangle_{j+1}$, so that $(p_0', \ldots, p_{l-1}), (q_0, \ldots, q_{m-1})$ is a decomposition of $\left(\langle s_0 \rangle^{i-\delta} s \langle s_0 \rangle_{i+1}, \ldots, s_{n-1} \right)$.

To illustrate Proposition 2, let us consider the example $l := \left(bcaabb, \overline{bbba}cba, \overline{ab}\overline{a} \right)$. It is bal-decomposable: $\left(aa, \overline{b}, \overline{ab}\overline{a} \right)$ and $\left(bc, bb, \overline{bba}cba \right)$ are two balanced components. Proposition 2 says that we can replace any letter in l with a string equivalent to that letter (case $\delta = 1$) without losing the bal-decomposable property. So let us replace the last b in l with $\overline{c}db\overline{d}\overline{b}dcb\overline{d} \equiv b$: the triple $l' := \left(bcaabb, \overline{bbba}cba, \overline{ac}db\overline{d}\overline{b}dcb\overline{d}\overline{a} \right)$ is still bal-decomposable, as witnessed by the components $\left(aa, \overline{b}, \overline{ac}db\overline{d}\overline{b}dcb\overline{a} \right)$ and $\left(bc, bb, \overline{bba}cba \right)$. For the case $\delta = 0$, Proposition 2 says that we can inject in l any string equivalent to ε without losing the bal-decomposable property. Taking again l as above, we can inject the string $\overline{c}db\overline{d}\overline{b}\overline{b}dcb\overline{d} \equiv \varepsilon$ at any position in l to obtain, for example $l'' := \left(bca\overline{c}db\overline{d}\overline{b}\overline{b}dcb\overline{d}abb, \overline{bbba}cba, \overline{ab}\overline{a} \right)$ which is still bal-decomposable via $\left(a\overline{c}db\overline{d}\overline{b}\overline{b}dcb\overline{d}a, \overline{b}, \overline{ab}\overline{a} \right)$ and $\left(bc, bb, \overline{bba}cba \right)$.

Corollary 3. *Assume that non-empty, the proper factorisation* $(s_0, \ldots, s_{n-1})$ *is non-balanced-decomposable in* Σ_n^*, *and that all the ultra-proper factorisations of any string in* Σ_n^* *of length strictly smaller than* $|s_0 \ldots s_{n-1}|$ *are balanced-decomposable. Then no* s_i *contains any ultra-proper factor equivalent to* ε *and longer than 0. Similarly, no* s_i *contains any ultra-proper factor equivalent to a letter and longer than 1.*

Proof. Immediate from Proposition 2.

The following lemma expresses formally the intuitive "sliding" argument of Sect. 3.1. More precisely, it states that it can only fail when the "canceled" bump is a left or right factor of a factor of a balanced word.

Lemma 2. *Let* s_0, s_1 *be non-empty strings of* Σ_2^*. *Assume that*

1. $\mathcal{B} := [i, 1 + i + k] \in B^\alpha(s_0) \cap \operatorname{argmin}_{\text{length}} \left(B^{\{\alpha, \overline{\alpha}\}}(s_0) \cup B^{\{\alpha, \overline{\alpha}\}}(s_1) \right)$,
2. (s_0, s_1) *is balanced (recall that this implies* $s_0 s_1 \in O_2$) *but not bal-decomposable,*
3. *for any* p_0 *and* p_1 *being substrings of* s_0 *and of* s_1, *respectively, such that* $p_0 p_1 \in O_2$ *and* $|p_0 p_1| < |s_0 s_1|$, *it holds that* (p_0, p_1) *is balanced-decomposable, and*
4. $k < |s_0| - 1$.

Then $i = 0$, $k \geq 2$, and $\overline{s_0 (i + 1)}^j \alpha \preceq x$ for some $x \in \{\underline{s_0}, s_1, \underline{s_1}\}$ and $1 \leq j < k$.

Proof. Note that Proposition 2, together with hypotheses 3 and 2, allows us to infer that

$$\forall [i', i' + 1 + j'] \in B(s_0) \cup B(s_1) . j' \geq 2, \tag{4}$$

and, in particular, $k \geq 2$; then, s_0 must be of the form $s_2 \alpha \beta^{1+l} \beta^{1+m} \overline{\alpha} s_3$, with $\beta :=$
$s_0 (i + 1) \notin \{\alpha, \overline{\alpha}\}$ and $s_3 \neq \emptyset$ thanks to hypothesis 4. Hypothesis 2 tells us that
$|s_0 s_1| > 4$ by a straightforward check. Therefore, using hypothesis 3, the set D of
bal-decompositions of $(s_0' := s_0 - \{i, 1 + i + k\}, s_1)$ (where we used the $-$ notation
introduced in Note 2) is not empty. Additionally, any decomposition in that set will be
made up of two components each of length 2 since, if that were not the case, we could
use Corollary 3 and hypothesis 3 to violate hypothesis 2;[2] to recapitulate:

$$D \neq \emptyset \text{ and } \forall x \in D. \, x = ((p_0, q_0), (p_1, q_1)) \text{ with } p_0, q_0, p_1, q_1 \neq \varepsilon. \tag{5}$$

Furthermore, any bal-decomposition in D must factor s_0 into at least two non-empty
strings "straddling" the bump $\mathcal{B}$, otherwise, again, we would easily violate hypothesis 2.

Let us then pick a $((p_0, q_0), (p_1, q_1)) \in D$. Without loss of generality, there are
only two cases: 1. $s_0 - \{i, 1 + i + k\} = p_0 p_1 q_0$, and 2. $s_0 - \{i, 1 + i + k\} = p_0 p_1$.
Since we have already shown that $k \geq 2$, we are left with proving that $i = 0$ and
that some extremal factor of s_0 or s_1 (depending on the case) has the wanted form
(i.e., $\overline{s_0 (i + 1)}^j \alpha$ or its reversal, again depending on the case). The proofs in the two
cases are similar; let us do the second,[3] so that (maybe after one reversal of s_1, which
would pose no problem since reversion does not interfere with bal-decomposability)
$s_1 = q_0 q_1$. We can assume that our q_0 has minimal length among all the choices of
p_0, q_0, p_1, q_1 satisfying case 2 above. Due to the above observation about straddling,
$p_0 = s_2 \beta^{1+l}$ with $|s_2| = i$. By (5), we can consider the rightmost letter of q_0 and
the leftmost one of q_1, calling them y and z, respectively. By minimality of $|q_0|$, and
since s_3 is non-empty, it cannot be $y = \beta$. Moreover, $y \neq \alpha$, otherwise we could
remove y from q_0 and reinstate α in p_0, obtaining a decomposition of (s_0, s_1) against
hypothesis 2. We now show that $y \neq \overline{\alpha}$:

proof that $y \neq \overline{\alpha}$: if y were $\overline{\alpha}$, then z could not be α due to (4), and could not be $\overline{\alpha}$ due
to hypothesis 2.
  case $z = \beta$: then we could write q_1 as $\beta^{1+l'} q_1'$ with $l' < l$ due to hypothesis 2, and
  q_1' not having β as its first letter. Note that

$$\left(\left(s_2 \beta^{l-l'}, q_0 \beta^{1+l'} \right), \left(\beta^{2+m+l'} s_3, q_1' \right) \right) \in D, \tag{6}$$

  so that q_1' cannot be empty due to (5). Its first letter cannot be α due to $l' < l$
  and hypothesis 3, and it cannot be β due to (4). It cannot be $\overline{\alpha}$ due to (6) and
  hypothesis 2. Contradiction.

---

[2] Assume there were a non-empty factor of s_0' having a zero balance. Then the corresponding
factor of s_0 obtained by reinstating α and $\overline{\alpha}$ would have balance zero or one, contradicting
Corollary 3.

[3] The reader is reminded that the first case was informally illustrated in Sect. 3.1.

case $z = \overline{\beta}$: then we could write q_1 as $\overline{\beta}^{1+m'} q_1'$ (with $m' < m$ due to hypotheses 2 and 4), and q_1' not having $\overline{\beta}$ as its first letter. Note that

$$\left(\left(s_2 \beta^{2+l+m'}, q_0 \overline{\beta}^{1+m'} \right), \left(\beta^{m-m'} s_3, q_1' \right) \right) \in D, \tag{7}$$

so that $q_1' \neq \varepsilon$ by (5). Its first letter cannot be α due to $m' < m$ and hyopthesis (3), and cannot be β due to (4). It cannot be $\overline{\alpha}$ due to (7) and hypothesis 2. Contradiction.

We conclude that $y = \overline{\beta}$ and hence $q_0 = \overline{\beta}$ by minimality of $|q_0|$, from which it follows that $l = 0$ and $s_2 = \varepsilon$ by (5): otherwise, we would have that $\left(\left(s_2 \beta^l, \beta^{2+m} s_3 \right), (q_0', y q_1) \right)$, where $q_0 = q_0' y$, is a bal-decomposition of (s_0', s_1), against the minimality of $|q_0|$. It is now easy to check that q_1 must contain at least one letter different from $\overline{\beta}$, so that we can write $q_1 = \overline{\beta}^h \gamma q_1'$ for some $h \in \mathbb{N}$, $\gamma \in \{\alpha, \overline{\alpha}, \beta\}$, and $q_1' \in \Sigma_2^*$. As usual, $\gamma \neq \beta$ by (4). Moreover, $h < m + 1 < k \Rightarrow h + 1 < k$: otherwise, since $s_3 \neq \varepsilon$, we could violate hypothesis 2. Therefore, $\left(\left(\beta^{1+h}, \overline{\beta}^{1+h} \right), \left(\beta^{m-h+1} s_3, \gamma q_1' \right) \right) \in D$. It follows that γ cannot be $\overline{\alpha}$, otherwise we would obtain a bal-decomposition of (s_0, s_1); hence, $\gamma = \alpha$, terminating the proof.

We are now in a position to perform our final proof.

Proof (of Theorem 1). Assume that the set in the statement of Theorem 1 is non-empty, and choose in it some (q_0, q_1) such that $|q_0 q_1|$ is minimal. q_0 and q_1 cannot both be short by virtue of Corollary 2. Therefore, by Lemma 2 and without loss of generality, we can assume that $ab^{2+m'} \overline{a} \preceq q_0$ and that $\overline{b}^{1+m} a \preceq x$ for some $0 \leq m \leq m'$, $x \in \{q_0, q_1\}$.

Case $x = q_0$: then there is a minimal β-bump either in q_0 or q_1, where $\beta \in \{b, \overline{b}\}$. This is because the shortest separation between any b and $\overline{b}$ in q_0 must be filled entirely by a's or $\overline{a}$'s, but not both, otherwise we would contradict Corollary 3, so that there is a β-bump in q_0, and hence a minimal one across q_0 and q_1. It is easy to check that minimal β-bump cannot be in q_0 by reapplying Lemma 2. Therefore, it must be in q_1 and, again by Lemma 2, we can assume it is a left factor of q_1, so that $\beta \overline{a}^{1+i} \overline{\beta} \preceq q_1$, with $\beta \in \{b, \overline{b}\}$: note that the second letter of q_1 cannot be a, otherwise Lemma 2 would make (q_0, q_1) bal-decomposable. But now, whichever value we choose for β, (q_0, q_1) would be bal-decomposable.

Case $x = q_1$: it is easy to check that, by Lemma 2, neither q_0 nor q_1 can have any β-bump, where $\beta \in \{b, \overline{b}\}$. This implies, by the same argument used in the previous case to show that q_0, there, had a β bump, that the last letter of q_0 cannot be $\overline{b}$; it therefore must be a to avoid decomposability. We can pick the minimal m making (q_0, q_1) bal-indecomposable. Then, by calling q_1' the string obtained from q_1 by swapping the $(1 + m)$-th $\overline{b}$ and the contiguous a in q_1, (q_0, q_1') will become bal-decomposable, and any bal-decomposition will split q_1' immediately after the leftmost a of q_1'. Since $m \leq m'$ and the occurrences of $\overline{b}$ in q_0 and of b in q_1', respectively, are zero, the only possibility is that there is a right factor q_2 of q_0 such that $q_2 \overline{b}^m a$ is balanced. But this would imply q_2 containing an a or $\overline{a}$-bump, and therefore at least $1 + m' > m$ occurrences of b, which is clearly impossible, since q_2 contains no $\overline{b}$.

4   Comparison with the Existing Proofs

In the peer-reviewed literature, there are four proofs of results relating to Theorem 1, with various degrees of generality. Contrary to the present proof, they are all derivations or improvements of the original proof in [13], from which we start our comparison. The two main criteria guiding our comparison are: 1. the feasibility of a concrete formalisation of the proof in a modern proof assistance language (e.g., Isabelle/HOL) and 2. the performance of a corresponding parser obtained from the definitions involved in the formalisation. To help assessing the first criterion, we will consider the requirements of each proof in terms of the mathematical definitions and results assumed as given, and the size in bytes of the textual proof. The second aspect has some elements of arbitrariness and discretion, given the somewhat difficult task of deciding what to extract from a paper to consider as part of a proof and what not. However, although crude, this byte counting approach is a well-established method when quantitatively analysing mathematical proofs, for example to establish their de Bruijn factor in the context of mechanised proving [2, 11]. In measuring the byte-size of paper proofs, the explanatory notes and the parts addressed to the reader, (as opposed to the proofs, statements, and definitions) were not considered. Technically, the size was measured by selecting the relevant text in the papers' pdf files. For the proof contained in this paper, this gives a result of around 10kB.

Regarding criterion (2), the mere existence of results, such as the one in the present paper, linking O_2 and some MCFG makes the membership problem for the given grammar trivial. The parsing problem remains more complicated: since we know that any string in O_2 is derivable in G_2, we can brute-force our way through a given string to build a parse tree, or use MCFG-dedicated algorithms to generate a parse tree [10, Section 7]. However, these algorithms tend to be complicated to implement and hence hard to verify, with complexities typically exceeding $O\left(n^6\right)$. On the other hand, Lemma 1 provides highly useful information to avoid brute forcing: it tells us that we can cancel a minimal bump and use the decomposition of the reduced string, instead, which gives the basis for a recursive parsing. Thus, the parsing problem can be recursively reduced to the simple problem of finding a list of bumps for a given string and book-keeping it as long as they are canceled one by one. We will now analyse whether the existing proofs provide any similar aspect.

The Original Proof by Salvati. The proof in [13] is geometrical and involves non-elementary notions such as that of homotopy, winding numbers, fundamental groups, covering spaces. Some facts, such as the unique path-lifting property and the homotopy-lifting property, are stated without proof. The remaining proofs and definitions (excluding MCFG-related definitions and proofs not strictly needed for the main proof, such as Lemma 1) amount to about 83kB. For these reasons, the proof looks very hard to formalise. The theorem on Jordan curves proved in Sect. 4 is highly non-constructive, for example resorting to Zorn's lemma to prove Lemma 13, thereby providing little insight from the parsing point of view.

The Proof by Nederhof. In [12], Nederhof provides a shorter proof than that in [13]. This avoids the need of problematic (from a formalisation point of view) dependencies such as homotopy, fundamental groups, covering spaces. However, many ideas are shared between the two proofs, including representing words as curves in the plane, using them to pass to the continuous geometrical view, and then finding a way to tame the extremely complicated curve's self-intersections which may arise. This implies that similar difficulties would arise with respect to the formalisation. For the same reason, building an efficient algorithm from this proof looks quite unfeasible. Another issue is that the proof given is quite unstructured (no separation into lemmas, theorems, propositions, etc.) and, mostly important, quite informal, with abundance of appeals to geometrical intuition (e.g., "it is clear that m, Q'_1 and Q'_2 satisfying these requirements must exist", or "The truncation consists in changing $sub_C(Q'_1, Q'_2)$ to become $seg(Q'_1, Q'_2)$, as illustrated by Fig. 7."). While this kind of style is likely readily accepted and appreciated by a human reader, it usually turns out to be problematic to formalise. This also makes the byte-size of the proof (about 24kB) unfairly underestimated in comparison to the others.

The Proof by Ho. In [9], the first generalisation to the case $n > 2$ was given. The proof is remarkably short at around 11kB. However, it crucially depends on a geometric, nonconstructive result ([3] and [1]) not present in proof assistants' libraries. Additionally, although more general than the previous and the present proofs, it pays for its simplicity by using grammars of higher fanout than necessary, even in the known $n = 2$ case. This makes this approach not feasible for a formalisation.

The Final Proof in [6]. This proof builds on that by Ho and improves it by using grammars of the lowest possible fanout for all values of n. The price to pay for that is a longer proof, of around 22kB. Additionally, this proof uses the Hobby-Rice theorem, not present in the Isabelle/HOL library. Interestingly, there is an extended version of the paper providing a Sect. 5 where the more geometric parts of the original proof are replaced by combinatorial equivalents, based on Tucker's and Ky Fan's lemmas. Unfortunately, neither is present in the Isabelle/HOL library as well, and any of them would probably require a substantial amount of work. This variation is also interesting because potentially impacting on criterion (2), since there are paper proofs of Tucker's lemma which are constructive. For both variants, the nature of the dependencies and the size of the proofs would make their formalisation quite a complex task.

5   Conclusions

A new proof for the fact that O_2 is a 2-MCFL has been presented. Contrary to the existing proofs, the present proof has no geometrical part whatsoever, and uses only elementary notions and methods. This impacts both on its formalisability, being such proof comparably much more amenable to being formalised than existing ones, and on its computational interest, in that Lemma 2 (expressing the informal argument of Sect. 3.1) gives an extremely simple construction for a recursive implementation of a corresponding parser. This paper therefore also serves as a starting point for a verified, efficient parser whose implementation is underway. This is both practically and theoretically important not only from the computational linguistics point of view, but also from the point of view of computational algebra, where the word problem for $\mathbb{Z}^2$

(tightly related to the parsing problem for O_2) has a prominent role. [8] While some of the existing proofs generalise beyond the $n = 2$ case, that case is important because it is connected to formal language theory, for example to the MIX language, and to a word problem in combinatorial group theory which has been open for decades [8]. Therefore, this work, besides its proof-theoretical and algorithmic interest, provides a first step towards extending proof assistants' libraries in those two directions. Work is already underway to provide an executable formalisation in Isabelle/HOL of the proposed proof, and a paper explaining the formalisation itself. Another venue for future work is the possibility of adapting the present proof to values of n bigger than 2: the main issue to tackle, in this direction, is the fact that the "sliding" argument presented in Sect. 3.1 works by cases on the possible characters occurring in particular spots of the considered decomposition, which is not trivial to extend to higher values of n. Lastly, it is hoped that a new proof and the novel concept involved, such as that of bump, can broaden the possible venues to tackle related open problems, such as the relationship between indexed languages [5] and the word problem [8].

Acknowledgments. The author is grateful to all the anonymous referees for their helpful feedback, to Mark Jan Nederhof for introducing the problem to him, and to Matthew Barnes at Lancaster University Library for his meticulous and extremely helpful support in providing difficult to access resources. He would also like to thank Mark Brewer at Heilbronn and Constantin Blome for their academic mentorship.

References

1. Alon, N., West, D.B.: The Borsuk-Ulam theorem and bisection of necklaces. Proc. Am. Math. Soc. **98**(4), 623–628 (1986)
2. Asperti, A., Sacerdoti Coen, C.: Some considerations on the usability of interactive provers. In: Intelligent Computer Mathematics: 10th International Conference, Aisc 2010, 17th Symposium, Calculemus 2010, and 9th International Conference, Mkm 2010, Paris, France, 5–10 July 2010. Proceedings, p. 147 (2010)
3. Burago, D.: Periodic metrics. In: Kuksin, S., Lazutkin, V., Poschel, J. (eds.) Seminar on Dynamical Systems: Euler International Mathematical Institute, St. Petersburg, pp. 90–95. Springer, Heidelberg (1994). https://doi.org/10.1007/978-3-0348-7515-8_7
4. Clark, A.: An introduction to multiple context free grammars for linguists. Linguist. Philos. **8**, 333–343 (1985)
5. Fischer, M.J.: Grammars with macro-like productions. Doctoral dissertation, Harvard University (1968)
6. Gebhardt, K., Meunier, F., Salvati, S.: O_n is an n-MCFL. J. Comput. Syst. Sci. **127**, 41–52 (2022)
7. Gilman, R.H.: Formal languages and their application to combinatorial group theory. Contemp. Math. **378**, 1–36 (2005)
8. Gilman, R.H., Kropholler, R.P., Schleimer, S.: Groups whose word problems are not semilinear. Groups Complex. Cryptol. **10**(2), 53–62 (2018)
9. Ho, M.C.: The word problem of $\mathbb{Z}_n$ is a multiple context-free language. Groups Complex. Cryptol. **10**(1), 9–15 (2018)
10. Kallmeyer, L.: Parsing Beyond Context-Free Grammars. Springer, Heidelberg (2010). https://doi.org/10.1007/978-3-642-14846-0

11. Naumowicz, A.: An example of formalizing recent mathematical results in Mizar. J. Appl. Log. **4**(4), 396–413 (2006)
12. Nederhof, M.J.: A short proof that O_2 is an MCFL. In: Proceedings of the 54th Annual Meeting of the Association for Computational Linguistics, vol. 1: Long Papers, pp. 1117–1126 (2016)
13. Salvati, S.: MIX is a 2-MCFL and the word problem in $\mathbb{Z}_2$ is captured by the IO and the OI hierarchies. J. Comput. Syst. Sci. **81**(7), 1252–1277 (2015)

Cyclic Operator Precedence Grammars for Improved Parallel Parsing

Michele Chiari[1]([✉])[iD], Dino Mandrioli[2][iD], and Matteo Pradella[2,3][iD]

[1] Institute of Computer Engineering, TU Wien, Treitlstraße 3, 1040 Vienna, Austria
michele.chiari@tuwien.ac.at
[2] DEIB, Politecnico di Milano, Piazza Leonardo Da Vinci 32, 20133 Milan, Italy
{dino.mandrioli,matteo.pradella}@polimi.it
[3] IEIIT, Consiglio Nazionale delle Ricerche, via Ponzio 34/5, 20133 Milan, Italy

Abstract. Operator precedence languages (OPL) enjoy the *local parsability property*, which essentially means that a code fragment enclosed within a pair of markers—playing the role of parentheses—can be compiled with no knowledge of its external context. Such a property has been exploited to build parallel compilers for languages formalized as OPLs. It has been observed, however, that when the syntax trees of the sentences have a linear substructure, its parsing must necessarily proceed sequentially making it impossible to split such a subtree into chunks to be processed in parallel. Such an inconvenience is due to the fact that so far much literature on OPLs has assumed the hypothesis that the equality precedence relation cannot be cyclic. We present an enriched version of operator precedence grammars which allows to remove the above hypothesis, therefore providing a little more expressive generality, and to further optimize parallel compilation.

Keywords: Operator Precedence Languages · Cyclic Precedence Relations · Parallel Parsing

1 Introduction

Operator precedence languages (OPL) are a "historical" family of languages invented by R. Floyd [10] to support fast deterministic parsing. Together with their *operator precedence grammars (OPG)*, they are still used within modern compilers to parse expressions with operators ranked by priority. The key feature that makes them well amenable for efficient parsing and compilation is that the syntax tree of a sentence is determined exclusively by three binary precedence relations over the terminal alphabet that are easily pre-computed from the grammar productions. For example: the arithmetic sentence $a + b \times c$ does not make manifest the natural structure $(a + (b \times c))$, but the latter is implied by the fact that the plus operator yields precedence to the times.

After Floyd's pioneering contribution much subsequent research discovered further important properties of OPLs, not necessarily related to deterministic parsing, that are

---

The extended version of this paper, available in [6], contains an omitted correctness proof, explanatory examples and figures developed in more depth.

© The Author(s), under exclusive license to Springer Nature Switzerland AG 2024
J. D. Day and F. Manea (Eds.): DLT 2024, LNCS 14791, pp. 98–113, 2024.
https://doi.org/10.1007/978-3-031-66159-4_8

typical of the much less powerful family of regular languages and therefore enabled applications in fields such as automatic verification. In synthesis, they are: OPLs are a boolean algebra [8]; they are also closed under concatenation and Kleene's * [7]; they are characterized, besides original OPGs, by a special and simple family of push-down automata, named *Operator Precedence automata (OPA)*, a *monadic second-order (MSO) logic* that naturally extends the classic one for regular languages [15], *operator precedence expressions (OPE)* which similarly extend traditional regular expressions [17], and in terms of a *syntactic congruence* with a finite number of equivalence classes (OPSC) [13]. Finally, an FO-complete temporal logic equivalent to *aperiodic OPLs* [17] has been defined which enabled the construction of a first model checker to verify properties of OPLs with a complexity comparable with that of model checkers for regular languages [5]. To the best of our knowledge, OPLs are the largest subclass of context-free languages that enjoys the same properties of regular ones.

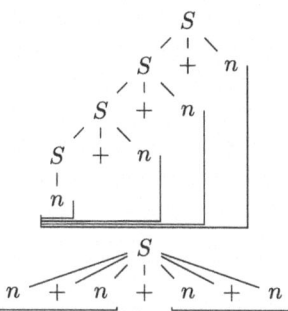

Fig. 1. Left-associative syntax tree (top) vs equal-level one (bottom) of the plus operator. The top syntax tree imposes a sequential left-to-right parsing and semantic processing whereas the bottom one can be split onto several branches to be partially processed independently and further aggregated.

It must be pointed out, however, that many—not all—of the algebraic properties discovered during such a research activity were proved by assuming a hypothesis on the precedence relations defined on the input alphabet. Although from a theoretical point of view this hypothesis slightly affects the generative power of OPGs—but is not necessary, e.g., for OPAs and MSO logic so that these two formalisms are a little more powerful than OPGs—, so far no practical limitation due to it was discovered in terms of formalizing the syntax of real-life programming and data description languages. Thus, it has been constantly adopted in the various developments to avoid making the mathematical notation and technical details too cumbersome.

Another distinguishing property of OPLs is their *local parsability*, i.e. the fact that a chunk included within a pair of symmetric precedence relations can be deterministically parsed even without knowing its context. This feature was exploited to produce a parallel parser generator which exhibited high performances w.r.t. traditional parsers [3,4].

The recent contribution [14], however, pointed out a weakness of the parallel compilation algorithm described in [3] which in some cases hampers partitioning the input to be parsed in well-balanced chunks, so that the benefits of parallelism are affected. Intuitively, the weakness is due to the fact that the "normal" precedence relation on the arithmetic operator + compels to parse a sequence thereof by associating them either to the left or to right so that parsing becomes necessarily sequential in this case. The authors also proposed a special technique to overtake this difficulty by allowing for an acceptable level of ambiguity, which in the case of OPGs determines a conflict for some precedence relations on the terminal alphabet.

Such normal precedence relation, however, which either lets + yield precedence to, or take precedence over itself, has no correspondence with the semantics of the operation, whose result does not depend on the order of association. So, why not giving the various occurrences of the $+$ operator the same precedence level as suggested by arithmetic's laws? Fig. 1 gives an intuitive idea of the different structures given to a sequence of $+$ operators by traditional OPGs and the natural semantics of the sum operation.

The answer to this question comes exactly from the above mentioned hypothesis: it forbids cyclic sequences of symbols that are at the same level of precedence; so that $+$ cannot be at the same level of itself. Thus, the discovery of this practical restriction "compelled" us to finally remove this relatively disturbing hypothesis: this is the object of the present paper. The solution we devised consists in allowing grammar *right hand sides (rhs) to include the Kleene * operator*. Thus, we introduce the *Cyclic operator-precedence grammars (C-OPG)* which include the above feature: whereas such a feature is often used in general context-free grammars to make the syntax of programming languages more compact but does not increase their expressive power, we show that C-OPGs are now fully equivalent to OPAs and other formalisms to define OPLs, such as the MSO logic. We also show that all results previously obtained under the above hypothesis still hold by using C-OPGs instead of traditional OPGs. Although the goal of this paper is not to develop parallel compilation algorithms rooted in C-OPGs, we show how they naturally overtake the difficulty pointed out by [14] and would allow to revisit their techniques, or to improve the efficiency of our previous parallel parser [3].

2   Background

We assume some familiarity with the classical literature on formal language and automata theory, e.g., [12, 19]. Here, we just list and explain our notations for the basic concepts we use from this theory. The terminal alphabet is usually denoted by Σ, and the empty string is ε. The character $\# \notin \Sigma$ is used as *delimiter*, and we define $\Sigma_\# = \Sigma \cup \{\#\}$.

A *context-free (CF) grammar* is a tuple $G = (\Sigma, V_N, P, S)$ where Σ and V_N, with $\Sigma \cap V_N = \emptyset$, are resp. the terminal and the nonterminal alphabets, the total alphabet is $V = \Sigma \cup V_N$, $P \subseteq V_N \times V^*$ is the rule (or production) set, and $S \subseteq V_N$, $S \neq \emptyset$, is the axiom set. For a generic rule, denoted as $A \to \alpha$, where A and α are resp. called the left/right hand sides (lhs / rhs), the following forms are relevant: *axiomatic* $A \in S$; *terminal* $\alpha \in \Sigma^+$; *empty* $\alpha = \varepsilon$; *renaming* $\alpha \in V_N$; *operator* $\alpha \notin V^* V_N V_N V^*$, i.e., at least one terminal is interposed between any two nonterminals occurring in α.

A grammar is *backward deterministic (BD)* if $(B \to \alpha, C \to \alpha \in P)$ implies $B = C$. If all rules of a grammar are in operator form, it is called an *operator grammar* or *O-grammar*. We give for granted the usual definition of *derivation* denoted by the symbols $\underset{G}{\Longrightarrow}$ (immediate derivation), $\underset{G}{\overset{*}{\Longrightarrow}}$ (reflexive and transitive closure of $\underset{G}{\Longrightarrow}$), $\underset{G}{\overset{+}{\Longrightarrow}}$ (transitive closure of $\underset{G}{\Longrightarrow}$); the subscript G will be omitted whenever clear from the context. We give also for granted the notion of *syntax tree (ST)*. As usual, the *frontier* of a syntax tree is the ordered left-to-right sequence of the leaves of the tree.

The *language* defined by G, said $L(G)$, is $\{w \mid w \in \Sigma^*, A \underset{G}{\overset{*}{\Longrightarrow}} w \wedge A \in S\}$. Two grammars defining the same language are *equivalent*. Two grammars generating the same set of syntax trees, up to a renaming of internal nodes, are *structurally equivalent*.

From now on, w.l.o.g., we exclusively deal with O-grammars without renaming and empty rules with the only exception that, if ε is part of the language, there is a unique empty rule whose lhs is an axiom that does not appear in the rhs of any production. In fact, this is a well-known normal form for CF grammars [2, 12].

We now define operator precedence grammars (OPGs) following primarily [16]. Intuitively, OPGs are O-grammars whose parsing is driven by three *precedence relations*, called *equal*, *yield* and *take*, included in $\Sigma_{\#} \times \Sigma_{\#}$. They are defined in such a way that two consecutive terminals of a grammar's rhs—ignoring possible nonterminals in between—are in the equal relation, while the two extreme ones—again, whether or not preceded or followed by a nonterminal—are preceded by a yield and followed by a take relation, respectively; in this way a complete rhs of a grammar rule is identified and can be *reduced* to a corresponding lhs by a typical bottom-up parsing.

Definition 1 ([10]). *Let $G = (\Sigma, V_N, P, S)$ be an O-grammar. Let a, b denote elements in Σ, A, B in V_N, C either an element of V_N or the empty string ε, and α, β range over V^*. The left and right terminal sets of nonterminals are respectively:*
$$\mathcal{L}_G(A) = \left\{ a \in \Sigma \mid \exists C : A \xRightarrow[G]{*} Ca\alpha \right\} \text{ and } \mathcal{R}_G(A) = \left\{ a \in \Sigma \mid \exists C : A \xRightarrow[G]{*} \alpha aC \right\}.$$
The operator precedence (OP) relations are defined over $\Sigma_{\#} \times \Sigma_{\#}$ as follows:
Equal in precedence $a \doteq b \Leftrightarrow \exists A \rightarrow \alpha a C b \beta \in P$.
Takes precedence $a \gtrdot b \Leftrightarrow \exists A \rightarrow \alpha B b \beta \in P, a \in \mathcal{R}(B); a \gtrdot \# \Leftrightarrow a \in \mathcal{R}(B), B \in S$.
Yields precedence $a \lessdot b \Leftrightarrow \exists A \rightarrow \alpha a B \beta \in P, b \in \mathcal{L}(B); \# \lessdot b \Leftrightarrow b \in \mathcal{L}(B), B \in S$.

The OP relations can be collected into a $|\Sigma_{\#}| \times |\Sigma_{\#}|$ array, called the operator precedence matrix *of the grammar, $OPM(G)$: for each (ordered) pair $(a, b) \in \Sigma_{\#} \times \Sigma_{\#}$, $OPM_{a,b}(G)$ contains the OP relations holding between a and b.*

An OPM is said conflict-free *iff $\forall a, b \in \Sigma_{\#}$, $0 \leq |M_{a,b}| \leq 1$. A conflict-free OPM is* total *or* complete *iff $\forall a, b \in \Sigma_{\#}$, $|M_{a,b}| = 1$. If $M_{\#,\#}$ is not empty, $M_{\#,\#} = \{\doteq\}$. An OPM is $\doteq$-acyclic if the transitive closure of the $\doteq$ relation over $\Sigma \times \Sigma$ is irreflexive.*

We extend the set inclusion relations and the Boolean operations in the obvious cell-by-cell way, to any two matrices having the same terminal alphabet. Two matrices are *compatible* iff their union is conflict-free.

Definition 2 (Operator precedence grammar). *A grammar G is an* operator precedence grammar *(OPG) iff the matrix $OPM(G)$ is conflict-free. An OPG is $\doteq$-acyclic if $OPM(G)$ is so. An* operator precedence language *(OPL) is a language generated by an OPG.*

Figure 2 (left) displays an OPG, G_{AE}, which generates simple, unparenthesized arithmetic expressions and its OPM (center). The left and right terminal sets of G_{AE}'s nonterminals E, T and F are, respectively: $\mathcal{L}(E) = \{+, \times, n\}$, $\mathcal{L}(T) = \{\times, n\}$, $\mathcal{L}(F) = \{n\}$, $\mathcal{R}(E) = \{+, \times, n\}$, $\mathcal{R}(T) = \{\times, n\}$, and $\mathcal{R}(F) = \{n\}$.
Remark. If the relation $\doteq$ is acyclic, then the length of the rhs of any rule of G is bounded by the length of the longest $\doteq$-chain in $OPM(G)$.

The key feature of OPLs is that a conflict-free OPM M defines a universe of *strings compatible with M* and associates to each of them a unique *syntax tree* whose internal nodes are unlabeled and whose leaves are elements of Σ. We illustrate such a feature through a simple example and refer the reader to previous literature for a thorough description of OP parsing [11, 16].

$$
G_{AE} : S = \{E\}
$$
$$
E \to E + T \mid T \times F \mid n
$$
$$
T \to T \times F \mid n
$$
$$
F \to n
$$

	$+$	$\times$	n	$\#$
$+$	$\gtrdot$	$\lessdot$	$\lessdot$	$\gtrdot$
$\times$	$\gtrdot$	$\gtrdot$	$\lessdot$	$\gtrdot$
n	$\gtrdot$	$\gtrdot$		$\gtrdot$
$\#$	$\lessdot$	$\lessdot$	$\lessdot$	

```
            N
          / | \
         #  N  #
          / | \
        N  +  N
       /|\     |
      N + N    n
      |   /|\
      n  N × N
         |   |
         n   n
```

Fig. 2. G_{AE} (left), its OPM (center), and the syntax tree of $n + n \times n + n$ according to the OPM (right).

Example 3. Consider the $OPM(G_{AE})$ of Fig. 2 and the string $n + n \times n + n$. Display all precedence relations holding between consecutive terminal characters, *including the relations with the delimiters $\#$* as shown here:

$$
\# \lessdot n \gtrdot + \lessdot n \gtrdot \times \lessdot n \gtrdot + \lessdot n \gtrdot \#
$$

Each pair $\lessdot, \gtrdot$ (with no further $\lessdot, \gtrdot$ in between) includes a *possible* rhs of a production of *any OPG* sharing the OPM with G_{AE}, not necessarily a G_{AE} rhs. Thus, as it happens in typical bottom-up parsing, we replace—*possibly in parallel*—each string included within the pair $\lessdot, \gtrdot$ with a *dummy nonterminal N*; this is because nonterminals are irrelevant for OPMs. The result is the string $\#N + N \times N + N\#$. Next, we compute again the precedence relation between consecutive terminal characters by *ignoring nonterminals*: the result is $\# \lessdot N + \lessdot N \times N \gtrdot +N \gtrdot \#$.

This time, there is only one pair $\lessdot, \gtrdot$ including a potential rhs determined by the OPM (the fact that the external $\lessdot$ and $\gtrdot$ "look matched" is coincidental as it can be easily verified by repeating the previous procedure with the string $n + n \times n + n + n$). Again, we replace the pattern $N \times N$, with the dummy nonterminal N; notice that there is no doubt about associating the two N to the $\times$ rather than to one of the adjacent $+$ symbols: if we replaced, say, just the $\times$ with an N we would obtain the string $N + NNN + N$ which cannot be derived by an O-grammar. By recomputing the precedence relations we obtain the string $\# \lessdot N + N \gtrdot +N \gtrdot \#$. Finally, by applying twice the replacing of $N + N$ by N we obtain $\#N\#$.

The result of the whole bottom-up reduction procedure is synthetically represented by the *syntax tree* of Fig. 2 (right) which shows the precedence of the multiplication operation over the additive one in traditional arithmetics. It also suggests a natural association to the left of both operations: if we reverted the order of the rhs of the rules rewriting E and T, the structure of the tree would have suggested associativity to the right of both operations which would not have altered the semantics of the two operations which can indifferently be associated to the left and to the right; not so is we dealt with, say, subtraction or division which instead *impose* association to the left.

Notice that the tree of Fig. 2 has been obtained—uniquely and deterministically— by using exclusively the OPM, not the grammar G_{AE} although the string $n+n\times n+n \in L(G_{AE})^1$.

Obviously, all sentences of $L(G_{AE})$ can be given a syntax tree by $OPM(G_{AE})$, but there are also strings in Σ^* that can be parsed according to the same OPM but are not in $L(G_{AE})$. E.g., the string $+ + +$ is parsed according to the $OPM(G_{AE})$ as a ST that associates the $+$ characters to the left. Notice also that, in general, not every string in Σ^* is assigned a syntax tree by an OPM; e.g., in the case of $OPM(G_{AE})$ the parsing procedure applied to nn is immediately blocked since there is no precedence relation between n and itself.

Definition 4 (OP-alphabet and Maxlanguage). *A string in Σ^* is compatible with an OPM M iff the procedure described in Example 3 terminates by producing the pattern $\#N\#$. The set of all strings compatible with an OPM M is called the* maxlanguage *or the* universe *of M and is simply denoted as $L(M)$.*

Let M be a conflict-free OPM over $\Sigma_\# \times \Sigma_\#$. We use the same identifier M to denote the—partial—function M that assigns to strings in Σ^* their unique ST as informally illustrated in Example 3.

The pair (Σ, M) where M is a conflict-free OPM over $\Sigma_\# \times \Sigma_\#$, is called an OP-alphabet. We introduce the concept of OP-alphabet as a pair to emphasize that it defines a universe of strings on the alphabet Σ—not necessarily covering the whole Σ^*—and implicitly assigns them a structure univocally determined by the OPM, or, equivalently, by the function M. The class of (Σ, M)-compatible OPGs and OPLs are respectively: $\mathscr{G}_M = \{G \mid G \text{ is an OPG and } OPM(G) \subseteq M\}$, $\mathscr{L}_M = \{L(G) \mid G \in \mathscr{G}_M\}$.

Various formal properties of OPGs and OPLs are documented in the literature, e.g., in [7, 8, 16]. The next proposition recalls those that are relevant for this article.

Proposition 5 (Algebraic properties of OPGs and OPLs). *If an OPM M is total, then the corresponding homonymous function is total as well, i.e., $L(M) = \Sigma^*$.*

Let (Σ, M) be an OP-alphabet where M is $\doteq$-acyclic. The class $\mathscr{G}_M$ contains an OPG, called the maxgrammar *of M, denoted by $G_{max,M}$, which generates the maxlanguage $L(M)$. For all grammars $G \in \mathscr{G}_M$, $L(G) \subseteq L(M)$.*

The closure properties of the family $\mathscr{L}_M$ of (Σ, M)-compatible OPLs defined by a total OPM are the following:

- *$\mathscr{L}_M$ is closed under union, intersection and set-difference, therefore also under complement (if a maxgrammar of M exists).*
- *$\mathscr{L}_M$ is closed under concatenation; if M is $\doteq$-acyclic, $\mathscr{L}_M$ is closed under *.*

In terms of expressive power, OPLs are strictly in-between Visibly Pushdown Languages [1] and deterministic context-free languages.

---

[1] The above procedure that led to the syntax tree of Fig. 2 could be easily adapted to become an algorithm that produces a new syntax tree whose internal nodes are labeled by G_{AE}'s nonterminals. Such an algorithm could be made deterministic by transforming G_{AE} into a structurally equivalent BD grammar sharing the same OPM [2, 12].

Remark. Thanks to the fact that a conflict-free OPM assigns to each string at most one ST—and exactly one if the OPM is complete—the above closure properties of OPLs w.r.t. Boolean operations automatically extend to sets of their STs. The same does not apply to the case of concatenation which in general may produce significant reshaping of the original STs [7]. Furthermore, any complete, conflict-free, $\doteq$-acyclic OPM defines a *universe of STs* whose frontiers are Σ^*.

The notion of *chain* introduced next is an alternative way to represent STs where internal nodes are irrelevant and "anonymized".

Definition 6 (Chains). *Let* (Σ, M) *be an OP-alphabet. A* simple chain *is a word* $a_0 a_1 a_2 \ldots a_n a_{n+1}$, *written as* $^{a_0}[a_1 a_2 \ldots a_n]^{a_{n+1}}$, *such that:* $a_0, a_{n+1} \in \Sigma \cup \{\#\}$, $a_i \in \Sigma$ *for every* $i : 1 \leq i \leq n$, $M_{a_0 a_{n+1}} \neq \emptyset$, *and* $a_0 \lessdot a_1 \doteq a_2 \ldots a_{n-1} \doteq a_n \gtrdot a_{n+1}$.

A composed chain *is a word* $a_0 x_0 a_1 x_1 a_2 \ldots a_n x_n a_{n+1}$, *with* $x_i \in \Sigma^*$, *where* $^{a_0}[a_1 a_2 \ldots a_n]^{a_{n+1}}$ *is a simple chain, and either* $x_i = \varepsilon$ *or* $^{a_i}[x_i]^{a_{i+1}}$ *is a chain (simple or composed), for every* $i : 0 \leq i \leq n$. *Such a composed chain will be written as* $^{a_0}[x_0 a_1 x_1 a_2 \ldots a_n x_n]^{a_{n+1}}$.

The body of a chain $^a[x]^b$, *simple or composed, is the word* x. *Given a chain* $^a[x]^b$ *the* depth $d(x)$ *of its body* x *is defined recursively:* $d(x) = 1$ *if the chain is simple, whereas* $d(x_0 a_1 x_1 \ldots a_n x_n) = 1 + \max_i d(x_i)$. *The depth of a chain is the depth of its body.*

For instance, the ST of Fig. 2 (right) is biunivocally represented by the composed chain $^\#[x_0 + x_1]^\#$, where, in turn x_0 is the body of the composed chain $^\#[y_0 + y_1]^+$, y_0 is the body of the simple chain $^\#[n]^+$, y_1 is the body of the composed chain $^+[z_0 \times z_1]^+$, etc. The depth of the main chain is 3.

As well as an OPG selects a set of STs within the universe defined by its OPM, an *operator precedence automaton (OPA)* selects a set of chains within the universe defined by an OP-alphabet.

Definition 7 (Operator precedence automaton (OPA)). *A nondeterministic OPA is given by a tuple:* $\mathcal{A} = \langle \Sigma, M, Q, I, F, \delta \rangle$ *where:* (Σ, M) *is an operator precedence alphabet,* Q *is a set of states (disjoint from* Σ), $I \subseteq Q$ *is a set of initial states,* $F \subseteq Q$ *is a set of final states,* δ, *the transition function, is a triple of functions* $\delta_{shift} : Q \times \Sigma \rightarrow \wp(Q), \delta_{push} : Q \times \Sigma \rightarrow \wp(Q), \delta_{pop} : Q \times Q \rightarrow \wp(Q)$.

We represent a nondeterministic OPA by a graph with Q as the set of vertices and $\Sigma \cup Q$ as the set of edge labelings. We write $p \xrightarrow{a} q$ iff $q \in \delta_{push}(p, a)$, $p \dashrightarrow{a} q$ iff $q \in \delta_{shift}(p, a)$, and $q \xRightarrow{p} r$ iff $r \in \delta_{pop}(q, p)$.

To define the semantics of the automaton, we introduce some notations. We use letters $p, q, p_i, q_i, \ldots$ to denote states in Q. Let Γ be $\Sigma \times Q$ and let Γ' be $\Gamma \cup \{\bot\}$; we denote symbols in Γ' as $[a, q]$ or $\bot$. We set $symbol([a, q]) = a$, $symbol(\bot) = \#$, and $state([a, q]) = q$. Given a string $\Pi = \bot \pi_1 \pi_2 \ldots \pi_n$, with $\pi_i \in \Gamma$, $n \geq 0$, we set $symbol(\Pi) = symbol(\pi_n)$, including the particular case $symbol(\bot) = \#$.

A *configuration* of an OPA is a triple $C = \langle \Pi, q, w \rangle$, where $\Pi \in \bot \Gamma^*, q \in Q$ and $w \in \Sigma^* \#$. The first component represents the contents of the stack, the second component represents the current state of the automaton, while the third component is the part of input still to be read.

A *computation* or *run* of the automaton is a finite sequence of *moves* or *transitions* $C_1 \vdash C_2$; there are three kinds of moves, depending on the precedence relation between the symbol on top of the stack and the next symbol to read:

push move: if $symbol(\Pi) \lessdot a$ then $\langle \Pi, p, ax \rangle \vdash \langle \Pi[a, p], q, x \rangle$, with $q \in \delta_{\text{push}}(p, a)$;

shift move: if $a \doteq b$ then $\langle \Pi[a, p], q, bx \rangle \vdash \langle \Pi[b, p], r, x \rangle$, with $r \in \delta_{\text{shift}}(q, b)$;

pop move: if $a \gtrdot b$ then $\langle \Pi[a, p], q, bx \rangle \vdash \langle \Pi, r, bx \rangle$, with $r \in \delta_{\text{pop}}(q, p)$.

Shift and pop moves are never performed when the stack contains only $\bot$.

Push and shift moves update the current state of the automaton according to the transition functions δ_{push} and δ_{shift}, respectively: push moves put a new element on the top of the stack consisting of the input symbol together with the current state of the automaton, whereas shift moves update the top element of the stack by changing its input symbol only. The pop move removes the symbol on the top of the stack, and the state of the automaton is updated by δ_{pop} on the basis of the pair of states consisting of the current state of the automaton and the state of the removed stack symbol; notice that in this move the input symbol is used only to establish the $\gtrdot$ relation and it remains available for the following move.

A configuration $\langle \bot, q_I, x\# \rangle$ is *initial* if $q_I \in I$; a configuration $\langle \bot, q_F, \# \rangle$ is *accepting* if $q_F \in F$. The language accepted by the automaton is:
$L(\mathcal{A}) = \{x \mid \langle \bot, q_I, x\# \rangle \vdash^* \langle \bot, q_F, \# \rangle, q_I \in I, q_F \in F\}$.

Example 8. The OPA depicted in Fig. 3 (top, left) based on the OPM at the (top, right) accepts the language of arithmetic expressions enriched w.r.t $L(G_{AE})$ in that it introduces the use of explicit parentheses to alter the natural precedence of arithmetic operations. The same figure (bottom) also shows an accepting computation on input $n + n \times (\!| n + n |\!)$.

Definition 9. *Let $\mathcal{A}$ be an OPA. A support for a simple chain $^{a_0}[a_1 a_2 \dots a_n]^{a_{n+1}}$ is any path in $\mathcal{A}$ of the form $q_0 \xrightarrow{a_1} q_1 \dashrightarrow \dots \dashrightarrow q_{n-1} \xrightarrow{a_n} q_n \xRightarrow{q_0} q_{n+1}$.*
Notice that the label of the last (and only) pop is exactly q_0, i.e. the first state of the path; this support is built due to the relations $a_0 \lessdot a_1$ and $a_n \gtrdot a_{n+1}$.
A support for the composed chain $^{a_0}[x_0 a_1 x_1 a_2 \dots a_n x_n]^{a_{n+1}}$ is any path in $\mathcal{A}$ of the form $q_0 \overset{x_0}{\rightsquigarrow} q_0' \xrightarrow{a_1} q_1 \overset{x_1}{\rightsquigarrow} q_1' \dashrightarrow \dots \dashrightarrow q_n \overset{x_n}{\rightsquigarrow} q_n' \xRightarrow{q_0'} q_{n+1}$, where for every $i, 0 \leq i \leq n$: if $x_i \neq \varepsilon$, then $q_i \overset{x_i}{\rightsquigarrow} q_i'$ is a support for the (simple or composed) chain $^{a_i}[x_i]^{a_{i+1}}$; if $x_i = \varepsilon$, then $q_i' = q_i$. Notice that the label of the last pop is exactly q_0'. The support of a chain with body x will be denoted by $q_0 \overset{x}{\rightsquigarrow} q_{n+1}$.

The context a, b of a chain $^a[x]^b$ is used by the automaton to build its support only because $a \lessdot x$ and $x \gtrdot b$; thus, the chain's body contains all information needed by the automaton to build the subtree whose frontier is that string, once it is understood that its first move is a push and its last one is pop. This is a distinguishing feature of OPLs, not shared by other deterministic languages: we call it their *locality principle*, which has been exploited to build parallel and/or incremental OP parsers [4].

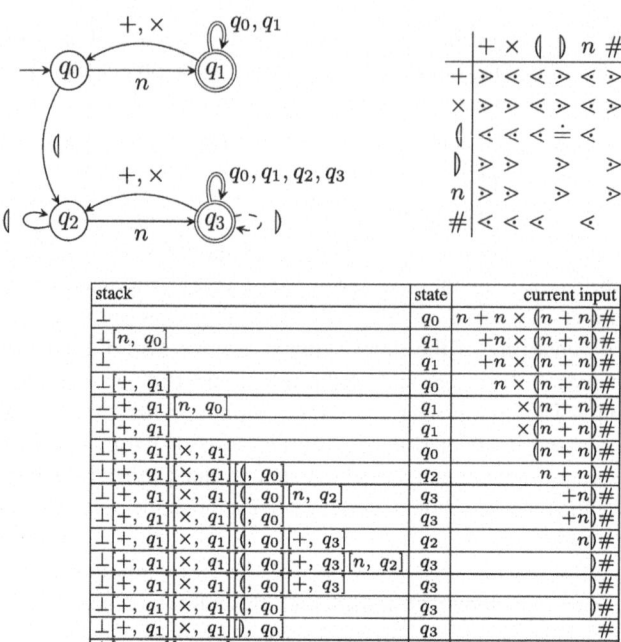

Fig. 3. An OPA (top, left), its OPM (top, right) and an example of computation for the language of Example 8 (bottom). Arrows $\longrightarrow$, $\dashrightarrow$ and $\Longrightarrow$ denote push, shift and pop transitions, respectively. To avoid confusion with the overloaded parenthesis symbols, the parentheses used as terminal symbols are denoted as $(\!|$ and $|\!)$.

3   Cyclic Operator Precedence Grammars (C-OPGs)

Proposition 5 shows that *some, but not all,* of the algebraic properties of OPLs depend critically on the $\doteq$-acyclicity hypothesis. This is due to the fact that without such a hypothesis the rhs of an OPG have an unbounded length but cannot be infinite: e.g., no OPG can generate the language $\{a, b\}^*$ if $a \doteq b$ and $b \doteq a$. In most cases cycles of this type can be "broken" as it has been done up to now, e.g., to avoid the $+ \doteq +$ relation in arithmetic expressions by associating the operator indifferently to the right or to left. From a theoretical point of view, the $\doteq$-acyclicity hypothesis affects the expressive power of OPGs; thus, the OPL familily as generated by OPGs is strictly included within the languages accepted by OPAs.[2] We assumed so far the $\doteq$-acyclicity hypothesis to keep the notation as simple as possible so that the two formalisms are equivalent.

Recently, however, it has been observed [14] that such a restriction may hamper the benefits achievable by the parallel compilation techniques that exploit the local parsability property of OPLs [3]. Thus, it is time to introduce the necessary extension of OPGs

---

[2] The language $\{a^n(bc)^n\} \cup \{b^n(ca)^n\} \cup \{c^n(ab)^n\} \cup (abc)^+$ cannot be generated by an OPG because the $a \doteq b \doteq c \doteq a$ relations are necessary [9], but it is accepted by OPAs.

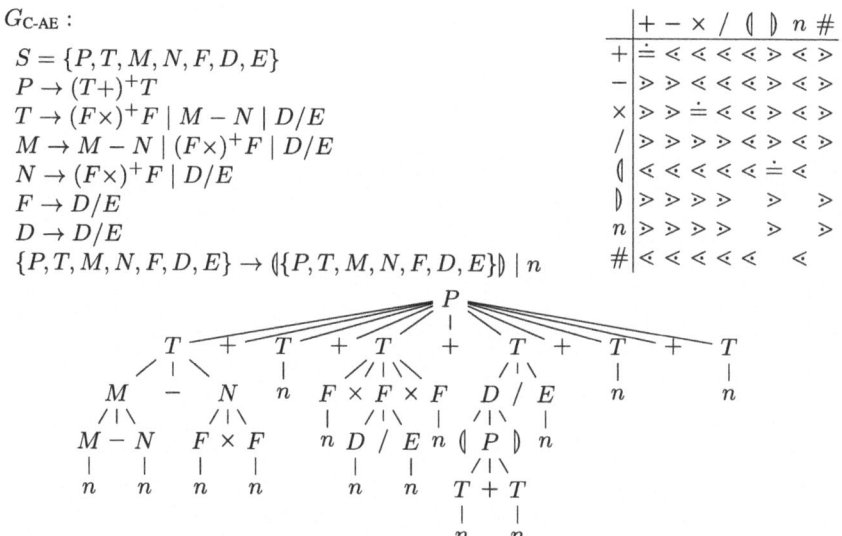

Fig. 4. A C-OPG (top left), its OPM (top right), and a ST generated by them. The notation $\{P, T, M, F, D, E\} \rightarrow (\!|\{P, T, M, N, F, D, E\}\!|)$ means that anyone of the nonterminals at the left can be rewritten as a pair of parentheses enclosing anyone of the same nonterminals.

so that the $\doteq$-acyclicity hypothesis can be avoided and they become fully equivalent to other formalisms to define OPLs.

Definition 10 (Cyclic Operator Precedence Grammar (C-OPG)). *A $^+$-O-expression on V^* is an expression obtained from the elements of V by iterative application of concatenation and the $^+$ operator[3], provided that any substring thereof has no two adjacent nonterminals; for convenience, and w.l.o.g., we assume that all subexpressions that are argument of the $^+$ operator are terminated by a terminal character.*

A Cyclic O-grammar (C-OG) is an O-grammar whose production rhs are $^+$-O-expressions. For a rule $A \rightarrow \alpha$ of a C-OG, the $\underset{G}{\Longrightarrow}$ (immediate derivation) relation is defined as $\beta A \gamma \Longrightarrow \beta \zeta \gamma$ iff ζ is a string belonging to the language defined by the $^+$-O-expression α, $L(\alpha)$. The $\doteq$ relation is redefined as $a \doteq b$ iff $\exists A \rightarrow \alpha \wedge \exists \zeta = \eta a C b \theta \mid (C \in V_N \cup \{\varepsilon\} \wedge \zeta \in L(\alpha))$. The other relations remain defined as for non-cyclic O-grammars. A C-OG is a C-OPG iff its OPM is conflict-free.

As a consequence of the definition of the immediate derivation relation for C-OPGs the STs derived therefrom can be unranked, i.e., their internal nodes may have an unbounded number of children.

Example 11. The C-OPG shown in Fig. 4 with its OPM generates a fairly complete language of parenthesized arithmetic expressions involving the four basic operations: as usual the multiplicative operations take precedence over the additive ones; subtraction takes precedence over sum and division over multiplication. The key novelty w.r.t.

---

[3] For our purposes $^+$ is more convenient than * without affecting the generality.

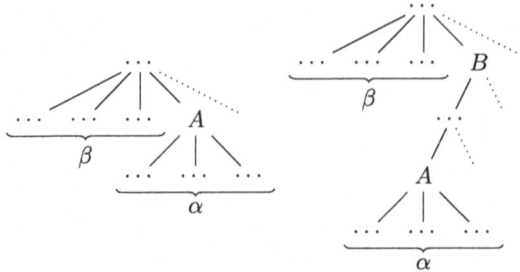

Fig. 5. When parsing α, the prefix previously under construction is β.

the traditional way of formalizing arithmetic expressions by means of OPGs are the $+ \doteq +$ and $\times \doteq \times$ OP relations; on the contrary we kept the structure that associates subtraction and division to the left, so that the grammar's STs—an example thereof is given at the bottom of Fig. 4—now fully reflect the semantics of arithmetic operations.

By looking at the ST of Fig. 4 and comparing it with the original Fig. 1, one can better envision why the introduction of cyclic $\doteq$ can support more effective parallel parsing algorithms for OPLs. Parallel parsing for OPLs is rooted in the *local parsability property* of this family: thanks to this property any fragment of input string enclosed within a pair of corresponding $\lessdot$ and $\gtrdot$ OP relations can be processed in parallel with other similar fragments. However, if, say, the $+$ operator is associated to the left (or right) as in the case of the upper ST of Fig. 1 the parsing of a sequence of $+$ must necessarily proceed sequentially from left to right. Conversely, if the ST has a structure like that of the lower tree of Fig. 1 the sequence of $+$—whether intermixed or not with other subtrees—can be arbitrarily split into several branches which can be parsed in parallel and, after that, can be joined into a unique subtree as imposed by the $\doteq$ OP relation between the corresponding extreme terminals of contiguous branches.

3.1 Equivalence Between C-OPGs and OPAs

The equivalence is obtained by adapting the analogous proof given in [15] where the additional hypothesis of M being $\doteq$-acyclic was exploited. First, we describe a procedure to build an OPA equivalent to a C-OPG. Then, we provide the converse construction.

Theorem 12 (From C-OPGs to OPAs). *Let (Σ, M) be an OP-alphabet. For any C-OPG defined thereon an equivalent OPA can be effectively built.*

A nondeterministic OPA[4] $\mathcal{A} = \langle \Sigma, M, Q, I, F, \delta \rangle$ from a given C-OPG G with the same precedence matrix M as G is built in such a way that a successful computation thereof corresponds to building bottom-up a syntax tree of G: the automaton performs a push transition when it reads the first terminal of a new rhs; it performs

---

[4] Any nondeterministic OPA can be transformed into a deterministic one at the cost of quadratic exponential increase in the size of the state space [15].

a shift transition when it reads a terminal symbol inside a rhs, i.e. a leaf with some left sibling leaf. It performs a pop transition when it completes the recognition of a rhs, then it guesses (nondeterministically) the nonterminal at the lhs. Each state contains two pieces of information: the first component is the prefix of the rhs under construction, whereas the second component is used to recover the rhs *previously under construction* whenever all rhs nested below have been completed (see Fig. 5).

Let $\hat{P}$ be the set of rhs γ where all $^+$ and related parentheses have been erased. Let $\tilde{P}$ be the set of strings $\tilde{\gamma} \in V^+$ belonging to the language of some rhs γ of P that is inductively defined as follows: if $(\eta)^+$ is a subexpression of γ such that η is a single string $\in V^+$ then $\tilde{\eta} = \{\eta, \eta\eta\}$; if $\eta = \alpha_1(\beta_1)^+\alpha_2(\beta_2)^+ \ldots \alpha_n$ where $\alpha_i \in V^*$, then $\tilde{\eta} = \{\eta_1, \eta_1\eta_1\}$ where $\eta_1 = \alpha_1\tilde{\beta}_1\alpha_2\tilde{\beta}_2 \ldots \alpha_n$.
E.g., let η be $(Ba(bc)^+)^+$; then $\hat{\eta} = \{Babc\}$ and $\tilde{\eta} = \{Babc, Babcbc, BabcBabc, BabcbcBabc, BabcBabcbc, BabcbcBabcbc\}$.

Let $\mathbb{P} = \{\alpha \in V^*\Sigma \mid \exists A \to \eta \in P \wedge \exists\beta(\alpha\beta \in \tilde{\eta})\}$ be the set of prefixes, ending with a terminal symbol, of strings $\in \tilde{P}$; define $\mathbb{Q} = \{\varepsilon\} \cup \mathbb{P} \cup N$, $Q = \mathbb{Q} \times (\{\varepsilon\} \cup \mathbb{P})$, $I = \{\langle\varepsilon, \varepsilon\rangle\}$, and $F = S \times \{\varepsilon\} \cup \{\langle\varepsilon, \varepsilon\rangle$ if $\varepsilon \in L(G)\}$. Note that $|\mathbb{Q}| = 1 + |\mathbb{P}| + |N|$ is $O(m^h)$ where m is the maximum length of the rhs in P, and h is the maximum nesting level of $^+$ operators in rhs; therefore $|Q|$ is $O(m^{2h})$.

The transition functions are defined by the following formulas, for $a \in \Sigma$ and $\alpha, \alpha_1, \alpha_2 \in \mathbb{Q}$, $\beta, \beta_1, \beta_2 \in \{\varepsilon\} \cup \mathbb{P}$, and where for any expression ξ, $\bar{\xi}$ is obtained from ξ by erasing parentheses and $^+$ operators:

$$
- \delta_{\text{shift}}(\langle\alpha, \beta\rangle, a) \ni
\begin{cases}
\text{if } \alpha \notin N: & \begin{cases} \text{if} \left(\begin{array}{l}\exists A \to \gamma \mid \gamma = \eta(\zeta)^+\theta \wedge \\ \alpha a = \bar{\eta}\bar{\zeta}\bar{\zeta} \wedge \alpha a\bar{\theta} \in L(\gamma) \cap \tilde{P}\end{array}\right) \\ \text{then } \langle\bar{\eta}\bar{\zeta}, \beta\rangle \text{ else } \langle\alpha a, \beta\rangle \end{cases} \\
\text{if } \alpha \in N: & \begin{cases} \text{if} \left(\begin{array}{l}\exists A \to \gamma \mid \gamma = \eta(\zeta)^+\theta \wedge \\ \beta\alpha a = \bar{\eta}\bar{\zeta}\bar{\zeta} \wedge \beta\alpha a\bar{\theta} \in L(\gamma) \cap \tilde{P}\end{array}\right) \\ \text{then } \langle\bar{\eta}\bar{\zeta}, \beta\rangle \text{ else } \langle\beta\alpha a, \beta\rangle \end{cases}
\end{cases}
$$

$$
- \delta_{\text{push}}(\langle\alpha, \beta\rangle, a) \ni \begin{cases} \langle a, \alpha\rangle & \text{if } \alpha \notin N \\ \langle\alpha a, \beta\rangle & \text{if } \alpha \in N \end{cases}
$$

$$
- \delta_{\text{pop}}(\langle\alpha_1, \beta_1\rangle, \langle\alpha_2, \beta_2\rangle) \ni \langle A, \gamma\rangle
$$

$$
\forall A: \begin{cases} \text{if } \alpha_1 \notin N: A \to \alpha \in P \wedge \alpha_1 \in L(\alpha) \cap \hat{P} \\ \text{if } \alpha_1 \in N: A \to \delta \in P \wedge \beta_1\alpha_1 \in L(\delta) \cap \hat{P} \end{cases} \text{ and } \gamma = \begin{cases} \alpha_2 \text{ if } \alpha_2 \notin N \\ \text{if } \alpha_2 \in N. \end{cases}
$$

The states reached by push and shift transitions have the first component in $\mathbb{P}$. If state $\langle\alpha, \beta\rangle$ is reached after a push transition, then α is the prefix of the rhs (deprived of the $^+$ operators) that is currently under construction and β is the prefix previously under construction; in this case α is either a terminal or a nonterminal followed by a terminal.

If the state is reached after a shift transition, and the α component of the previous state was not a single nonterminal, then the new α is the concatenation of the first component of the previous state with the read character. If, instead, the α component of the previous state was a single nonterminal—which was produced by a pop transition—then the new α also includes the previous β and β is not changed from the previous state. However, if the new α becomes such that a suffix thereof is a double occurrence of a string $\zeta \in L((\zeta)^+)$—hence $\alpha \in \mathbb{P}$—then the second occurrence of ζ is cut from the new α, which therefore becomes a prefix of an element of $\hat{P}$.

stack	state	current input
$\perp$	$\langle \varepsilon, \varepsilon \rangle$	$n + n + n/n/n + n + n\#$
$\perp[n,\ \langle \varepsilon, \varepsilon \rangle]$	$\langle n, \varepsilon \rangle$	$+n + n/n/n + n + n\#$
$\perp$	$\langle T, \varepsilon \rangle$	$+n + n/n/n + n + n\#$
$\perp[+,\ \langle T, \varepsilon \rangle]$	$\langle T+, \varepsilon \rangle$	$n + n/n/n + n + n\#$
$\perp[+,\ \langle T, \varepsilon \rangle][n,\ \langle T+, \varepsilon \rangle]$	$\langle n, T+ \rangle$	$+n/n/n + n + n\#$
$\perp[+,\ \langle T, \varepsilon \rangle]$	$\langle T, T+ \rangle$	$+n/n/n + n + n\#$
$\perp[+,\ \langle T, \varepsilon \rangle]$	$\langle \mathbf{T+, T+} \rangle$	$n/n/n + n + n\#$
$\perp[+,\ \langle T, \varepsilon \rangle][n,\ \langle T+, T+ \rangle]$	$\langle n, T+ \rangle$	$/n/n + n + n\#$
$\perp[+,\ \langle T, \varepsilon \rangle]$	$\langle D, T+ \rangle$	$/n/n + n + n\#$
$\perp[+,\ \langle T, \varepsilon \rangle][/,\ \langle D, T+ \rangle]$	$\langle D/, T+ \rangle$	$n/n + n + n\#$
$\perp[+,\ \langle T, \varepsilon \rangle][/,\ \langle D, T+ \rangle][n,\ \langle /, D \rangle]$	$\langle n, D/ \rangle$	$/n + n + n\#$
$\perp[+,\ \langle T, \varepsilon \rangle][/,\ \langle D, T+ \rangle]$	$\langle E, D/ \rangle$	$/n + n + n\#$
$\perp[+,\ \langle T, \varepsilon \rangle]$	$\langle D, T+ \rangle$	$/n + n + n\#$
$\perp[+,\ \langle T, \varepsilon \rangle][/,\ \langle D, T+ \rangle]$	$\langle D/, T+ \rangle$	$n + n + n\#$
$\perp[+,\ \langle T, \varepsilon \rangle][/,\ \langle D, T+ \rangle][n,\ \langle D/, T+ \rangle]$	$\langle n, D/ \rangle$	$+n + n\#$
$\perp[+,\ \langle T, \varepsilon \rangle][/,\ \langle D, T+ \rangle]$	$\langle E, D/ \rangle$	$+n + n\#$
$\perp[+,\ \langle T, \varepsilon \rangle]$	$\langle T, T+ \rangle$	$+n + n\#$
$\perp[+,\ \langle T, \varepsilon \rangle]$	$\langle \mathbf{T+, T+} \rangle$	$n + n\#$
$\perp[+,\ \langle T, \varepsilon \rangle][n,\ \langle T+, T+ \rangle]$	$\langle n, T+ \rangle$	$+n\#$
$\perp[+,\ \langle T, \varepsilon \rangle]$	$\langle T, T+ \rangle$	$+n\#$
$\perp[+,\ \langle T, \varepsilon \rangle]$	$\langle \mathbf{T+, T+} \rangle$	$n\#$
$\perp[+,\ \langle T, \varepsilon \rangle][n,\ \langle T+, T+ \rangle]$	$\langle n, T+ \rangle$	$\#$
$\perp[+,\ \langle T, \varepsilon \rangle]$	$\langle T, T+ \rangle$	$\#$
$\perp$	$\langle P, \varepsilon \rangle$	$\#$

Fig. 6. A run of the OPA built from the C-OPG of Fig. 4 accepting the sentence $n + n + n/n/n + n + n$. The states truncated by erasing a repeated suffix ζ occurring under the scope of a $^+$ operator are emphasized in boldface.

The states reached by a pop transition have the first component in N: if $\langle A, \gamma \rangle$ is such a state, then A is the corresponding lhs, and γ is the prefix previously under construction.

For instance, imagine that a C-OPG contains the rules $A \to (Ba(bc)^+)^+a$ and $B \to h$ and that the corresponding OPA $\mathcal{A}$ parses the string $habcbchabca$: after scanning the prefix $habcb$ $\mathcal{A}$ has reduced h to B and has $Babcb$ as the first component of its state; after reading the new c it recognizes that the suffix of the first state component would become a second instance of bc belonging to $(bc)^+$; thus, it goes back to $Babc$. Then, it proceeds with a new reduction of h to B and, when reading with a shift the second a appends Ba to its current β which was produced by the previous pop so that the new α becomes $BabcBa$; after shifting b it reads c and realizes that its new α would become $BabcBabc$, i.e., an element of $(Ba(bc)^+)^+$ and therefore "cuts" it to the single instance thereof, i.e., $Babc$. Finally, after having shifted the last a it is ready for the last pop.

The result of δ_{shift} and δ_{push} is a singleton, whereas δ_{pop} may produce several states, in case of repeated rhs. Thus, if G is BD, the corresponding $\mathcal{A}$ is deterministic.

Example 13. Figure 6 displays a run of the OPA obtained from the C-OPG of Fig. 4 accepting the sentence $n + n + n/n/n + n + n$.

The construction of a C-OPG equivalent to a given OPA is far simpler than the converse one, thanks to the explicit structure associated to words by the precedence matrix. The key difference w.r.t. the analogous construction given in [15] is that even simple chains can have unbounded length because the $\doteq$ relation may be circular.

In analogy with the definition of $\tilde{P}$, we define *essential supports* as those supports where possible cyclic behaviors of the OPA along a sequence of terminals—whether with interposing nonterminals or not—occur exactly twice.

Definition 14. *An essential support of a simple chain* $^{a_0}[a_1 a_2 \ldots a_n]^{a_{n+1}}$ *is any path in* $\mathcal{A}$ *of the form* $q_0 \xrightarrow{a_1} q_1 \dashrightarrow \ldots \dashrightarrow q_{n-1} \xdashrightarrow{a_n} q_n \xRightarrow{q_0} q_{n+1}$, *where any cycle* $q_i a_{i1} \ldots a_{ik} q_i$ *is repeated exactly twice. Composed chains are treated similarly.*

E.g., with reference to the OPA built from the C-OPG of Fig. 4 an essential support of the chain $^{\#}[n + n + n + n + n]^{\#}$ is: $\langle \varepsilon, \varepsilon \rangle \xrightarrow{n} \langle T, \varepsilon \rangle \xrightarrow{+} \langle T+, \varepsilon \rangle \xrightarrow{n} \langle T, T+ \rangle \dashrightarrow^{+}$ $\langle T+, T+ \rangle \xrightarrow{n} \langle T, T+ \rangle \dashrightarrow^{+} \langle T+, T+ \rangle \xrightarrow{n} \langle T, T+ \rangle \xRightarrow{\langle T, \varepsilon \rangle} \langle P, \varepsilon \rangle$.

Lemma 15. *The essential supports of simple chains of any OPA have an effectively computable bounded length.*

Theorem 16 (From OPAs to C-OPGs). *Let* (Σ, M) *be an OP-alphabet. For any OPA defined thereon an equivalent C-OPG can be effectively built.*

Proof Given an OPA $\mathcal{A} = \langle \Sigma, M, Q, I, F, \delta \rangle$, we build an equivalent C-OPG G having OPM M. The equivalence between $\mathcal{A}$ and G is then rather obvious.

G's nonterminals are the 4-tuples $(a, q, p, b) \in \Sigma \times Q \times Q \times \Sigma$, written as $\langle ^{a}p, q^{b} \rangle$. G's rules are built as follows:

For every essential support of a simple chain, P contains $\langle ^{a_0}q_0, q_{n+1}{}^{a_{n+1}} \rangle \to a_1 a_2 \ldots a_n$, where every double sequence $a_{i1} \ldots a_{ik} a_{i1} \ldots a_{ik}$ is recursively replaced by $(a_{i1} \ldots a_{ik})^{+}$ by proceeding from the innermost cycles to the outermost ones. Furthermore, if $a_0 = a_{n+1} = \#$, q_0 is initial, and q_{n+1} is final, then $\langle ^{\#}q_0, q_{n+1}{}^{\#} \rangle$ is in S.

For every essential support of a composed chain $^{a_0}[x_0 a_1 x_1 a_2 \ldots a_n x_n]^{a_{n+1}}$, P contains the rule $\langle ^{a_0}q_0, q_{n+1}{}^{a_{n+1}} \rangle \to \Lambda_0 a_1 \Lambda_1 a_2 \ldots a_n \Lambda_n$, where, for every $i = 0, 1, \ldots, n$, $\Lambda_i = \langle ^{a_i}q_i, q'_i{}^{a_{i+1}} \rangle$ if $x_i \neq \varepsilon$ and $\Lambda_i = \varepsilon$ otherwise, and double cyclic sequences $\alpha_i \alpha_i$ are replaced by $(\alpha_i)^{+}$ in the same way as for simple chains. If $a_0 = a_{n+1} = \#$, q_0 is initial, and q_{n+1} is final, then $\langle ^{\#}q_0, q_{n+1}{}^{\#} \rangle$ is in S; if ε is accepted by $\mathcal{A}$, $A \to \varepsilon$ is in P, A being a new axiom not otherwise occurring in any other rule.

Notice that the above construction is effective thanks to Lemma 15 and to the fact that subchains of composed chains are replaced by nonterminals Λ_i.

4   Equivalences and Closure Properties

Previous literature proved the equivalence of defining OPLs through OPGs, OPAs, MSO logic, OPEs and OPSC. The reciprocal inclusions between MSO and OPA, between OPA and the finite equivalence classes of OPL syntactic congruence, and the inclusion of MSO in OPE have been proved without the hypothesis of non-circularity of the $\doteq$ relation; the reciprocal inclusions between C-OPG and OPA have been restated in Sect. 3; it remains the inclusion of OPE in OPG which was proved in [17] under the restrictive hypothesis. The proof used the claim that deriving an OPG from an OPE may exploit the closure properties of OPLs, in particular, w.r.t. *; such a closure, however, was proved in [7] by using OPGs, again, under the restrictive hypothesis.

In OPEs the * operator is applied to subexpressions independently on the OP relations between their last and first terminal character. Thus, it is first convenient to rewrite the OPE in a normal form using the $^+$ operator instead of the * one to avoid having to deal explicitly with the case of the ε string. Then, subexpressions of type $(\alpha)^+$ where the last terminal of α is not in relation $\doteq$ with the first one are replaced by the same procedure defined in [7] to prove effectively the closure w.r.t. the * operator. The new rules will produce a right or left-linear subtree of the occurrences of α depending on the OP relation between the two extreme terminals of α and will avoid the use of the * and $^+$ operators which are not permitted in the original OPGs.

The remaining substrings including the $^+$ operator are the new rhs of the C-OPG. The other technicalities of the construction of an OPG equivalent to an OPE are identical to those given in [17] and are not repeated here.

All major closure properties of OPLs have been originally proved by referring to their generating OPGs and some of them, in particular the closure w.r.t. *, required the $\doteq$-acyclicity hypothesis. Thus, it is necessary to prove them again. However, since some of those proofs are technically rather involved, here we simply observe that it is easier to restate the same properties by exploiting OPAs which are now fully equivalent to C-OPGs. Thanks to the determinization of nondeterministic OPAs, closure w.r.t. boolean operations is "for free". Closure w.r.t. concatenation can be seen as a corollary of the closure proved in [15] of the concatenation between an OPL whose strings have finite length and an ω-OPL, i.e., a language of infinite strings. The construction is based on a nondeterministic guess of the position where a string of the first language could end—and a nontrivial technique to decide whether it could be accepted even in the absence of the # delimiter—. Then, the closure w.r.t. * is obtained simply by allowing the OPA to repeat such a guess any number of times until the real # is found.

5   Conclusion

We have filled up a longstanding "hole" in the theory of OPLs under the pressure of recent applications in the field of parallel compilation that showed how such a hole could hamper the benefits of parallelism [14]. The new formalism of C-OPGs, fully equivalent to OPAs, MSO-logic, and OPEs, can be exploited for the parallel compilation techniques of [14] or to improve the efficiency of techniques based on the less powerful OPGs [3].

Other algebraic and logic properties of OPLs, e.g., aperiodicity, star-freeness, first-order definability [17, 18] can be re-investigated in the light of this generalization.

 Acknowledgments. This work was partially funded by the EU Commission in the Horizon Europe research and innovation programme under grant agreement No. 101107303 (MSCA-PF CORPORA).

References

1. Alur, R., Madhusudan, P.: Adding nesting structure to words. J. ACM **56**(3) (2009). https://doi.org/10.1145/1516512.1516518
2. Autebert, J., Berstel, J., Boasson, L.: Context-free languages and pushdown automata. In: Handbook of Formal Languages, no. 1, pp. 111–174 (1997). https://doi.org/10.1007/978-3-642-59136-5_3
3. Barenghi, A., Crespi Reghizzi, S., Mandrioli, D., Panella, F., Pradella, M.: Parallel parsing made practical. Sci. Comput. Program. **112**(3), 195–226 (2015). https://doi.org/10.1016/j.scico.2015.09.002
4. Barenghi, A., Crespi Reghizzi, S., Mandrioli, D., Pradella, M.: Parallel parsing of operator precedence grammars. Inf. Process. Lett. **113**(7), 245–249 (2013). https://doi.org/10.1016/j.ipl.2013.01.008
5. Chiari, M., Mandrioli, D., Pontiggia, F., Pradella, M.: A model checker for operator precedence languages. ACM Trans. Program. Lang. Syst. **45**(3), 19:1–19:66 (2023). https://doi.org/10.1145/3608443
6. Chiari, M., Mandrioli, D., Pradella, M.: Cyclic operator precedence grammars for parallel parsing. CoRR arxiv:2309.04200 (2023)
7. Crespi Reghizzi, S., Mandrioli, D.: Operator precedence and the visibly pushdown property. J. Comput. Syst. Sci. **78**(6), 1837–1867 (2012). https://doi.org/10.1016/j.jcss.2011.12.006
8. Crespi Reghizzi, S., Mandrioli, D., Martin, D.F.: Algebraic properties of operator precedence languages. Inf. Control **37**(2), 115–133 (1978)
9. Crespi Reghizzi, S., Pradella, M.: Beyond operator-precedence grammars and languages. J. Comput. Syst. Sci. **113**, 18–41 (2020). https://doi.org/10.1016/j.jcss.2020.04.006
10. Floyd, R.W.: Syntactic analysis and operator precedence. J. ACM **10**(3), 316–333 (1963)
11. Grune, D., Jacobs, C.J.: Parsing Techniques: A Practical Guide. Springer, New York (2008). https://doi.org/10.1007/978-0-387-68954-8
12. Harrison, M.A.: Introduction to Formal Language Theory. Addison Wesley, Boston (1978)
13. Henzinger, T.A., Kebis, P., Mazzocchi, N., Saraç, N.E.: Regular methods for operator precedence languages. In: Etessami, K., Feige, U., Puppis, G. (eds.) 50th International Colloquium on Automata, Languages, and Programming, ICALP 2023, Paderborn, Germany, 10–14 July 2023. LIPIcs, vol. 261, pp. 129:1–129:20. Schloss Dagstuhl - Leibniz-Zentrum für Informatik (2023). https://doi.org/10.4230/LIPIcs.ICALP.2023.129
14. Li, L., Taura, K.: Associative operator precedence parsing: a method to increase data parsing parallelism. In: Proceedings of the International Conference on High Performance Computing in Asia-Pacific Region, HPC Asia 2023, Singapore, 27 February–2 March 2023, pp. 75–87. ACM (2023).https://doi.org/10.1145/3578178.3578233
15. Lonati, V., Mandrioli, D., Panella, F., Pradella, M.: Operator precedence languages: their automata-theoretic and logic characterization. SIAM J. Comput. **44**(4), 1026–1088 (2015). https://doi.org/10.1137/140978818
16. Mandrioli, D., Pradella, M.: Generalizing input-driven languages: theoretical and practical benefits. Comput. Sci. Rev. **27**, 61–87 (2018). https://doi.org/10.1016/j.cosrev.2017.12.001
17. Mandrioli, D., Pradella, M., Crespi Reghizzi, S.: Aperiodicity, star-freeness, and first-order definability of structured context-free languages. Logical Methods Comput. Sci. **19** (2023).https://doi.org/10.46298/lmcs-19(4:12)2023
18. McNaughton, R., Papert, S.: Counter-Free Automata. MIT Press, Cambridge (1971)
19. Salomaa, A.K.: Formal Languages. Academic Press, New York (1973)

On the Complexity and Approximability of Bounded Access Lempel Ziv Coding

Ferdinando Cicalese$^{(\boxtimes)}$ and Francesca Ugazio

Department of Computer Science, University of Verona, Strada le Grazie 15,
37134 Verona, Italy
ferdinando.cicalese@univr.it, francesca.ugazio@studenti.univr.it

Abstract. We study the complexity of constructing an optimal parsing φ of a string $\mathbf{s} = s_1 \dots s_n$ under the constraint that given a position p in the original text, and the LZ76 (also known as LZ77 or simply Lempel-Ziv) encoding of T based on φ, it is possible to identify/decompress the character s_p by performing at most c accesses to the LZ encoding, for a given integer c. We refer to such a parsing φ as a c-bounded access LZ parsing or c-BLZ parsing of $\mathbf{s}$. We show that for any constant c the problem of computing the optimal c-BLZ parsing of a string, i.e., the one with the minimum number of phrases, is NP-hard and also APX hard, i.e., no PTAS can exist under the standard complexity assumption $P \neq NP$. We also study the ratio between the sizes of an optimal c-BLZ parsing of a string $\mathbf{s}$ and an optimal LZ76 parsing of $\mathbf{s}$ (which can be greedily computed in polynomial time).

Keywords: Lempel Ziv parsing · bounded access encoding · NP-hardness · approximation

1 Introduction

LZ76 [14][1] parses a text $\mathbf{s}$ from left to right by defining each new phrase as the longest substring already appeared in the text plus a new character. It is known that such a parsing is the one with the minimum number of phrases among all parsing where each phrase must be chosen to be either a single character or a previously occurred substring plus an additional character. LZ76 parsing is the basis of the LZ76 encoding [21], where each phrase ϕ is encoded as either (i) the triple (p, ℓ, c) where p and ℓ are the position and length of the previously

---

[1] This parsing appeared originally in 1976 in [14] in the context of defining a measure of string complexity. In 1977, in [21], the same parsing was presented by the same authors as the basis of a compression algorithm. In the literature, both the parsing and the compression algorithm are referred to as LZ77. Here, we decided to refer both to the original idea, hence we choose to use the name LZ76.

Ferdinando Cicalese is member of the Gruppo Nazionale Calcolo Scientifico-Istituto Nazionale di Alta Matematica (GNCS-INdAM).

J. D. Day and F. Manea (Eds.): DLT 2024, LNCS 14791, pp. 114–130, 2024.
https://doi.org/10.1007/978-3-031-66159-4_9

occurring substring equal to the longest proper prefix of ϕ and c is the last character of ϕ; or (ii) as the triple $(0, 0, c)$ when ϕ has exactly one character. Remarkably, LZ76 parsing and LZ76 encoding of a string $\mathbf{s}$ can be done in linear time as well as the decompression of an encoding can also be done in linear time, reconstructing the original string from left to right [17].

However, when an encoding is employed for compressed data structures, additional properties are desirable. A property that has recently gained attention [3,4,8–10,15] is the following: given the encoding of a string $\mathbf{s}$ and a position p (or positions p, q) we would like to be able to return the character s_p (or the substring $\mathbf{s}[p, q]$) without having to decompress the whole string, and possibly in time $O(1)$ (resp. $O(q - p)$), i.e. proportional to the length of the extracted substring and not of the encoding or the text.

To clarify this issue, let us consider the basic LZ76 encoding of a string $\mathbf{s}$. If we want to extract the character s_p we can compute the index i_1 of the phrase φ_{i_1} that contains the pth character. However, only if s_p is the last character of φ_{i_1} we would find it directly in the encoding. Otherwise, if s_p is in the proper prefix of φ_{i_1}, from the encoding, we can compute the index i_2 of the phrase φ_{i_2} and the position q such that $s_q = s_p$ is part of φ_{i_2}. Again, if s_q is the ending character of φ_{i_2} we have access to it, otherwise additional *hops* will be necessary until we reach a phrase φ_{i_t} such that s_p is its end character. In such case we say that position p requires t hops or access to extract s_p.

It should be clear that given a parsing φ (or equivalently the associated encoding) of a string $\mathbf{s}$ we can compute for each position $p \in [|\mathbf{s}|]$ the value $hop^{\varphi}(p)$ corresponding to the number of hops necessary to identify s_p. We are interested in parsings/encodings that guarantees $hop^{\varphi}_{\max}(\mathbf{s}) = \max_i hop^{\varphi}(i)$ to be small. The problem is that in general for the LZ76 greedy parsing/encoding $\varphi^* = \varphi_{LZ76}$ the value $hop^{\varphi^*}_{\max}(\mathbf{s})$ appears to be large [15].

In this paper we study the complexity of the following problem, which was originally considered in [12] and recently experimentally analyzed in [2,15]:

BOUNDED ACCESS LEMPEL ZIV PARSING (BLZ)
Input: A strings $\mathbf{s}$ and an integer c.
Output: A parsing φ^*_c of $\mathbf{s}$ with the minimum number of phrases among all LZ-parsings φ_c of $\mathbf{s}$ satisfying $hop^{\varphi_c}_{\max}(\mathbf{s}) \leq c$

Our Results. We show that the above problem is NP-hard for any constant c. Moreover, we show that, in terms of approximation, the problem is *APX*-hard, which, under the standard complexity assumption $NP \neq P$ implies that no PTAS is expected to exist. Finally we give some results regarding the ratio between the size of an optimal LZ76 parsing and the size of an optimal c-BLZ parsing (i.e., guaranteeing $hop^{\varphi_c}_{\max}(\mathbf{s}) \leq c$) for the same string $\mathbf{s}$.

Related Work. The BLZ problem was originally introduced in [12] where it was called LZ-COST. In the same paper the authors introduced and focused on another variant of the LZ76 algorithm, called LZ-END, where the constraint is that the longest proper prefix of any phrase must have a previous occurrence whose last character is the last character of another phrase. Both LZ-END and

LZ-COST were motivated by the possibility of providing fast access to substrings from the compressed text. In [12], the authors stated not to have found any efficient parsing algorithm for LZ-COST. In [15] several algorithms for constructing c-BLZ parsing are presented with extensive empirical evaluation with respect to an optimal LZ76 greedy parsing. The authors show that, with some tolerable cost in terms of additional space, their parsing algorithms can guarantee much better access cost with respect to the original LZ76 parsing. To the best of our knowledge, ours is the first result on the complexity of computing or approximating an optimal c-BLZ parsing. Extensive literature exists on LZ-End parsing. Finding an optimal LZ-End parsing was proved NP-hard in [3]. Previously, the focus had been on comparing the greedy LZ-End parsing to the greedy LZ76 parsing. In [11] it is shown that the optimal LZ-End parsing is bounded by $O(z \log^2 n)$ where z is the size of the LZ76 parsing. The upper bound was later improved by a factor of $\log \log(n/z)$ in [8].

Computing an optimal 1-BLZ parsing of $\mathbf{s}$ (our problem in the particular case $c = 1$) is equivalent to finding an optimal compression for $\mathbf{s}$ in the *unidirectional non-recursive OPM scheme* as defined and studied by Storer in [18,19]. Remarkably, the NP-hardness of this problem, which we prove here, was conjectured in [18] and has remained open until now.

2   Notation and Basic Facts

Definitions for Strings. We denote by Σ^* the set of all strings of finite length over a finite alphabet Σ. For a string $\mathbf{s} \in \Sigma^*$ we denote by $|\mathbf{s}|$ the length of $\mathbf{s}$, i.e., the number of its characters. We refer to the unique string of length 0 as the null-string, and denote it by ϵ. We assume that $\epsilon \in \Sigma^*$. For any integer $n \geq 0$, let $\Sigma^n = \{\mathbf{s} \in \Sigma^* \mid |\mathbf{s}| = n\}$. For a string $\mathbf{s} \in \Sigma^n$ and any integers $1 \leq i \leq j \leq n$, we denote by s_i the ith character of $\mathbf{s}$ and by $\mathbf{s}[i,j]$ the sequence of characters $s_i s_{i+1} \ldots s_j$. We say that $\mathbf{s}[i,j]$ is a substring of $\mathbf{s}$. If $i = 1$ (resp. $j = n$) we say that $\mathbf{s}[i,j]$ is a *prefix* (resp. *suffix*) of $\mathbf{s}$. A square of a string $\mathbf{s}$ is a substring $\mathbf{t}$ of $\mathbf{s}$ which occurs twice consecutively in $\mathbf{s}$, i.e., $\mathbf{s} = \mathbf{s'tts''}$, for some (possibly empty) strings $\mathbf{s'}, \mathbf{s''}$. If no square is present in $\mathbf{s}$ we say that $\mathbf{s}$ is *square-free*.

Definition 1 (Parsing). *A parsing φ of a string $\mathbf{s}$ is a partition of $\mathbf{s}$ into consecutive substrings, $\mathbf{s}[1, i_1]\mathbf{s}[i_1 + 1, i_2] \cdots \mathbf{s}[i_{t-1} + 1, i_t]$. For each $j = 1, \ldots, t$, the substring $\mathbf{s}[i_{j-1} + 1, i_j]$ is called a* phrase *of the parsing φ and will be denoted by φ_j. The size of the parsing $\varphi = \mid \varphi_1 \mid \cdots \mid \varphi_t \mid$, is the number t of its phrases, and will be denoted by $|\varphi(\mathbf{s})|$ (or also $|\varphi|$ whenever $\mathbf{s}$ is clear from the context).*

For a substring $\mathbf{w}$ of $\mathbf{s}$, we use $\varphi(\mathbf{w})$ to denote the parsing induced by φ on $\mathbf{w}$ and with $|\varphi(\mathbf{w})|$ the number of its phrases[2].

Definition 2 (LZ-parsing and encoding). *We say that φ is an LZ-parsing if for each $j = 1, \ldots, t$ the phrase $\varphi_j = \mathbf{s}[i_{j-1} + 1, i_j]$ is either a single character*

---

[2] We will generally consider situations where $\mathbf{w}$ starts and ends at the boundary of some phrase of φ, i.e., $\mathbf{w} = \mathbf{s}[i_j + 1, i_{j'}]$. In such a case, we have $\varphi(\mathbf{w}) = \mid \varphi_{j+1} \mid \varphi_{j+2} \mid \cdots \mid \varphi_{j'} \mid$ and $|\varphi(\mathbf{w})| = j' - j..$

or $s[i_{j-1}+1, i_j - 1]$ is a substring of s occurring in s strictly before the starting position $i_{j-1}+1$ of the phrase φ_j, i.e., there exists $\ell \leq i_{j-1}$, such that $s[\ell, \ell + (i_j - i_{j-1}) - 2] = s[i_{j-1}+1, i_j - 1]$. We refer to $s[\ell, \ell + (i_j - i_{j-1}) - 2] = s[i_{j-1}+1, i_j-1]$ as the source of φ_j. Analogously, for each position $p = i_{j-1}+t$, for $t = 1, \ldots, i_j - 2 - i_{j-1}$ we say that position $q = \ell + t - 1$ is the source of p in φ.

If $\ell + (i_j - i_{j-1}) - 2 \geq i_{j-1} + 1$ i.e., the substrings $s[\ell, \ell + (i_j - i_{j-1}) - 2]$ and φ_j overlap, it is easy to see that φ_j must be a power of $s[\ell, i_{j-1}]$, i.e., $\varphi_j = s[\ell, i_{j-1}]s[\ell, i_{j-1}] \ldots s[\ell, i_{j-1}]s[\ell, \ell-1+(i_j-i_{j-1}) \mod (i_{j-1}-\ell+1)]$. In this case, for each $t = 1, \ldots, i_j - i_{j-1} - 2$ the source of $p = i_{j-1}+t$ is the position $q = \ell+((t-1) \mod (i_{j-1} - \ell + 2))$, i.e., the position of the corresponding character in the repeated substring $s[\ell, i_{j-1}]$.

For an LZ-parsing we assume that for any non-singleton phrase, $\varphi_j = s[i_{j-1}+1, i_j]$, the source substring's starting position ℓ is also chosen together with the parsing[3].

Given an LZ-parsing φ of a string s an LZ76 encoding of s is obtained by replacing each phrase of φ_j by

- the triple $(0, 0, s_{i_{j-1}+1})$ if φ_j is a singleton phrase, i.e., $i_j = i_{j-1}+1$.
- the triple $(\ell, i_j - i_{j-1} - 1, s_{i_j})$ if $|\varphi_j| > 1$ and $\varphi_j = s[i_{j-1}+1, i_j] = s[\ell, \ell + (i_j - i_{j-1}) - 2]s_{i_j}$.

Definition 3 (Access time - hop-number). Let φ be an LZ-parsing of a string s. We define the access time function $hop^\varphi : [|s|] \mapsto \mathbb{N}$ mapping position p to the number of accesses necessary to identify s_p given the LZ-parsing φ:

$$hop^\varphi(p) = \begin{cases} 0 & \text{if } p \text{ is the position of the last character of a phrase of } \varphi \\ hop^\varphi(q) + 1 & \text{if position } q \text{ is the source of position } p \text{ in } \varphi \end{cases}$$

We will also refer to $hop^\varphi(x)$ as the hop-number of position x.
Given an LZ parsing φ of a string s, we define

$$hop^\varphi_{\max}(s) = \max_{p=1,\ldots,|s|} hop^\psi(p).$$

Definition 4 (c-BLZ parsing). Fix an LZ-parsing φ of a string s. We say that φ is a c-BLZ parsing of s if $hop^\varphi_{\max}(s) \leq c$, i.e., any character of s can be recovered from the parsing φ with at most c hops.

Definitions for Graphs. All graphs considered in this paper are undirected. A *graph* is a pair $G = (V, E)$ where $V = V(G)$ is a finite set of *vertices* and $E = E(G)$ is a set of 2-element subsets of V called *edges*. Two vertices u and v in a graph $G = (V, E)$ are *adjacent* if $\{u, v\} \in E$. We also say that edge $\{u, v\}$

---

[3] Note that in general for a given phrase there could be more than one possible preceding substring which coincide to its longest proper prefix. The assumption that sources are chosen with the parsing φ implies that the encoding is uniquely defined by φ and allows us to focus only on the parsing.

is incident in vertices u and v. We say that G is B-regular if for each vertex v there are exactly B edges incident in v.

A *vertex cover* in G is a set C of vertices such that every edge of G is incident in at least one vertex of C. The minimum size of a vertex cover of G is denoted by $\tau(G)$. The MINIMUM VERTEX COVER (MIN-VC) problem is the following minimization problem: Given a graph G find a vertex cover of G of size $\tau(G)$. We will use MIN-VC-4 for the restriction of MIN-VC to 4-regular graphs.

3   The BLZ Problem is NP-hard

For the hardness of the BLZ PROBLEM in the introduction, it would be enough to prove the existence of a single c for which the problem is NP-hard. In fact, we are able to show a stronger result, namely, that the hardness of the problem holds for any possible choice of c (constant w.r.t. $\mathbf{s}$). For this, we consider the following variant where the parameter c is fixed (not part of the input).

c-BOUNDED ACCESS LEMPEL ZIV PARSING (c-BLZ)
Input: A strings $\mathbf{s}$.
Output: An LZ-parsing φ_c^* of $\mathbf{s}$ with the minimum number of phrases among all LZ-parsings φ_c of $\mathbf{s}$ satisfying $hop_{\max}^{\varphi_c}(\mathbf{s}) \leq c$

We show that for any integer $c \geq 1$ the c-BLZ problem is NP-hard. This implies the hardness of the BLZ PROBLEM in the introduction.

3.1   The Reduction

Let $G = (V, E)$ be a graph. We are going to show how to build a string $\mathbf{s} = \mathbf{s}_G$ over an alphabet of size $O(m + n)$, where $n = |V|$ and $m = |E|$. The string $\mathbf{s}_G$ will be constructed in order to guarantee that G has a vertex cover of size k iff the string $\mathbf{s}_G$ has a c-BLZ parsing of size $4n + 6m + k + \ell(c - 1)$, where ℓ is an integer[4] $> k$. Since finding a minimum vertex cover is NP-hard, from our reduction the NP-hardness of the c-BLZ problem follows.

Let $v_1, \ldots, v_n$ denote the vertices of V and $e_1, \ldots, e_m$ denote the edges in E. The string $\mathbf{s}$ will be the concatenation of several pieces that we are now going to describe.

Let P be the following string that contains a distinct character for each vertex and a distinct character for each edge of G separated by additional and distinct *separator*-characters $\#_1^p, \ldots, \#_{n+m}^p$.

$$P = v_1 \#_1^p v_2 \#_2^p \ldots v_n \#_n^p e_1 \#_{n+1}^p e_2 \#_{n+2}^p \cdots e_m \#_{n+m}^p$$

For each $i = 1, \ldots n$, we have a (*vertex*) string X_i associated to vertex v_i, which is defined as follows:

$$X_i = v_i' v_i \#_i^v.$$

---
[4] The parameter ℓ is important only for the case $c > 1$. For the proof it is crucial that $\ell > k$, hence setting or assuming $\ell = n$ is sufficient in all following arguments.

Moreover, we define X to be the concatenation of these vertex strings:

$$X = X_1 X_2 \ldots X_n.$$

For each $i = 1, \ldots m$, we have an (*edge*) string Y_i associated to edge $e_i = (v_p, v_q)$, which is defined as follows:

$$Y_i = v'_p \, v_p \, e_i \, \$_i \, v'_q \, v_q \, e_i \, \$_i \, \#_i^e. \tag{1}$$

Moreover, we define Y to be the concatenation of these edge strings:

$$Y = Y_1 Y_2 \ldots Y_m.$$

Let $\mathbf{s}^{(1)} = PXY$. For any integer $\ell > k$, we recursively define the strings $\alpha^{(i)}$, for $i = 1, 2, \ldots$

$$\alpha^{(i)} = \begin{cases} \mathbf{s}^{(1)} = PXY & i = 1 \\ \alpha^{(i-1)} \, \beta^{(i-1)} & i > 1 \end{cases}$$

where for $i \geq 1$

$$\beta^{(i)} = \alpha^{(i)} \, \#_1^{(i)} \ldots \alpha^{(i)} \, \#_\ell^{(i)},$$

Then, we let the string $\mathbf{s} = \mathbf{s}_G$ be defined by:

$$\mathbf{s} = \alpha^{(c)} = \alpha^{(1)} \, \beta^{(1)} \, \beta^{(2)} \ldots \beta^{(c-1)}$$

$$= \underbrace{\underbrace{\underbrace{\alpha^{(1)} \beta^{(1)}}_{\alpha^{(2)}} \, \underbrace{\alpha^{(2)} \, \#_1^{(2)} \ldots \alpha^{(2)} \, \#_\ell^{(2)}}_{\beta^{(2)}}}_{\alpha^{(3)}} \, \underbrace{\alpha^{(3)} \#_1^{(3)} \ldots \alpha^{(3)} \#_\ell^{(3)}}_{\beta^{(3)}} \ldots \underbrace{\alpha^{(c-1)} \#_1^{(c-1)} \ldots \alpha^{(c-1)} \#_\ell^{(c-1)}}_{\beta^{(c-1)}}}_{\alpha^{(4)}}$$

For each $i = 1, \ldots, c-1$, and for $\gamma = 1, \ldots, \ell$, we denote by $\alpha_\gamma^{(i)}$ the γth copy of $\alpha^{(i)}$ that appears as a substring in $\beta^{(i)}$, i.e., the substring of $\beta^{(i)}$ of size $|\alpha^{(i)}|$ immediately preceding the unique occurrence in $\beta^{(i)}$ of the character $\#_\gamma^{(i)}$.

The Intuition Behind the Reduction. The basic idea of the reduction is based on the fact that for any parsing of the string $\mathbf{s}^{(1)} = PX_1 \ldots X_n Y_1 \ldots Y_m$ the following holds:

– there is a unique way to parse P,
– for each X_i there are two ways to parse it, into 3 or 2 phrases; which will, respectively, encode the inclusion or not of the vertex v_i into a vertex cover for G
– for a Y_j corresponding to edge $e_j = (v_i, v_{i'})$, Y_j can be parsed into 4 phrases if and only if at least one between X_i and $X_{i'}$ has been parsed into 3 phrases, which encodes the fact that e_j is covered only if one of its incident vertices are in the vertex cover. Otherwise Y_j must be parsed into ≥ 5 phrases.

The bound on the parsing set by the reduction can then be attained only if each Y_j is parsed into 4 phrases, which by the above properties, implies that the set of X_i's parsed into 3 phrases must be a vertex cover of G.

The next theorem formalizes the relationship between minimum vertex covers of G and shortest 1-BLZ parsing of the prefix $\mathbf{s}^{(1)}$ of $\mathbf{s}_G$. We note that this theorem is already enough to establish that 1-BLZ is NP-hard, which implies also the hardness of the BLZ problem in the introduction.

Theorem 1. *Fix a graph $G = (V, E)$ and a non-negative integer k, and let $\mathbf{s}$ be the string produced by the construction above. Then, G has a vertex cover of size k if and only if there exists a 1-BLZ parsing φ for $\mathbf{s}^{(1)}$ of size $4n + 6m + k$.*

Proof. String P can be parsed in a unique way with each character being a distinct phrase, since each character is different from any other one, hence

$$\varphi(P) = \mid v_1 \mid \#_1^p \mid v_2 \mid \#_2^p \mid \cdots \mid v_n \mid \#_n^p \mid e_1 \mid \#_1^p \mid e_2 \mid \#_1^p \mid \cdots \mid e_m \mid \#_m^p \mid .$$

For each $X_i = v_i' v_i \#_i^v$ there are only two ways to parse it, satisfying the bound that the max-hop-number is ≤ 1, namely

$$v_i' \mid v_i \mid \#_i^v \mid \tag{2}$$

$$v_i' \mid v_i \#_i^v \mid \tag{3}$$

To see this, observe the occurrence of v_i' in X_i is the first occurrence of this character in $\mathbf{s}^{(1)}$, while v_i already occurred in P.

We refer to the case (2) by saying that the parsing *selects* the vertex v_i. Moreover, we also say that φ selects vertex v_i if the pair of adjacent phrases $\mid v_i' \mid v_i \mid$ appears somewhere in $\varphi(\mathbf{s}^{(1)})$, with result that the hop-numbers of the positions of these occurrences of v_i and v_i' are both 0.

Conversely, if the pair of consecutive phrases $\mid v_i' \mid v_i \mid$ does not appear in φ we say that the parsing *does not select* the vertex v_i.

Claim 1. Without loss of generality, we can assume that if a parsing selects v_p then, it parses the substring X_p (in $\mathbf{s}^{(1)}$) into three phrases as in (2)

$$\varphi(X_p) = \mid v_p' \mid v_p \mid \#_p^v \mid$$

Proof. It is enough to observe that if a parsing φ selects v_p and does not parse X_p into three phrases, then there is an edge substring Y_i, corresponding to some edge e_i incident to v_p, where φ produces the two consecutive phrases $\mid v_p' \mid v_p \mid$. Consider the parsing φ' that: (i) parses X_p into three phrases and uses a single phrase for the substring $Y_i' = v_p' v_p$ in Y_i; and (ii) for any phrase ϕ that uses the substring $Y_i' = v_p' v_p$, of Y_i, as a source (where both positions have $hop = 0$ in φ) we make the substring $v_p' v_p$ of X_i to be the source of ϕ in φ'. It follows that $|\varphi'| = |\varphi|$ and

$$hop_{\max}^{\varphi'}(\mathbf{s}) = hop_{\max}^{\varphi}(\mathbf{s})$$

and φ' satisfies the claim with respect to the way v_p is selected.   □

Let us now consider a substring Y_i corresponding to the edge $e_i = \{v_p, v_q\}$. Recall the definition of Y_i from (1).

Claim 2. Every 1-BLZ parsing of $\mathbf{s}^{(1)}$ parses Y_i into ≥ 4 phrases.

Proof. Since before its first occurrence in Y_i the character e_i only appeared in P and it was not preceded by v_p no phrase containing v_p can contain also $\$_i$. Moreover since the first occurrence of $\$_i$ is the first occurrence of such character in $\mathbf{s}^{(1)}$ every parsing will have a phrase ending here. Therefore, the substring $v_p' v_p e_i \$_i$ must be parsed in at least two phrases. Analogously, no phrase can contain both v_q and $\$_i$, hence also the substring $v_q' v_q e_i \$_i \#_i^e$ requires at least 2 phrases. $\qquad\square$

Claim 3. A 1-BLZ parsing φ of $\mathbf{s}^{(1)}$ can parse Y_i into 4 phrases if and only if φ selects at least one of the vertices v_p, v_q.

Proof. If both vertices v_p and v_q are selected a valid 1-BLZ parsing of Y_i into exactly 4 phrases is:

$$\varphi(Y_i) = \underset{1\ \ 1\ \ 0\ \ \ 0}{v_p' v_p e_i \mid \$_i} \mid \underset{1\ \ 1\ \ 0\ \ \ 1\ \ 0}{v_q' v_q e_i \mid \$_i \#_i^e},$$

where the number underneath each character is the hop-number for the corresponding position. These follow since $\varphi(X_p) = \mid v_p' \mid v_p \mid \#_p^v \mid$, $\varphi(X_p) = \mid v_p' \mid v_p \mid \#_p^v \mid$ provide sources $v_p' v_p$ and $v_q' v_q$ where all positions have hop-number 0.

If only vertex v_p is selected, from $\varphi(X_p)$ and $\varphi(X_q)$ we have a source $v_p' v_p$ where both positions have hop-number 0 and the source v_q' with hop-number 0. Hence, a valid 1-BLZ parsing for Y_i is:

$$\varphi(Y_i) = \underset{1\ \ 1\ \ 0\ \ \ 0}{v_p' v_p e_i \mid \$_i} \mid \underset{1\ \ 0\ \ \ 1\ \ 1\ \ 0}{v_q' v_q \mid e_i \$_i \#_i^e},$$

where the source of $\mid e_i \$_i \#_i^e \mid$ is the previous occurrence of $e_i \$_i$ in Y_i. Analogously, if only the vertex v_q is selected a valid 1-BLZ parsing for Y_i is:

$$\varphi(Y_i) = \underset{1\ \ 0\ \ \ 1\ \ 0}{v_p' v_p \mid e_i \$_i} \mid \underset{1\ \ 1\ \ 0\ \ \ 1\ \ 0}{v_q' v_q e_i \mid \$_i \#_i^e},$$

with the sources: for $\mid e_i \$_i \mid$ from P and for $\mid \$_i \#_i^e \mid$ from previous $\$_i$ in Y_i.

It remains to be shown that if the 1-BLZ parsing φ does not select any of the vertices v_p, v_q, then it must parse Y_i into ≥ 5 phrases. Since the vertex p is not selected there is no substring $v_p' v_p$ where both corresponding positions have hop-number 0. Because of this in the first half of Y_i, that encodes the incidence in the the vertex p, namely, the substring $v_p' v_p e_i \$_i$, no phrase containing $v_p' v_p$ can contain also e_i, for otherwise the position of v_p' or the position of v_p would have hop-number 2. Analogously since also the vertex q is not selected no phrase containing $v_q' v_q$ can contain also e_i. Additionally, as reported in the previous claim, the first occurrence of $\$_i$ and $\#_i^e$ need to be positioned at the end of a

phrase. These observations imply that for any valid 1-BLZ parsing of Y_i there
are only two options left: (i) the parsing splits the two adjacent characters $e_i \, \$_i$
into two singleton phrases giving each one of them hop-number 0 and uses a
single phrase for the suffix $| \, e_i \$_i \#_i^e \, |$. The resulting parsing is then

$$v_p' \, v_p \mid e_i \mid \$_i \mid v_q' \, v_q \mid e_i \$_i \, \#_i^e$$
$$\;\; 1 \;\; 0 \quad\; 0 \quad\; 0 \quad\;\; 1 \;\; 0 \quad 1 \;\, 1 \;\, 0$$

with 5 phrases. (ii) Alternatively, the substring $e_i \, \$_i$ in the first half is parsed
as a single phrase. In this case the two positions have hop-number 1 and 0,
respectively, as the only occurrence of e_i is in P and it is not followed by $\$_i$. In
this case the second occurrence of e_i cannot be parsed in the same phrase of $\#_i^e$
and two different phrases are needed for the last three character of Y_i, namely
we have the following two possible parsings:

$$v_p' \, v_p \mid e_i \$_i \mid v_q' \, v_q \mid e_i \mid \$_i \, \#_i^e \qquad\qquad v_p' \, v_p \mid e_i \$_i \mid v_q' \, v_q \mid e_i \$_i \mid \#_i^e$$
$$\;\; 1 \;\, 0 \quad 1 \;\, 0 \quad\;\; 1 \;\, 0 \quad\; 0 \;\; 1 \;\, 0 \qquad\qquad\quad 1 \;\, 0 \quad 1 \;\, 0 \quad\;\; 1 \;\, 0 \quad 1 \;\, 0 \quad\; 0$$

both leading to a total number of phrases equal to 5.

$\square$

We can now complete the proof of the theorem by considering the two directions
of the equivalence in the statement.

1. *If $\tau(G) \leq k$ then there exists a 1-BLZ parsing of* s *of size* $\leq 4n + 6m + k$.
Let us now assume that there is a vertex cover C of G of size $\leq k$. Consider the
parsing φ_C that in X *selects* each vertex in C. Such parsing, by Claim 2, will
have: 4 phrases for each edge substring Y_i; 2 phrases for each X_i corresponding
to a non-selected vertex v_i; and (by Claim 1) 3 phrases for each X_i corresponding
to a selected vertex v_i. Considering also the $2n + 2m = |P|$ phrases necessary to
parse P, in total, the parsing φ_C will have size

$$|\varphi_C| = |P| + 2V| + |C| + 4|E| \leq 4n + 6m + k,$$

where for the last inequality we use $|C| \leq k$.

2. *If $\tau(G) > k$ then any 1-BLZ parsing of* s *has size* $> 4n + 6m + k$. Recall
that we say that a parsing φ selects a vertex v_p if it produces the sequence of
two adjacent phrases $| \, v_p' \mid v_p \, |$, hence with both positions having hop-number
0, either in the substring X_p or in some substring Y_i corresponding to an edge
incident to v_p. By Claim 1 we can assume that selection of a vertex v_p is only
done in the string X_p.

Let $k' = \tau(G) > k$. Given φ there are two possible scenarios:

Case 1. φ selects more than k vertices. Let h be the number of vertices selected
by φ. By Claim 1, for each selected vertex v_i the corresponding X_i is parsed in
3 phrases so the number of phrases used by φ over all X_i is $|\varphi(X)| = 2n + h$.
By Claim 2, the number of phrases used to parse all the Y_i is at least $4m$. Then,

$|\varphi| = |\varphi(P)| + |\varphi(X)| + |\varphi(Y)| \geq 4n + 6m + h > 4n + 6m + k$, where for the last inequality we use the hypothesis $h > k$.

Case 2. φ selects $\leq k$ vertices. Let C be the set of such selected vertices. Then there exist $t \geq k' - |C|$ edges not covered by C. We can affirm this because, it is possible to build a vertex cover from C adding a vertex for each edge that is not covered in C, this vertex cover would have a size equal to $|C| + t$ that need to be $\geq k'$ since k' is the size of a minimum vertex cover in G. Each one of the t edges not covered corresponds to a Y_i that is parsed by φ in at least 5 phrases, by Claim 3. The total number of phrases used to parse all these Y_is is then $\geq 5t$, hence the number of phrases used to parse $Y = Y_1 \cdots Y_m$ is at least $4m + t$. Moreover, the number of phrases used to parse all of $X = X_1 \cdots X_n$ is at least $2n + |C|$. Putting together these two bound and the fact that $|\varphi(P)| = 2n + 2m$, we conclude that

$$|\varphi| = |\varphi(P)| + |\varphi(X)| + |\varphi(Y)| \geq 4n + 6m + t + |C| \geq 4n + 6m + k' > 4n + 6m + k,$$

where we are using $t + |C| \geq k'$ and $k' > k$.   □

The following theorem extends the result of Theorem 1 to the case of $c > 1$. The idea is to show that any c-BLZ φ parsing for $\mathbf{s}_G$ that has size $\leq 4n + 6m + k + (c-1)\ell$ when restricted to $\mathbf{s}^{(1)}$ has to be a 1-BLZ parsing of $\mathbf{s}^{(1)}$ of size $4n + 6m + k$, hence encoding a vertex cover of G of size k. The main steps of the proof are summarized in Sect. 3.2. Due to the space limitation, the proof of the technical lemmas are deferred to an extended version of the paper.

Theorem 2. *Fix a graph $G = (V, E)$ and let $n = |V|, m = |E|$. Fix integers $c \geq 1$ and $0 \leq k \leq n$, and $\ell > k$. Let $\mathbf{s}_G$ be the string produced by the procedure in Sect. 3.1. Then G has a vertex cover of size k if and only if there is a c-BLZ parsing of $\mathbf{s}_G$ of size $4n + 6m + k + \ell(c-1)$.*

We are ready to state the main result of this section.

Theorem 3. *For any integer $c > 1$ the c-BLZ problem is NP-hard.*

Proof. The problem of deciding for a given graph G and a given k whether G has a vertex cover of size k is well known to be NP-hard (see, e.g., [5]). By Theorem 2 the polynomial time function that maps a graph G and integers k and $\ell > k$ to the string $\mathbf{s}_G$ is a polynomial time reduction between the problem of deciding whether G has a vertex cover of size k and the problem of deciding whether $\mathbf{s}_G$ has a c-BLZ parsing of size $4n + 6m + (c-1)\ell$. It follows that the latter problem is also NP-hard.   □

3.2   The Structure of the Proof of Theorem 2

The proof is based on several lemmas.

Lemma 1. *If G has a vertex cover of size k then there exists a c-BLZ parsing of $\mathbf{s}$ of size $4n + 6m + k + (c-1)\ell$.*

Lemma 2. *Assume that $\tau(G) = k' > k$ and let φ be a c-BLZ parsing of s that on $\alpha^{(1)}$ uses $4n+6m+t$ phrases for some $t \leq k$. Then, there are at least $z = k'-t$ positions in distinct Y_j substrings of $\alpha^{(1)}$ with hop-number ≥ 2.*

Lemma 3. *Let $W = \{x_1, \ldots, x_z\}$ be a set of indices of characters in $\alpha^{(1)}$, where*

- *for each $j = 1, \ldots, z$, $hop(x_j) \geq 2$,*
- *for $1 \leq j < j' \leq z$ there exist $i_j \neq i_{j'}$ such that x_j is in Y_{i_j}, $x_{j'}$ is in $Y_{i_{j'}}$.*

For each $x_j \in W$ let us denote by x_j^γ the corresponding position in $\alpha_\gamma^{(1)}$. Moreover, we denote by x_{z+1}^γ the unique position in $\beta^{(1)}$ of the character $\#_\gamma^{(1)}$.
If there exists $j \in \{1, \ldots, z\}$ such that for every $\gamma = 1, \ldots, \ell$, there is no phrase of φ ending between x_j^γ and x_{j+1}^γ then in each one of the substrings $\alpha_\gamma^{(1)}$ there is a position x with $hop(x) \geq 3$.

Corollary 1. *Recall the definition of W in the previous lemma. If G has no vertex cover of size k and φ uses $\leq 4n + 6m + k$ phrases in $\alpha^{(1)}$ and $< \ell + z$ phrases in $\beta^{(1)}$ then on $\alpha^{(2)}$ it produces ℓ positions $x_1^{(1)}, \ldots, x_\ell^{(1)}$ within $\beta^{(1)}$ with hop ≥ 3 such that $\#_j^{(1)}$ is between positions $x_j^{(1)}$ and position $x_{j+1}^{(1)}$ for each $j = 1, \ldots, \ell - 1$.*

Lemma 4. *Let φ be a parsing of s such that for some $i \in \{1, 2, \ldots, c - 2\}$, and for each $j = 1, 2, \ldots, \ell$ in the substring $\alpha_j^{(i)}$ of $\beta^{(i)}$ there is a position x_j such that $hop(x_j) > i + 1$. For each $\gamma = 1, \ldots, \ell$ let x_j^γ be the position in $\alpha_\gamma^{(i+1)}$ corresponding to x_j. If φ uses strictly less than 2ℓ phrases in $\beta^{(i+1)}$ then there is a $j \in \{1, 2, \ldots, \ell\}$ such that for each $\gamma = 1, \ldots, \ell$ the position x_j^γ in $\alpha_\gamma^{(i+1)}$ satisfies $hop(x_j^\gamma) > i + 2$.*

Lemma 5. *If $\tau(G) > k$ then any c-BLZ parsing φ of s has size $> 4n + 6m + k + (c - 1)\ell$.*

Proof. Let φ be a c-BLZ parsing of s.

Claim 1. for each $i = 1, \ldots, c - 1$, $|\varphi(\beta^{(i)})| \geq \ell$. The claim easily follows from the fact that in $\beta^{(i)}$ there is the first occurrence of the characters $\#_1^{(i)}, \ldots, \#_\ell^{(i)}$, hence each one of these characters must be the end of a distinct phrase.

We now argue by cases.

Case 1. $|\varphi(\alpha^{(1)})| > 4n + 6m + k$. Then, by Claim 1, we have that:

$$|\varphi(\mathbf{s})| = |\varphi(\alpha^{(1)})| + \sum_{i=1}^{c-1} |\varphi(\beta^{(i)})| > 4n + 6m + k + \ell(c - 1).$$

Case 2. $|\varphi(\alpha^{(1)})| = 4n + 6m + t$, for some $t \leq k$.

Since $\tau(G) > k$, by Lemma 2 we have that for some integer $z \geq \tau(G) - t$ and for each $i = 1, \ldots, z$ there is a position x_i in a distinct substring Y_i of $\alpha^{(1)}$ such that $hop(x_i) \geq 2$.

Subcase 2.1 $|\varphi(\beta^{(1)})| \geq \ell + z$.

Then, again by Claim 1, we have that in total $|\varphi|$ satisfies the desired bound:

$$|\varphi(\mathbf{s})| = |\varphi(\alpha^{(1)})| + |\varphi(\beta^{(1)})| + \sum_{i=2}^{c-1} |\varphi(\beta^{(i)})| \geq 4n + 6m + t + \ell + z + \ell(c-2)$$

$$\geq 4n + 6m + t + \ell(c-1) + \tau(G) - t > 4n + 6m + k + \ell(c-1). \quad (4)$$

Subcase 2.2 $|\varphi(\beta^{(1)})| < \ell + z$.

Then, by Corollary 1, it follows, that for each $\gamma = 1, \ldots, \ell$ in the substring $\alpha_\gamma^{(2)}$ there is a position $x_\gamma^{(2)}$ such that $hop(x_\gamma^{(2)}) \geq 3$.

We now observe that under the standing hypothesis there must exist an integer $j^* \in \{2, 3, \ldots, c-1\}$ such that $|\varphi(\beta^{(j^*)})| \geq 2\ell$. For otherwise, by repeated application of Lemma 4 we have that for $i = 2, \ldots, c-1$ there are ℓ position with hop-number $\geq i + 2$. This in particular means that in $\beta^{(c-1)}$ there are positions with hop-number $c + 1$ contradicting the hypothesis that φ is a c-BLZ parsing.

Therefore, we can again bound the number of phrases of φ as follows:

$$|\varphi(\mathbf{s})| = |\varphi(\alpha^{(1)})| + |\varphi(\beta^{(j^*)})| + \sum_{\substack{i=1 \\ i \neq j^*}}^{c-1} |\varphi(\beta^{(i)})| \geq 4n + 6m + t + 2\ell + \ell(c-2)$$

$$\geq 4n + 6m + \ell + \ell(c-1) > 4n + 6m + k + \ell(c-1), \quad (5)$$

where we have used the fact that in each $\beta^{(i)}$ any parsing must use at least ℓ phrases and the assumption $\ell > k$.

We have shown that in all possible cases the size of φ satisfied the desired bound. □

The Proof of Theorem 2. The result is an immediate corollary of Lemma 1 and Lemma 5. □

4   APX Hardness

In this section we focus on the existence of polynomial time approximation algorithms for the problem of finding an optimal c-BLZ parsing of a string. We show that the problem is APX-hard, hence there exists a constant κ such that no polynomial time κ-approximation algorithm can exists for the BLZ problem unless $P = NP$.

For the proof of APX hardness we will employ the concept of an L-reduction: Let $\mathbb{A}, \mathbb{B}$ be two minimization problems. $\mathbb{A}$ is said to be L-reducible to $\mathbb{B}$, denoted by $\mathbb{A} \leq_L \mathbb{B}$, if two polynomial time computable functions f, g and two constants a, b exist such that for any instance χ of $\mathbb{A}$:

1. $f(\chi)$ is an instance of $\mathbb{B}$ such that $OPT_{\mathbb{B}}(f(\chi)) \leq a\,OPT_{\mathbb{A}}(\chi)$
2. for any solution $s_{\mathbb{B}}(f(\chi))$ to the instance $f(\chi)$ of $\mathbb{B}$, $g(s_{\mathbb{B}}(f(\chi)))$ is a solution to the instance χ of $\mathbb{A}$ such that

$$cost(g(s_{\mathbb{B}}(f(\chi)))) - OPT_{\mathbb{A}}(\chi) \leq b\,(cost(s_{\mathbb{B}}(f(\chi)))) - OPT_{\mathbb{B}}(f(\chi)),$$

where $OPT_{\mathbb{P}}(x)$ denotes the value of the optimal solution to instance x of problem $\mathbb{P}$ and $cost_{\mathbb{P}}(s)$ denotes the cost in problem $\mathbb{P}$ of a solution s. When conditions 1-2 are satisfied, the quadruple (f, g, a, b) is an L-reduction from $\mathbb{A}$ a to $\mathbb{B}$.

Theorem 4. *For any fixed constant $c \geq 1$, the problem c-BLZ is APX-hard.*

Proof. We prove that there exists an L-reduction (f, g, a, b) from MIN-VC-4, the problem of finding a minimum vertex cover in a 4-regular graph $G = (V, E)$ to the minimization problem c-BLZ on the string s_G. Since MIN-VC-4 is APX-hard [5], the existence of the L-reduction implies that c-BLZ is also APX-hard [7] (also see [1, 16]).

The function f, that associates instances of the vertex cover problem to instances of c-BLZ problem, in this L-reduction is going to be the same one used for the proof of NP-hardness.

Since we are working with 4-regular graph we know that $m = |E| = 2|V| = 2n$ and $\frac{n}{2} \leq \tau(G)$, where $\tau(G)$ is the size of a minimum vertex cover of G.

Let $k^* = \tau(G) = OPT_{VC}(G)$ and φ^* denote a c-BLZ parsing of s_G of minimum size, i.e., $|\varphi^*| = OPT_{BLZ}(s_G, c)$. By Theorem 2, with $\ell = n > k^*$, we have $|\varphi^*| = 4n + 6m + k^* + \ell(c-1) = 16n + k^* + \ell(c-1)$.

This holds for each $\ell \geq k$ where k is the vertex cover's size. In particular, it holds for $\ell = n$. Using $\ell = n$ and the property of $|\tau(G)| \geq \frac{n}{2}$ from the previous inequality we get $|\varphi^*| = 15n + nc + k^* \leq 30k^* + k^* + 2k^*c = (31 + 2c)k^*$. Since c is a constant, setting $a = (31 + 2c)$, we have that our reduction satisfies the first item in the definition of an L-reduction.

We define the function g that maps a parsing φ of the string s_G to a vertex cover C_φ of G by providing the following algorithm to compute it:

Start with C_φ being the empty set. Given φ, a parsing of the string s_G, for each X_i of $\alpha^{(1)}$, if the parsing φ selects X_i, i.e., it parses it into 3 phrases, then v_i is added to C_φ. Let d be the number of X_i selected in φ, hence at this point we have $|C_\varphi| = d$.

If C_φ is a vertex cover for G then the computation of g terminates. Otherwise, we look for every Y_i in $\alpha^{(1)}$ which is either parsed into ≥ 5 phrases or it includes a position x with $hop(x) = 2$ and 4 phrases. We let p be the number of Y_is that in φ is parsed into ≥ 5 phrases and z be the number of Y_is that in φ are parsed into 4 phrases and include a position x with $hop(x) = 2$. For each one of these $p + z$ many Y_i, let e_i be the corresponding edge of G and add to C_φ one of the vertices incident to e_i. It is possible that for some of these the same vertex is chosen. The resulting C_φ is a vertex cover for G, as we know that each Y_i not considered in the last step is parsed into 4 phrases and does not have a position with hop-number > 1. Let the edge of G corresponding to Y_i be $e_i = \{v_p, v_q\}$. Then by Claim 3 in Theorem 1, one of v_p and v_q is selected by φ, hence in C_φ.

Then, we set $g(\varphi) = C_\varphi$.

We have $|C_\varphi| \le d+p+z$. We also have that $|\varphi| \ge 4n+6m+d+p+z+\ell(c-1)$. This follows by Lemma 3, Corollary 1 and Lemma 4 implying that for each one of the z distinct Y_is parsed into 4 phrases and including a position x with $hop(x) = 2$ there must exists in $\varphi(\beta^{(2)}\beta^{(3)}\ldots\beta^{(c-1)})$ an additional phrase besides the $\ell(c-1)$ necessary to accommodate the end characters of each $\beta^{(i)}$. For otherwise φ would not satisfy the bound on the maximum hop-number (see also the argument in Lemma 5 (Case 2)).

Hence $|\varphi| \ge |C_\varphi| + 4n + 6m + \ell(c - 1)$. From Theorem 2, we also have $|\tau(G)| = |\varphi^*| - (4n + 6m + \ell(c - 1))$. Thus, $|C_\varphi| - |\tau(G)| \le |\varphi| - |\varphi^*|$, which implies that the function g defined above and $b = 1$ satisfy property 2 of an L-reduction.

We conclude that the quadruple (f, g, a, b) is the desired L-reduction from MIN-VC-4 to c-BLZ. It follows that the latter problem is APX-hard. $\quad\square$

5   On the Approximation of BLZ to LZ76

For a string $\mathbf{s}$, we denote by $OPT_{LZ}(s)$ the size of an optimal LZ parsing for $\mathbf{s}$ and by $OPT_{c-BLZ}(\mathbf{s})$ the size of an optimal c-BLZ parsing for $\mathbf{s}$.

Trivially we have $OPT_{LZ}(s) \le OPT_{c-BLZ}(\mathbf{s})$. Before our results, $OPT_{LZ}(s)$ was the only known and used lower bound on the size of $OPT_{c-BLZ}(\mathbf{s})$ [2,15].

In this section we derive a new non trivial lower bound on the size of an optimal c-BLZ parsing for a string $\mathbf{s}$. As a corollary, our lower bounds allows to show that in the worst case the ratio between $OPT_{c-BLZ}(\mathbf{s})$ and $OPT_{LZ}(\mathbf{s})$ is unbounded. For this, we exhibit a class of strings over a ternary alphabet for which the optimal LZ parsing is known to be logarithmic in the size of the string while the optimal c-BLZ parsing is $\Omega(n^{\frac{1}{c+1}})$.

Remarkably, the latter result provides the first non trivial lower bound on the optimal c-BLZ parsing of a string which might be significantly larger than the size of the LZ-parsing. More precisely, for the size of an optimal c-BLZ parsing of a square free string we prove the following lower bound.

Theorem 5. *Let us fix an integer $c \ge 1$ and a square free string $\mathbf{s}$, then for every c-BLZ parsing φ of s it holds that $|\varphi(\mathbf{s})| \ge \sqrt[c+1]{|\mathbf{s}|} - 1$.*

Proof. We start by observing that if the string is square free then the source of a phrase cannot overlap with it.

We argue by contradiction. We assume $|\varphi(\mathbf{s})| < \sqrt[c+1]{n} - 1$. Then there exists a phrase $\phi^{(1)}$, such that: $|\phi^{(1)}| > \dfrac{n}{\sqrt[c+1]{n} - 1} > n^{\frac{c}{c+1}} + 1$.

For $i = 1, \ldots, c$, we define strings $\phi^{(i)}$ and $\tilde{\phi}^{(i)}$ as follows: (i) $\tilde{\phi}^{(i)}$ is the source of $\phi^{(i)}$; and $\phi^{(i+1)}$ is the largest substring of $\tilde{\phi}^{(i)}$ which is contained in a phrase of φ. Then, inductively we can show that for each $i = 1, \ldots, c$ we have

$$|\phi^{(i)}| > n^{\frac{c-i+1}{c+1}} + 1, \qquad \text{hence} \qquad |\tilde{\phi}^{(i)}| > n^{\frac{c-i+1}{c+1}}.$$

Base case $i = 1$. From $|\phi^{(1)}| > n^{\frac{c}{c+1}} + 1$ above, we immediately also have $|\tilde{\phi}^{(1)}| > n^{\frac{c}{c+1}}$.

Inductive step $i > 1$. Since, by the induction hypothesis on $i - 1$, we have
$$|\phi^{(i)}| \geq \frac{|\tilde{\phi}^{(i-1)}|}{\sqrt[c+1]{n} - 1} > \frac{n^{\frac{c-(i-1)+1}{c+1}} + 1}{n^{\frac{1}{c}} - 1} > n^{\frac{c-i+1}{c+1}} + 1, \text{ it holds that:}$$

$$|\phi^{(i)}| > n^{\frac{c-i+1}{c+1}} + 1 \quad \text{hence} \quad |\tilde{\phi}^{(i)}| > n^{\frac{c-i+1}{c+1}}.$$

For each i, the maximum hop-number of the positions of $\phi^{(i)}$ is $\leq c - i + 1$ therefore for $i = c$, we have that the maximum hop-number of the positions of $\phi^{(c)}$ is ≤ 1. Then, the sources of these positions (i.e., the positions of $\tilde{\phi}^{(c)}$) must be consecutive positions whose hop-number is 0. This can only be the case if they are the positions of single character phrases. Hence:

$$|\varphi| \geq |\tilde{\phi}^{(c)}| > n^{\frac{c-c+1}{c+1}} = n^{\frac{1}{c+1}}.$$

This contradicts the initial (absurdum) hypothesis and completes the proof. □

From the results in [6,13] we have the following upper bound on $OPT_{LZ}(\mathbf{s})$.

Fact 1. *There are infinitely many string $s \in \{1, 2, 3\}^*$, which are square free and such that: $OPT_{LZ}(s) = O(\log |s|)$.*

This fact and the lower bound in Theorem 5 leads to the following result on the worst case ratio between $OPT_{c-BLZ}(\mathbf{s})$ and $OPT_{LZ}(\mathbf{s})$. We note here that the same conclusion can be also derived—although in less explicit way—from the main result of [20].

Corollary 2. *Fix an integer $c \geq 1$. Then, the worst case ratio between $OPT_{c-BLZ}(\mathbf{s})$ and $OPT_{LZ}(\mathbf{s})$ over all ternary strings $\mathbf{s}$ is unbounded, i.e.,*
$$\max_{\mathbf{s} \in \{1,2,3\}^*} \frac{OPT_{c-BLZ}(\mathbf{s})}{OPT_{LZ}(\mathbf{s})} \to \infty.$$

Proof. Let $\mathcal{S}$ be the set of square free ternary string. Then, by Fact 1 and Theorem 5, there exists a constant a such that:
$$\max_{\mathbf{s} \in \mathcal{S}} \frac{OPT_{c-BLZ}(\mathbf{s})}{OPT_{LZ}(\mathbf{s})} \geq \frac{|\mathbf{s}|^{\frac{1}{c+1}} - 1}{a \log |\mathbf{s}|} \to \infty.$$

Remark: Since there exist also square free strings $\mathbf{s}$ s.t. $OPT_{LZ}(\mathbf{s}) \geq |\mathbf{s}|^{\frac{1}{c+1}}$ by Theorem 5 we obtain the following general lower bound on the size of the optimal c-BLZ parsing for a square free string $\mathbf{s}$: $OPT_{c-BLZ}(\mathbf{s}) \geq \max\{OPT_{LZ}(\mathbf{s}), |\mathbf{s}|^{\frac{1}{c+1}}\}$.

6 Conclusion and Open Problems

We studied c-BLZ, a variant of the LZ76, that allows to decompress characters of the text in a bounded number of accesses to the encoding, i.e., without needing to decompress the whole text. We proved that for any constant c computing the optimal parsing that guarantees decompression of a character with at most c accesses is NP-hard and also APX hard. We also showed that the ratio to the size of the optimal LZ76 parsing is unbounded in the worst case, providing a first non-trivial lower bound on the size of the optimal c-BLZ parsing.

A main direction for future research is the investigation of approximation algorithms as well as parameterized algorithms for bounded parameters of practical importance, e.g., alphabet size. The unbounded ratio to the optimal LZ76 parsing leaves open the question of lower bounds on OPT_{c-BLZ}.

Acknowledgements. The authors would like to thank the anonymous reviewers. Their insightful comments and the pointers to relevant literature have helped us to significantly improve the presentation of our results.

References

1. Alimonti, P., Kann, V.: Hardness of approximating problems on cubic graphs. In: Bongiovanni, G., Bovet, D.P., Di Battista, G. (eds.) CIAC 1997. LNCS, vol. 1203, pp. 288–298. Springer, Heidelberg (1997). https://doi.org/10.1007/3-540-62592-5_80

2. Bannai, H., Funakoshi, M., Hendrian, D., Matsuda, M., Puglisi, S.J.: Height-bounded Lempel-Ziv encodings. arXiv preprint arXiv:2403.08209 (2024)

3. Bannai, H., Funakoshi, M., Kurita, K., Nakashima, Y., Seto, K., Uno, T.: Optimal LZ-end parsing is hard. In: Bulteau, L., Lipták, Z. (eds.) 34th Annual Symposium on Combinatorial Pattern Matching (CPM 2023). Leibniz International Proceedings in Informatics (LIPIcs), vol. 259, pp. 3:1–3:11 (2023)

4. Cenzato, D., Lipták, Zs.: A theoretical and experimental analysis of BWT variants for string collections. In: Bannai, H., Holub, J. (eds.) 33rd Annual Symposium on Combinatorial Pattern Matching, CPM 2022, 27–29 June 2022, Prague, Czech Republic. LIPIcs, vol. 223, pp. 25:1–25:18. Schloss Dagstuhl - Leibniz-Zentrum für Informatik (2022). https://doi.org/10.4230/LIPICS.CPM.2022.25

5. Chlebík, M., Chlebíková, J.: Complexity of approximating bounded variants of optimization problems. Theoret. Comput. Sci. **354**(3), 320–338 (2006)

6. Constantinescu, S., Ilie, L.: The Lempel-Ziv complexity of fixed points of morphisms. SIAM J. Discret. Math. **21**(2), 466–481 (2007)

7. Crescenzi, P.: A short guide to approximation preserving reductions. In: Proceedings of Computational Complexity. Twelfth Annual IEEE Conference, pp. 262–273 (1997). https://doi.org/10.1109/CCC.1997.612321

8. Gawrychowski, P., Kosche, M., Manea, F.: On the number of factors in the LZ-End factorization. In: Nardini, F.M., Pisanti, N., Venturini, R. (eds.) SPIRE 2023. LNCS, vol. 14240, pp. 253–259. Springer, Cham (2023). https://doi.org/10.1007/978-3-031-43980-3_20

9. Ideue, T., Mieno, T., Funakoshi, M., Nakashima, Y., Inenaga, S., Takeda, M.: On the approximation ratio of LZ-End to LZ77. In: Lecroq, T., Touzet, H. (eds.) SPIRE 2021. LNCS, vol. 12944, pp. 114–126. Springer, Cham (2021). https://doi.org/10.1007/978-3-030-86692-1_10

10. Kempa, D., Kosolobov, D.: LZ-End parsing in linear time. In: 25th Annual European Symposium on Algorithms (ESA 2017). Schloss Dagstuhl-Leibniz-Zentrum fuer Informatik (2017)

11. Kempa, D., Saha, B.: An Upper Bound and Linear-Space Queries on the LZ-End Parsing, pp. 2847–2866. https://doi.org/10.1137/1.9781611977073.111. https://epubs.siam.org/doi/abs/10.1137/1.9781611977073.111

12. Kreft, S., Navarro, G.: LZ77-like compression with fast random access. In: 2010 Data Compression Conference, pp. 239–248 (2010). https://doi.org/10.1109/DCC.2010.29

13. Leech, J.: 2726. a problem on strings of beads. Math. Gazette **41**(338), 277–278 (1957)

14. Lempel, A., Ziv, J.: On the complexity of finite sequences. IEEE Trans. Inf. Theory **22**(1), 75–81 (1976). https://doi.org/10.1109/TIT.1976.1055501

15. Lipták, Zs., Masillo, F., Navarro, G.: BAT-LZ out of hell. In: 35th Annual Symposium on Combinatorial Pattern Matching (CPM 2024). Leibniz International Proceedings in Informatics (LIPIcs) (2024, to appear)

16. Papadimitriou, C.H., Yannakakis, M.: Optimization, approximation, and complexity classes. J. Comput. Syst. Sci. **43**(3), 425–440 (1991). https://doi.org/10.1016/0022-0000(91)90023-X. https://www.sciencedirect.com/science/article/pii/002200009190023X

17. Rodeh, M., Pratt, V.R., Even, S.: Linear algorithm for data compression via string matching. J. ACM **28**(1), 16-24 (1981). https://doi.org/10.1145/322234.322237

18. Storer, J.: NP-completeness results concerning data compression. Technical Report 234 (1977)

19. Storer, J.A., Szymanski, T.G.: Data compression via textual substitution. J. ACM **29**(4), 928–951 (1982)

20. Verbin, E., Yu, W.: Data structure lower bounds on random access to grammar-compressed strings. In: Fischer, J., Sanders, P. (eds.) CPM 2013. LNCS, vol. 7922, pp. 247–258. Springer, Heidelberg (2013). https://doi.org/10.1007/978-3-642-38905-4_24

21. Ziv, J., Lempel, A.: A universal algorithm for sequential data compression. IEEE Trans. Inf. Theory **23**(3), 337–343 (1977)

How to Find Long Maximal Exact Matches and Ignore Short Ones

Travis Gagie[(✉)] [ID]

Faculty of Computer Science, Dalhousie University, Halifax, Canada
travis.gagie@dal.ca

Abstract. Finding maximal exact matches (MEMs) between strings is an important task in bioinformatics, but it is becoming increasingly challenging as geneticists switch to pangenomic references. Fortunately, we are usually interested only in the relatively few MEMs that are longer than we would expect by chance. In this paper we show that under reasonable assumptions we can find all MEMs of length at least L between a pattern of length m and a text of length n in $O(m)$ time plus extra $O(\log n)$ time only for each MEM of length at least nearly L using a compact index for the text, suitable for pangenomics.

Keywords: Maximal exact matches · pangenomics · Burrows-Wheeler Transform · grammar-based compression

1 Introduction

Finding maximal exact matches (MEMs) has been an important task at least since Li's introduction of BWA-MEM [11]. A MEM (in other contexts sometimes called a super-MEM or SMEM [10]) of a pattern $P[1..m]$ with respect to a text $T[1..n]$ is a non-empty substring $P[i..j]$ of P such that

- $P[i..j]$ occurs in T,
- $i = 1$ or $P[i-1..j]$ does not occur in T,
- $j = m$ or $P[i..j+1]$ does not occur in T.

If we have a suffix tree for T then we can find all the MEMs of P with respect to T in $O(m)$ time, but its $\Theta(n)$-word space bound is completely impractical in bioinformatics. The textbook compact solution (see [12,13,16]) uses a bidirectional FM-index to simulate a suffix tree, which is slightly slower—typically using $\Theta(\log n)$ extra time per MEM, and with significantly worse constant coefficients overall—but takes $\Theta(n)$ bits for DNA instead of $\Theta(n)$ words. As geneticists have started aligning against pangenomic references consisting of hundreds or thousands of genomes, however, even $\Theta(n)$ bits is unacceptable. Bioinformaticians have started designing indexes for MEM-finding that can work with such massive and highly repetitive datasets [1,4,6,9,14] but, although they have shown some practical promise [17], there is still definite room for improvement.

© The Author(s), under exclusive license to Springer Nature Switzerland AG 2024
J. D. Day and F. Manea (Eds.): DLT 2024, LNCS 14791, pp. 131–140, 2024.
https://doi.org/10.1007/978-3-031-66159-4_10

Fortunately, we are usually interested only in the relatively few MEMs that are longer than we would expect by chance. For example, consider the randomly chosen string over $\{A, C, G, T\}$ shown at the top of Fig. 1, with the highlighted substring copied below it and then edited by having each of its characters replaced with probability 1/4 by another character chosen uniformly at random from $\{A, C, G, T\}$ (so a character could be replaced by a copy of itself). The differences from the original substring are shown highlighted in the copy, with the lengths of the MEMs of the copy with respect to the whole string shown under the copy. The occurrences of the MEMs in the whole string are shown at the bottom of the figure, with the two reasonably long MEMs—of length 12 and 8—highlighted in red. These are the two interesting MEMs, and the others are really more trouble than they are worth. (The unhighlighted substrings with more than 6 characters are formed by consecutive or overlapping occurrences of MEMs with at most 6 characters.) The whole string contains 225 of 256 possible distinct 4-tuples (88%), 421 of 1024 possible distinct 5-tuples (41%), and 512 of 4096 possible distinct 6-tuples (13%), so even if the copied substring were completely scrambled we would still expect quite a lot of MEMs of these lengths—and we should ignore them, since they are mostly just noise.

In this paper we show how to find long, interesting MEMs without wasting time finding all the short, distracting ones. We show that under reasonable assumptions we can find all the MEMs of length at least L in time $O(m)$ time plus extra $O(\log n)$ time only for each MEM of length at least nearly L, using a compact index suitable for pangenomics. Specifically, suppose the size of the alphabet is polylogarithmic in n, ϵ is a constant strictly between 0 and 1, $L \in \Omega(\log n)$ and we are given a straight-line program with g rules for T. Then there is an $O(r + \bar{r} + g)$-space index for T, where r and $\bar{r}$ are the numbers of runs in the Burrows-Wheeler Transforms of T and of the reverse of T, with which when given P we can find all the MEMs of P with respect to T with length at least L correctly with high probability and in $O(m + \mu_{(1-\epsilon)L} \log n)$ time, where μ_x is the number of MEMs of length at least x.

The closest previous work to this paper is Li's [10] forward-backward algorithm for finding all MEMs, and a recent paper by Goga et al.'s [7] about how lazy evaluation of longest common prefix (LCP) queries can speed up finding long MEMs in practice. We review those results and some related background in Sect. 2 and then combine them in Sect. 3 to obtain our result.

2   Previous Work

As far as we know, the asymptotically fastest way to compute the MEMs of a pattern $P[1..m]$ with respect to an indexed text $T[1..n]$ using one of the indexes designed for massive and highly repetitive datasets, is to first compute the forward-match and backward-match pointers of P with respect to T. Figure 2 shows a small example of match pointers.

Definition 1. *Let* $\mathrm{MF}[1..m]$ *and* $\mathrm{MB}[1..m]$ *be arrays of positions in* T *such that* $T[\mathrm{MF}[i]..n]$ *has the longest common prefix with* $P[i..m]$ *of any suffix of* T *and*

```
TCTTAGCTGACGTTCGGGGCGGGTTAGGCCATCTTCTATAGATTTCTCAG
AGACATCCTAGCCGTGCTGAAGTTGTCACTCGCGGCCGTGTTTCCTAACG
CCACCTGATAGCGTGTTCCAAGCACTTGAGTGTCGGGCTGTAGGGGCTCA
CTCTGCGCAGGATCACGGCTGTTTGTACCTATATCGTTATCGTACTGAAT
AAGTAGAATATCCAAACTTTCAGATTCCGGTTTGGCTGCCAAAACTAGGT
GGGATGTGATGCGCGGCGAATTGTGATCTCGCATTGTATATTATCAATCT
CAGCTTAGCTTGACTTGCACAAAATGAACCCTACGGCGGTGGAGGATTAC
GACCGGAAGCGTCCTGCCTCGGAAAGCGTCCTCCTCAGAAGACGCGCGTG
AGGTCCGTCTTGTGGTCGCGACACAATACGCGACACGAACGACTGGTACC
GGATCAAGTTCTCGATAGGCTGAATTGGCTCTTGTATACATGATGATTGT
GGAATCTATACTGTGAACTTATAGGCAAATCCTATGCCACTACATTACGG


AAGTCTTATACCCAAACTTACGGATTCCGGTTTGTCTGCCGAAATTAGGT
4 556 5 44 8   6   6  55(12)    6 5 455 4 4444455


TCTTAGCTGACGTTCGGGGCGGGTTAGGCCATCTTCTATAGATTTCTCAG
AGACATCCTAGCCGTGCTGAAGTTGTCACTCGCGGCCGTGTTTCCTAACG
CCACCTGATAGCGTGTTCCAAGCACTTGAGTGTCGGGCTGTAGGGGCTCA
CTCTGCGCAGGATCACGGCTGTTTGTACCTATATCGTTATCGTACTGAAT
AAGTAGAATATCCAAACTTTCAGATTCCGGTTTGGCTGCCAAAACTAGGT
GGGATGTGATGCGCGGCGAATTGTGATCTCGCATTGTATATTATCAATCT
CAGCTTAGCTTGACTTGCACAAAATGAACCCTACGGCGGTGGAGGATTAC
GACCGGAAGCGTCCTGCCTCGGAAAGCGTCCTCCTCAGAAGACGCGCGTG
AGGTCCGTCTTGTGGTCGCGACACAATACGCGACACGAACGACTGGTACC
GGATCAAGTTCTCGATAGGCTGAATTGGCTCTTGTATACATGATGATTGT
GGAATCTATACTGTGAACTTATAGGCAAATCCTATGCCACTACATTACGG
```

Fig. 1. A randomly chosen string (**top**) over $\{A, C, G, T\}$ with the highlighted substring copied (**center**) and then edited. The differences from the original substring are shown highlighted in red in the copy, with the lengths of the MEMs of the copy with respect to the whole string shown under the copy; 12 is shown as (12) to distinguish it from 1 followed by 2. The occurrences of the MEMs in the whole string (**bottom**) are shown in black when they have lengths 4, 5 or 6, and in red when they have lengths 8 or 12. Substrings longer than 6 characters shown in black are formed by consecutive or overlapping occurrences of MEMs of length at most 6. (Color figure online)

$T[1..MB[i]]$ *has the longest common suffix with* $P[1..i]$ *of any prefix of* T, *for* $1 \leq i \leq m$. *We call* $MF[1..m]$ *and* $MB[1..m]$ *the* forward-match *and* back-ward match pointers *of* P *with respect to* T.

Bannai, Gagie and I [1] showed how to compute MF in $O(m(\log \log n + \log \sigma))$ time using an $O(r)$-space index for T, where σ is the size of the alphabet and r is the number of runs in the Burrows-Wheeler Transform (BWT) of T. Applying a speedup by Nishimoto and Tabei [15], their time bound becomes $O(m \log \sigma)$, or $O(m)$ when σ is polylogarithmic in n (using multiary wavelet trees [3]). If we apply the same ideas to the reverses of P and T, we can compute MB in the same time with an $O(\bar{r})$-space index, where $\bar{r}$ is the number of runs in the BWT of the reverse of T.

	1 2	3	4	5 6 7 8 9 10 11 12
$T =$	**G A**	**T**	**T**	**A G A T A** C A T
$P =$	**T A**	C	A	**T A G A T** T A G
MF =	8 9	10	7	4 5 1 2 3 4 5 1
MB =	3 5	10	11	12 9 6 7 8 4 5 6

Fig. 2. The forward-match and backward-match pointers MF[1..m] and MB[1..m] of $P = $ TACATAGATTAG with respect to $T = $ GATTAGATACAT. Since $T[5..12]$ has the longest common prefix AGAT with $P[6..12]$, MF[6] $= 5$ (**red**); since $T[1..12]$ has the longest common suffix CAT with $P[1..5]$, MB[5] $= 12$ (**blue**). (Color figure online)

Theorem 1. *There is an $O(r + \bar{r})$-space index for T, where r and $\bar{r}$ are the number of runs in the BWT of T and the reverse of T, with which when given P we can compute MF and MB in $O(m \log \sigma)$ time, or $O(m)$ time when σ is polylogarithmic in n.*

Suppose we have MF and MB and we can compute in $O(t(n))$ time both the length LCP $(P[i..m], T[\mathrm{MF}[i]..n])$ of the longest common prefix of $P[i..m]$ and $T[\mathrm{MF}[i]..n]$ and the length LCS $(P[1..i], T[1..\mathrm{MB}[i]])$ of the longest common suffix of $P[1..i]$ and $T[1..\mathrm{MB}[i]]$. Then we can use a version of Li's [10] forward-backward algorithm to find all MEMs. To see why, suppose we know that the kth MEM from the left starts at $P[i_k]$; then it ends at $P[j_k]$, where

$$j_k = i_k + \mathrm{LCP}\,(P[i_k..m], T[\mathrm{MF}[i_k]..n]) - 1\,,$$

which we can find in $O(t(n))$ time. Since MEMs cannot nest, the next character $P[j_k + 1]$ is in the $(k + 1)$st MEM from the left. (For simplicity and without loss of generality, we assume all the characters in P occur in T; otherwise, since MEMs cannot cross characters that do not occur in T, we split P into maximal subpatterns consisting only of characters that do.) That MEM starts at $P[i_{k+1}]$, where

$$i_{k+1} = (j_k + 1) - \mathrm{LCS}\,(P[1..j_k + 1], T[1..\mathrm{MB}[j_k + 1]]) + 1\,,$$

which we can also find in $O(t(n))$ time. If there are μ MEMs then, since the first from the left starts at $P[1]$, we can find them all in $O(\mu t(n))$ time.

Suppose we are given a straight-line program with g rules for T. By balancing it [5] and augmenting its symbols with the Karp-Rabin hashes of their expansions, we can build an $O(g)$-space data structure with which, given i and j and constant-time access to the Karp-Rabin hashes of the substrings of P— which we can support after $O(m)$-time preprocessing of P—we can compute LCP$(P[i..m], T[j..n])$ and LCS$(P[1..i], T[1..j])$ correctly with high probability and in $O(\log n)$ time; see [2, Appendix A] for more details of the implementation.

Verbin and Yu [18] showed that any data structure using space S polynomial in g needs $\Omega\left(\frac{\log^{1-\delta} n}{\log S}\right)$ time for random access to T in the worst case, for any positive constant δ, and Kempa and Kociumaka [8] showed that $r, \bar{r} \in O(g \log^2 n)$. Since $g \in \Omega(\log n)$ and we can use our LCP or LCS queries to support random access to T in $O(\sigma t(n))$ time, it follows that we cannot have $t(n)$ significantly sublogarithmic while still using space polynomial in $r + \bar{r} + g$ in the worst case.

Using our straight-line program to get $t(n) \in O(\log n)$ gives us the following result, which we believe to be the current state of the art.

Theorem 2. *There is an $O(r + \bar{r} + g)$ index for T with which, given P, we can find all the μ MEMs of P with respect to T correctly with high probability and in $O(m \log \sigma + \mu \log n)$ time, or $O(m + \mu \log n)$ time when σ is polylogarithmic in n.*

Goga et al. [7] recently noted that if a MEM starts at $P[i]$ and j is the next value at least $i + L - 1$ such that $j = m$ or $\mathrm{MB}[j+1] \neq \mathrm{MB}[j]+1$, then any MEM of length at least L starting in $P[i..j]$ includes $P[j]$. To see why, consider that a MEM of length at least L starting after $P[i]$ must end at or after $P[i + L - 1]$, and that a MEM cannot end at $P[j]$ if $\mathrm{MB}[j + 1] = \mathrm{MB}[j] + 1$. This means that if we are searching only for MEMs of length at least a given L and we have performed an LCS query and found that a MEM starts at $P[i]$, then we can wait to perform another LCS query until we reach $P[j]$.

To see how much faster Goga et al.'s approach can be than finding all MEMs with Theorem 2, suppose $L \in \omega(\log n)$, the longest common substring of P and T has length $O(\log n)$, and there are $\Theta(m)$ MEMs. Then when we evaluate LCS queries lazily we use only $O\left(m + \frac{m \log n}{L}\right) = O(m)$ time—dominated by the time to compute MB—but with Theorem 2 we use $\Theta(m \log n)$ time. (For consistency with the literature about the BWT and matching statistics, Goga et al. work right to left and so presented their approach as lazy LCP evaluation rather than lazy LCS evaluation.)

On the other hand, if every proper prefix $P[1..i]$ of P occurs in T, followed sometimes by $P[i + 1]$ and sometimes by some other character, then we could have $\mathrm{MB}[j + 1] \neq \mathrm{MB}[j] + 1$ for every $j < m$, even though every proper substring of P can be extended to a longer match in T and thus P itself is the only MEM. For example, if $P = \mathtt{GATTACAT}$ and

$$\begin{array}{cccccccccccccccccccccccccccccccccccccccccccc}
1&2&3&4&5&6&7&8&9&10&11&12&13&14&15&16&17&18&19&20&21&22&23&24&25&26&27&28&29&30&31&32&33&34&35&36&37&38&39&40&41&42&43
\end{array}$$
$$T = \mathtt{GCGAAGATAGATTCGATTAGGATTACCGATTACAAGATTACAT}$$

then we can have $\mathrm{MB}[1..8] = [1, 4, 8, 13, 19, 26, 34, 43]$. On an example like this, Goga et al.'s approach can use $\Omega(m \log n)$ time, while Theorem 2 says we can use $O(m \log \sigma + \log n)$ time.

The weakness of Theorem 2 is that it spends logarithmic time on each MEM, and the weakness of Goga et al.'s approach is that when it finds a very long MEM it can spend logarithmic time on each prefix longer than L of that MEM. When there are both many short MEMs and a few very long MEMs, neither approach may work well.

3   Result

Suppose again that we have MF and MB and we can compute in $O(t(n))$ time both LCP $(P[i..m], T[MF[i]..n])$ and LCS $(P[1..i], T[1..MB[i]])$ for any i, and we are interested only in MEMs of length at least a given threshold L. We now show how to modify the forward-backward algorithm to find only those MEMs, using something like Goga et al.'s approach.

Assume we have already found all MEMs of length at least L that start in $P[1..i_k - 1]$ and that $P[i_k]$ is the start of a MEM, for some $i_k \leq m - L + 1$. Notice that any MEMs of length at least L that start in $P[i_k..i_k + L - 1]$ include $P[i_k + L - 1]$. We set

$$b = LCS\,(P[1..i_k + L - 1], T[1..MB[i_k + L - 1]])$$

and consider two cases:

1. If $b \geq L$ then we set

$$f = LCP\,(P[i_k..m], T[MF[i_k]..n]) \geq L\,,$$

   so $P[i_k..i_k + f - 1]$ is the next MEM of length at least L. We report $P[i_k..i_k + f - 1]$ and—unless $i_k + f - 1 = m$ and we stop—set i_{k+1} to the starting position

$$i_k + f - LCS\,(P[1..i_k + f], T[1..MB[i_k + f]]) + 1$$

   of the next MEM from the left after $P[i_k..i_k + f - 1]$.
2. If $b < L$ then there is no MEM of length L starting in $P[i_k..i_k + L - b - 1]$, so we set $i_{k+1} = i_k + L - b$ which is the starting position of a MEM by our choice of b.

After this, we have either reported all MEMs of length at least L and stopped, or we have reported all MEMs of length at least L that start in $P[1..i_{k+1} - 1]$ with $i_{k+1} > i_k$ and $P[i_{k+1}]$ is the starting position of a MEM. Some readers may wonder whether we can ever have $b > L$; notice that $P[i_k - 1]$ could be the start of one MEM of length much more than L and $P[i_k]$ could be the start of another, in which case

$$LCS\,(P[1..i_k + L - 1], T[1..MB[i_k + L - 1]]) \geq L + 1\,.$$

Algorithm 1 shows our pseudocode, starting with i set to 1 and increasing it until it exceeds $m - L + 1$. Figure 3 shows a trace of how Algorithm 1 processes our example of $P = $ TACATAGATTAG and $T = $ GATTAGATACAT from Fig. 2, with $L = 4$. The reader may wonder why we do not follow Goga et al. more closely and set

$$b = LCS\,(P[1..j], T[1..MB[j]])\,,$$

where j is the next value at least $i + L - 1$ such that $j = m$ or MB$[j + 1] \neq$ MB$[j] + 1$, and adjust the rest of the algorithm accordingly. This could indeed line 1: $i \leftarrow 1$

line 2: $i \leq 9$

line 3: $\text{MB}[4] = 11$ so $b \leftarrow \text{LCS}(P[1..4], T[1..11]) = 4$

line 4: $b \geq 4$

line 5: $\text{MF}[1] = 8$ so $f \leftarrow \text{LCP}(P[1..12], T[8..12]) = 5$

line 6: we report $P[1..5]$

line 7: $i + f - 1 \neq 12$

line 10: $\text{MB}[6] = 9$ so $i \leftarrow 6 - \text{LCS}(P[1..6], T[1..9]) + 1 = 4$

line 2: $i \leq 9$

line 3: $\text{MB}[7] = 6$ so $b \leftarrow \text{LCS}(P[1..7], T[1..6]) = 3$

line 4: $b < 4$

line 12: $i \leftarrow 5$

line 2: $i \leq 9$

line 3: $\text{MB}[8] = 7$ so $b \leftarrow \text{LCS}(P[1..8], T[1..7]) = 4$

line 4: $b \geq 4$

line 5: $\text{MF}[5] = 4$ so $f \leftarrow \text{LCP}(P[5..12], T[4..12]) = 5$

line 6: we report $P[5..9]$

line 7: $i + f - 1 \neq 12$

line 10: $\text{MB}[10] = 4$ so $i \leftarrow 10 - \text{LCS}(P[1..10], T[1..4]) + 1 = 7$

line 2: $i \leq 9$

line 3: $\text{MB}[10] = 4$ so $b \leftarrow \text{LCS}(P[1..10], T[1..4]) = 4$

line 4: $b \geq 4$

line 5: $\text{MF}[7] = 1$ so $f \leftarrow \text{LCP}(P[7..12], T[1..12]) = 6$

line 6: we report $P[7..12]$

line 7: $i + f - 1 = 12$

line 8: we break

	1	2	3	4	5	6	7	8	9	10	11	12
$T =$	G	A	T	T	A	G	A	T	A	C	A	T
$P =$	T	A	C	A	T	A	G	A	T	T	A	G
MF $=$	8	9	10	7	4	5	1	2	3	4	5	1
MB $=$	3	5	10	11	12	9	6	7	8	4	5	6

Fig. 3. A trace (**top**) of how Algorithm 1 processes our example (**bottom**) of $P =$ TACATAGATTAG and $T =$ GATTAGATACAT from Fig. 2, with $L = 4$.

Algorithm 1. Pseudocode for our version of Li's forward-backward algorithm, modified to find only MEMs of length at least L.

```
1: i ← 1
2: while i ≤ m − L + 1 do
3:      b ← LCS (P[1..i + L − 1], T[1..MB[i + L − 1]])
4:      if b ≥ L then
5:          f ← LCP (P[i..m], T[MF[i]..n])
6:          report P[i..i + f − 1]
7:          if i + f − 1 = m then
8:              break
9:          end if
10:         i ← i + f − LCS (P[1..i + f], T[1..MB[i + f]]) + 1
11:     else
12:         i ← i + L − b
13:     end if
14: end while
```

be faster in some cases but we do not see that our worst-case bounds (and it complicates our pseudocode and trace).

We can charge the $O(t(n))$ time we spend in each first case ($b \geq L$ in line 4 of Algorithm 1) to the MEM $P[i_k..i_k + f − 1]$ that we then report, and get a bound of $O(\mu_L t(n))$ total time for all the first cases, where μ_L is the number of MEMs of length at least L. To bound the time we spend on the second cases ($b < L$ in line 4 of Algorithm 1), we observe that for each second case, we either find a MEM of length at least $(1 − \epsilon)L$—which may or may not be of length at least L, and so which we may or may not report later in a first case—or we advance at least ϵL characters. These two subcases are illustrated in Fig. 4.

Choose ϵ strictly between 0 and 1 and consider that when $(1 − \epsilon)L \leq b < L$, we can charge the $O(t(n))$ time for the second case to the MEM starting at $i_{k+1} = i_k + L − b$, which has length at least $b \geq (1 − \epsilon)L$. On the other hand, when $b < (1 − \epsilon)L$ we can charge a $(\frac{1}{\epsilon L})$-fraction of the $O(t(n))$ time for the second case to each of the

$$i_{k+1} − i_k = L − b > \epsilon L$$

characters in $P[i_k..i_{k+1} − 1]$.

Although we may charge the $O(t(n))$ time for a second case to a MEM of length at least $(1 − \epsilon)L$, and then right after charge the $O(t(n))$ time for a first case to the same MEM—because it also has length at least L—we do this at most once to each such MEM. In total we still charge $O(t(n))$ time to each MEM of length at least $(1−\epsilon)L$ and $O\left(\frac{t(n)}{\epsilon L}\right)$ time to each character in P. This means we use $O\left(\left(\frac{m}{\epsilon L} + \mu_{(1−\epsilon)L}\right) t(n)\right)$ time overall, where $\mu_{(1−\epsilon)L} \geq \mu_L$ is the number of MEMs of length at least $(1 − \epsilon)L$. Since our algorithm does not depend on ϵ, this bound holds for all ϵ strictly between 0 and 1 simultaneously.

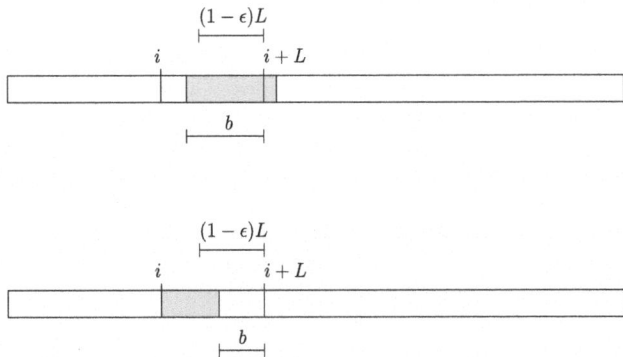

Fig. 4. The two subcases of second cases ($b < L$ in line 4 of Algorithm 1. When $(1 - \epsilon)L \leq b < L$ (**top**), there is a MEM of length at least L starting at $i_k + L - b$ (**shown in grey**). When $b < (1 - \epsilon)L$ (**bottom**), there are $L - b > \epsilon L$ characters between i and $i + L - b$ (**shown in grey**).

Theorem 3. *Suppose we have MF and MB and we can compute in $O(t(n))$ time LCP $(P[i..m], T[\mathrm{MF}[i]..n])$ and LCS $(P[1..i], T[1..\mathrm{MB}[i]])$ for any i. Then we can find all MEMs of P with respect to T with length at least a given threshold L in $O\left(\left(\frac{m}{\epsilon L} + \mu_{(1-\epsilon)L}\right) t(n)\right)$ time for all ϵ strictly between 0 and 1 simultaneously.*

Combining this result with those from Sect. 2 gives us something like Theorem 2 but with the query time depending on $\mu_{(1-\epsilon)L}$ instead of on μ. We note that our final result does not depend on the number of MEMs much shorter than L (the weakness of Theorem 2), nor on the length of MEMs much longer than L (the weakness of Goga et al.'s approach).

Theorem 4. *Suppose σ is polylogarithmic in n, ϵ is a constant strictly between 0 and 1 and $L \in \Omega(\log n)$. Then there is an $O(r + \bar{r} + g)$-space index for T with which, given P, we can find all the MEMs of P with respect to T with length at least L correctly with high probability and in $O(m + \mu_{(1-\epsilon)L} \log n)$ time.*

Acknowledgments. This work was done while the author visited Paola Bonizzoni's group at the University of Milano-Bicocca. Many thanks to them and Lore Depuydt for helpful discussions—especially to Luca Denti for pointing out Li's forward-backward algorithm—and to the anonymous reviewers for pointing out some mistakes in the submitted draft and for suggesting improvements to the presentation. This research was funded by NSERC Discovery Grant RGPIN-07185-2020 to the author and NIH grant R01HG011392 to Ben Langmead.

References

1. Bannai, H., Gagie, T., I, T.: Refining the r-index. Theor. Comput. Sci. **812**, 96–108 (2020)
2. Depuydt, L., et al.: r-indexing without backward searching. arXiv preprint arXiv:2312.01359v2 (2024)
3. Ferragina, P., Manzini, G., Mäkinen, V., Navarro, G.: Compressed representations of sequences and full-text indexes. ACM Trans. Algorithms **3**(2), article 20 (2007)
4. Gagie, T., Navarro, G., Prezza, N.: Fully functional suffix trees and optimal text searching in BWT-runs bounded space. J. ACM **67**(1), 1–54 (2020)
5. Ganardi, M., Jeż, A., Lohrey, M.: Balancing straight-line programs. J. ACM **68**(4), 1–40 (2021)
6. Gao, Y.: Computing matching statistics on repetitive texts. In: Data Compression Conference (DCC), pp. 73–82 (2022)
7. Goga, A., Depuydt, L., Brown, N.K., Fostier, J., Gagie, T., Navarro, G.: Faster maximal exact matches with lazy LCP evaluation. In: 2024 Data Compression Conference (DCC), pp. 123–132 (2024)
8. Kempa, D., Kociumaka, T.: Resolution of the Burrows-Wheeler transform conjecture. Commun. ACM **65**(6), 91–98 (2022)
9. Kempa, D., Kociumaka, T.: Collapsing the hierarchy of compressed data structures: suffix arrays in optimal compressed space. In: 64th Symposium on Foundations of Computer Science (FOCS), pp. 1877–1886 (2023)
10. Li, H.: Exploring single-sample SNP and INDEL calling with whole-genome de novo assembly. Bioinformatics **28**(14), 1838–1844 (2012)
11. Li, H.: Aligning sequence reads, clone sequences and assembly contigs with BWA-MEM. arXiv preprint arXiv:1303.3997 (2013)
12. Mäkinen, V., Belazzougui, D., Cunial, F., Tomescu, A.I.: Genome-Scale Algorithm Design: Bioinformatics in the Era of High-Throughput Sequencing, 2nd edn. Cambridge University Press, Cambridge (2023)
13. Navarro, G.: Compact Data Structures: A Practical Approach. Cambridge University Press, Cambridge (2016)
14. Navarro, G.: Computing MEMs on repetitive text collections. In: 34th Symposium on Combinatorial Pattern Matching (CPM) (2023)
15. Nishimoto, T., Tabei, Y.: Optimal-time queries on BWT-runs compressed indexes. In: 48th International Colloquium on Automata, Languages, and Programming (ICALP) (2021)
16. Ohlebusch, E.: Bioinformatics algorithms: sequence analysis, genome rearrangements, and phylogenetic reconstruction. Oldenbusch Verlag (2013)
17. Rossi, M., Oliva, M., Langmead, B., Gagie, T., Boucher, C.: MONI: a pangenomic index for finding maximal exact matches. J. Comput. Biol. **29**(2), 169–187 (2022)
18. Verbin, E., Yu, W.: Data structure lower bounds on random access to grammar-compressed strings. In: 24th Symposium on Combinatorial Pattern Matching (CPM), pp. 247–258 (2013)

The Pumping Lemma for Context-Free Languages is Undecidable

Hermann Gruber[1], Markus Holzer[2]([⊠]), and Christian Rauch[2]

[1] Planerio GmbH, Theresienhöhe 11A, 80339 München, Germany
h.gruber@planerio.de
[2] Institut für Informatik, Universität Giessen, Arndtstr. 2, 35392 Giessen, Germany
{holzer,christian.rauch}@informatik.uni-giessen.de

Abstract. Recently, the computational complexity of the PUMPING-PROBLEM, that is, for a given finite automaton A and a value p, deciding whether the language $L(A)$ satisfies a previously fixed regular pumping lemma w.r.t. the value p, was considered in [H. GRUBER and M. HOLZER and C. RAUCH. The Pumping Lemma for Regular Languages is Hard. *CIAA 2023*, pp. 128-140.]. Here we generalize the PUMPING-PROBLEM by investigating Bar-Hillel's context-free pumping lemma instead. It turns out that for context-free languages, the PUMPING-PROBLEM for Bar-Hillel's pumping lemma is undecidable. When restricted to regular languages, the problem under consideration becomes decidable.

1 Introduction

Since the beginning of automata and formal language theory, researchers have studied pumping and iteration properties of formal languages to gain better insights into the computational complexity and expressive power of various types of language accepting or generating mechanisms. It is well known that not all formal language families obey pumping properties as, e.g., context-sensitive or Type-0 languages. Hence, satisfying a particular pumping property gives certain information about the structure of the language family, and is very often used to show that a particular language does not belong to the language family in question. For instance, Bar-Hillel's lemma [3] applied to the language $L = \{ a^n b^n c^n \mid n \geq 0 \}$ shows that this language is *not* context free. In fact, the literature on pumping properties is far-reaching, with very different applications, see, e.g., [13], where language families are defined *via* pumping properties, [20] with a focus on pumping with the additional requirement that repeating a sub-word is allowed only if it is done a minimal number of times, or [15], which investigates regular pumping of Turing machine languages and learnability, just to mention a few.

We continue our research on the PUMPING-PROBLEM initiated in [7]. This is the problem of deciding for an automaton A (or a grammar G, respectively) and a value p, whether the language $L(A)$ (the set $L(G)$, respectively) satisfies a previously fixed pumping lemma w.r.t. the value p. For finite automata

J. D. Day and F. Manea (Eds.): DLT 2024, LNCS 14791, pp. 141–155, 2024.
https://doi.org/10.1007/978-3-031-66159-4_11

with Kozen's [16] or Jaffe's [14] pumping lemmata for regular languages, it has been shown that this problem is already intractable for DFAs, namely coNP-complete—this is the case for both pumping lemmata. This is quite remarkable, as it is a rare example of a finite automaton problem where the studied property becomes intractable for a single deterministic device. Jaffe's pumping property turns out to be more complex for NFAs, namely PSPACE-complete, while for Kozen's lemma it is shown to be coNP-hard and contained in Π_2^P for nondeterministic finite state devices. Furthermore, analysis of these problems has led to the conclusion that they are inapproximable unless the Exponential Time Hypothesis (ETH) fails. Since regular languages also satisfy context-free pumping lemmata, e.g., Bar-Hillel's pumping lemma [3], the natural question is how complicated it is to decide whether a regular or a context-free language satisfies a given context-free pumping lemma w.r.t. the pumping parameter involved? This is the starting point of the current paper.

Here we study the complexity of the PUMPING-PROBLEM for regular, linear context-free, and context-free languages w.r.t. Bar-Hillel's pumping lemma [3] and its variants for regular [18] and linear context-free language [12]. Before investigating these problems in detail, we present some basic properties of the minimal pumping constants w.r.t. these pumping lemmata. It is shown that in almost all cases the PUMPING-PROBLEM is undecidable, except when we consider regular languages represented by finite automata or right-linear grammars, where the problem becomes decidable. A more detailed analysis shows that the PUMPING-PROBLEM for context-free languages w.r.t. Bar Hillel's pumping lemma is complete for the Π_1^0-level of the arithmetical hierarchy. Observe that every language in Π_1^0 is the complement of a recursively enumerable language. A summary of the results obtained is given in Table 1.

Table 1. Decidability status of the PUMPING-PROBLEM for different language families and pumping lemmata.

Language family	Pumping w.r.t. value p		
	regular (Lem. 3)	linear context-free	context-free (Lem. 1)
REG	decidable		
LIN	undecidable (Π_1^0)		
CFL			

The paper is organized as follows: in the next section we introduce the necessary notations for context-free and regular languages and two special pumping lemmata for these language families. One of these pumping lemmata is the well known Bar-Hillel lemma [3] for context-free languages. The PUMPING-PROBLEM is then studied. First, the relation between the pumping constants induced by Bar-Hillel's lemma and its variants for regular and linear context-free languages is investigated. It is then shown that the PUMPING-PROBLEM for regular languages is decidable, for each of the pumping lemmata under consideration. We conclude with a summary and topics for further investigation.

2   Preliminaries

We assume the reader to be familiar with the basic notions on grammars and languages as contained in [12]. In particular, a *context-free grammar* (CFG) is a 4-tuple $G = (N, T, P, S)$, where N and T are disjoint alphabets of *nonterminals* and *terminals*, respectively, $S \in N$ is the *axiom*, and P is a finite set of *productions* of the form $A \to \alpha$, where $A \in N$ and $\alpha \in (N \cup T)^*$. As usual, the transitive closure of the derivation relation $\Rightarrow_G$ is written as $\Rightarrow_G^*$. If there is no danger of confusion, we simply write $\Rightarrow$ ($\Rightarrow^*$, respectively) instead of $\Rightarrow_G$ ($\Rightarrow_G^*$, respectively). The *language generated* by G is defined as

$$L(G) = \{ w \in T^* \mid S \Rightarrow_G^* w \}.$$

We also consider the following restrictions of context-free grammars: (i) a context-free grammar is said to be *linear context-free* (LIN) if the productions are of the form $A \to \alpha$, where $A \in N$ and $\alpha \in T^*(N \cup \{\varepsilon\})T^*$—here ε refers to the *empty word*, and (ii) a context-free grammar is said to be *right-linear* or *regular* (REG) if the productions are of the form $A \to \alpha$, where $A \in N$ and $\alpha \in T^*(N \cup \{\varepsilon\})$.

The following pumping lemma for context-free languages can be found in [12, page 125, Lemma 6.1] and is a variant of the well-known Bar-Hillel pumping lemma [3, page 154, Theorem 4.1], see also [6, page 84, Lemma 3.1.1]. Observe that the original Bar-Hillel pumping lemma uses two pumping constants, one for the length of the word and the other for the sub-word that can be pumped.

Lemma 1. *Let L be a context-free language over Σ. Then, there is a constant p (depending on L) such that the following holds: If $z \in L$ and $|z| \geq p$, then there are words $u, v, w, x, y \in \Sigma^*$ such that $z = uvwxy$, $|vx| \geq 1$, $|vwx| \leq p$, and $uv^t wx^t y \in L$ for $t \geq 0$—it is then said that v and x can be (simultaneously) pumped in z.*

For a context-free language L, let $\mathtt{mpcf}(L)$ denote the minimal number p satisfying the conditions of Lemma 1. Let us give a small example:

Example 2. Let $p \geq 2$. Consider the linear context-free grammar

$$G_p = (N, T, P_p, S)$$

with nonterminals $N = \{S\}$, terminals $T = \{a, b\}$, and the set of productions P_p containing the rules

$$S \to a^{p-1} Sb \mid \varepsilon.$$

It is easy to see that G_p generates the language $\{ a^{n(p-1)} b^n \mid n \geq 0 \}$. By inspection, the sub-words a^{p-1} and b can be pumped in any word of $L(G_p)$, but any shorter word cannot since otherwise the number of a's and b's is no longer well correlated anymore. Hence, the minimal pumping constant w.r.t. Lemma 1 is equal to p. Thus, $\mathtt{mpcf}(T_p) = p$, where T_p refers to the language generated by the linear context-free grammar G_p.   □

The previous example shows that already for linear context-free grammars, with a single nonterminal, the minimal pumping constant can be arbitrarily large w.r.t. Lemma 1. The pumping lemma for linear context-free languages reads like the pumping lemma for context-free languages, but with one exception: instead of $|vwx| \leq p$, now the condition $|uvxy| \leq p$ is required—see [12, page 143, Exercise 6.11]. For a linear context-free language L let $\mathtt{mplin}(L)$ denote the minimal number p satisfying the conditions of the pumping lemma for linear context-free languages. Note that $\mathtt{mplin}(T_p) = p$ holds.

When considering regular languages, we usually use finite automata instead of right-linear grammars. A *nondeterministic finite automaton* (NFA) is a quintuple $A = (Q, \Sigma, \cdot, q_0, F)$, where Q is the finite set of *states*, Σ is the finite set of *input symbols*, $q_0 \in Q$ is the *initial state*, $F \subseteq Q$ is the set of *accepting states*, and the *transition function* $\cdot$ maps $Q \times \Sigma$ to 2^Q. Here 2^Q refers to the powerset of Q. The *language accepted* by the NFA A is defined as

$$L(A) = \{\, w \in \Sigma^* \mid (q_0 \cdot w) \cap F \neq \emptyset \,\},$$

where the transition function is recursively extended to a mapping $Q \times \Sigma^* \to 2^Q$ in the usual way. An NFA A is said to be *deterministic* (DFA) if $|q \cdot a| = 1$ for all $q \in Q$ and $a \in \Sigma$. In this case we simply write $q \cdot a = p$ instead of $q \cdot a = \{p\}$.

The syntactic monoid for a given language $L \subseteq \Sigma^*$, is defined by the *syntactic congruence* $\equiv_L$ over Σ^* where $v_1 \equiv_L v_2$ if and only if $uv_1w \in L \iff uv_2w \in L$ for every $u, w \in \Sigma^*$. Then the *syntactic monoid* is the quotient monoid $M(L) = \Sigma^* / \equiv_L$, where the concatenation of equivalence classes $[u]_{\equiv_L} \cdot [v]_{\equiv_L} = [uv]_{\equiv_L}$ serves as the monoid operation. The syntactic monoid of a regular language L is the smallest monoid recognizing the language under consideration (with respect to the division relation) and it is isomorphic to the transformation monoid of the minimal deterministic finite automaton accepting L. Here a language $L \subseteq \Sigma^*$ is *recognizable* if and only if there exists a finite monoid M, a morphism $\varphi_L : \Sigma^* \to M$, and a subset $N \subseteq M$ such that $L = \varphi_L^{-1}(N)$, which in turn is equivalent to the regularity (acceptance by a finite state machine) of L. For an n-state DFA (NFA, respectively) the size of the syntactic monoid is at most n^n (2^{n^2}, respectively).

The pumping lemma for regular languages, which can be found in [18, page 119, Lemma 8], [4, page 252, Folgerung 5.4.10], and [12, page 56, Lemma 6.1], reads as follows:

Lemma 3. *Let L be a regular language over Σ. Then there is a constant p (depending on L) such that the following holds: If $w \in L$ and $|w| \geq p$, then there are words $x \in \Sigma^*$, $y \in \Sigma^+$, and $z \in \Sigma^*$ such that $w = xyz$, $|xy| \leq p$, and $xy^tz \in L$ for $t \geq 0$. It is said that y can be* pumped *in w.*

Similarly to context-free languages, for a regular language L let $\mathtt{mpl}(L)$ denote the smallest number p that satisfies the above statement. A more relaxed version of the previous lemma can be found in [16, page 70, Theorem 11.1], where the condition $|xy| \leq p$ is *not* required. We call this variant *Kozen's pumping lemma*. Obviously, Lemma 3 implies Kozen's pumping lemma. For a regular language L, the smallest value p that satisfies Kozen's pumping lemma is denoted $\mathtt{mpc}(L)$. Then we have

$$\mathtt{mpc}(L) \leq \mathtt{mpl}(L) \leq \mathtt{sc}(L),$$

where $\mathtt{sc}(L)$ is the number of states of the minimal deterministic finite automaton (DFA) accepting L, as shown in [5]. For further properties of $\mathtt{mpc}$ and $\mathtt{mpl}$, see [5,7,10,11].

3   The Language-Pumping-Problem

Recently in [7], the computational complexity of the pumping problem for regular languages w.r.t. Jaffe's [14] and Kozen's [16] pumping lemmata was investigated. It turned out that this problem is already intractable for DFAs and becomes PSPACE-complete for NFAs. Both pumping lemmata don't require any *upper* bound on the length of the pumped sub-word v or uv in $z = uvw$. The problem under consideration is defined as follows:

LANGUAGE-PUMPING-PROBLEM or for short PUMPING-PROBLEM:
  FIXED: Particular pumping lemma, such as Lemma 1.
  INPUT: an accepting or generating device A for a formal language family such as, e.g., the regular languages, and a natural number p, i.e., an encoding $\langle A, 1^p \rangle$.
  OUTPUT: Yes, if and only if the statement from a previously fixed pumping lemma holds for the language $L(A)$ w.r.t. the value p.

Thus, the LANGUAGE-PUMPING-PROBLEM is a natural host for different problem variants by considering different pumping lemmata and different formal language families. Here, we are particularly interested in the family of regular and context-free languages and some of their pumping lemmata as mentioned above.

Before investigating these problems in detail, we present some simple properties of the minimal pumping constants w.r.t. Lemma 1 and 3.

Theorem 4. *Let L be a regular language. Then $\mathtt{mpcf}(L) \leq \mathtt{mpl}(L)$ and moreover $\mathtt{mplin}(L) \leq \mathtt{mpl}(L)$.*

Proof. We only prove the first relation $\mathtt{mpcf}(L) \leq \mathtt{mpl}(L)$, which is immediate, because any pumpable decomposition of a word z w.r.t. Lemma 3 can be read as a pumpable decomposition w.r.t. Lemma 1. To this end consider a pumpable decomposition of $z = uvw$ with $|uv| \leq \mathtt{mpl}(L)$ and $|v| \geq 1$ according to Lemma 3. Then define $u' = u$, $v' = v$, $w' = \varepsilon$, $x' = \varepsilon$, and $y' = w$. Obviously $z = u'v'w'x'y'$ and is a valid pumpable decomposition w.r.t. Lemma 1. This proves the stated claim.   □

Recall, that from [5] it is known that $\mathtt{mpl}(L) \leq \mathtt{sc}(L)$, for any regular language L. Here $\mathtt{sc}(L)$ can be replaced by $\mathtt{nsc}(L)$ as mentioned in [7]. For context-free languages we find the following situation, which follows from the proof of the pumping lemma given in [8, Chapter 6]:

Theorem 5. *Let L be generated by the context-free grammar $G = (N, T, P, S)$. Set $n := |N|$ and $m := \max\{2, |\alpha| \mid A \to \alpha \in P\}$, then $\mathtt{mpcf}(L) \leq m^{2n+3}$.*   □

If the given context-free grammar is in Chomsky normal form,[1] the proof in [12] yields the bound $\mathtt{mpcf}(L) \leq 2^n$. Notice, however, that the conversion to normal form incurs a size blow-up in the worst case [17]. For linear context-free languages, the next theorem applies:

Theorem 6. *Let L be a linear context-free language generated by the linear context-free grammar $G = (N, T, P, S)$. Set $n := |N|$ and similarly as above define $m := \max\{|\alpha| \mid A \to \alpha \in P\}$, then $\mathtt{mplin}(L) \leq (m-1) \cdot n + 2$.*

Proof. Consider a derivation $S \Rightarrow^* z$ of G generating a word $z \in L$, with $|z| \geq (m-1) \cdot n + 2$. (In case there is no such word z, then the pumping condition is trivially satisfied). Observe that the derivation must have at least $n + 1$ steps: the last derivation step can generate at most m terminal symbols, and every other derivation step can generate at most $m - 1$ terminal symbols each. Within n steps, the grammar G can thus generate only terminal words of length at most $(m-1)n + 1 < |z|$.

We find a decomposition $z = uvwxy$ that meets the needs as follows. Let A denote the first variable in the derivation that appears twice. Then the derivation can be written as

$$\underbrace{S \Rightarrow^* uAy \Rightarrow^* uvAxy}_{\leq n+1 \text{steps}} \Rightarrow^* uvwxy.$$

Then $uvAxy$ is generated in at most $n + 1$ steps, each of which generates one variable and at most $m - 1$ terminal symbols, hence $|uvxy| \leq (m-1)(n+1)$. In case $|vx| = 0$, we can cut out the part $uAy \Rightarrow^* uvAxy$ from the derivation and recursively find a suitable decomposition for the shorter derivation. Finally, it is immediate that $A \Rightarrow^* v^i A x^i$ for all $i \geq 0$, so $z = uvwxy$ is a decomposition with the desired properties. This proves the stated upper bound.   □

The relation between $\mathtt{mpcf}(L)$ and $\mathtt{mplin}(L)$ for regular and linear context-free languages L is subject to further research.

3.1 Decidability of Context-Free Pumping for Regular Languages

The main result of this section is that the pumping problem for regular languages w.r.t. Lemma 1 is decidable. The statement reads as follows:

---

[1] A context-free grammar $G = (N, T, P, S)$ is in *Chomsky normal form* if every production is either of the form $A \to a$ or $A \to BC$ or $S \to \varepsilon$, for $A, B, C \in N$ and $a \in T$.

Theorem 7. *Given a finite automaton A and a natural number p, it is decidable whether for the language $L(A)$ the statement of Lemma 1 holds for the value p.*

Before we come to the proof of this result, let us take a closer look at the pumping lemma for regular languages, as stated in Lemma 3.

Theorem 8. *Let L be a regular language over Σ that is accepted by an n-state finite automaton and $p \leq n$. If for every $w \in L$ with $p \leq |w| \leq n + |M(L)|$, there are words $x \in \Sigma^*$, $y \in \Sigma^+$, and $z \in \Sigma^*$ such that $w = xyz$, $|xy| \leq p$, and $xy^t z \in L$ for $t \geq 0$, then Lemma 3 is satisfied w.r.t. the value p.*

Proof. It suffices to show that for every word w with $|w| > n + |M(L)|$ there are words $x \in \Sigma^*$, $y \in \Sigma^+$, and $z \in \Sigma^*$ such that $w = xyz$, $|xy| \leq p$, and $xy^t z \in L$ for $t \geq 0$. Since $|xy| \leq p$ and $p \leq n$, by assumption the pumping of y only appears in the prefix of length at most n of w. Hence we decompose w into $w = uv$ such that $|u| = n$ and replace v by the shortest word v' that is equivalent w.r.t. the syntactic monoid of L, i.e., $\varphi_L(v') = \varphi_L(v)$ for the syntactic morphism $\varphi_L : \Sigma^* \to M(L)$ and there is no shorter word than v' that satisfies the above equality. Then uv' is of length at most $n + |M(L)|$ and

$$w = uv \in L \iff uv' \in L.$$

By the precondition of the implication there are words $x \in \Sigma^*$, $y \in \Sigma^+$, and $z \in \Sigma^*$ such that $uv' = xyz$, $|xy| \leq p$, and $xy^t z \in L$ for $t \geq 0$. By construction xy entirely lies within u, while z may contain letters from the right end of u followed by the whole word v'. Thus, we can write $z = z_1 z_2$ with $z_1 = (xy)^{-1} u$ and $z_2 = v'$. But then we can use this decomposition to construct a y-pumpable decomposition for the original word w we started from, by using the words x, y, and $z_1 v$ without changing the acceptance of the word

$$xy^t z = xy^t z_1 z_2 \in L \iff xy^t z_1 v \in L,$$

for $t \geq 0$, because $z_2 = v'$ and v belong to the same equivalence class w.r.t. the syntactic congruence $\equiv_L$ of the language L.   □

With the help of the previous theorem we can now show that the regular pumping-problem for regular languages is decidable.

Theorem 9. *Given a finite automaton A and a natural number p, it is decidable whether for the language $L(A)$ the statement of Lemma 3 holds for the value p.*

Proof. Let n be the number of states of the automaton A. A Turing machine M that decides the problem in question works as follows: first M constructs a list of all words of length at least p and at most $n + 2^{n^2}$, that belong to the language $L(A)$. Observe that $n^n \leq 2^{n^2}$ and therefore $|M(L)| \leq 2^{n^2}$ holds. If this list is empty, then the Turing machine halts and accepts, because in this case $L(A)$ is a finite language, whose longest word is shorter than p. Next assume that the constructed list is *not* empty. Then M cycles through all words z in the list and tries to find a valid decomposition $z = uvw$ according to Lemma 3 with value p such that v can be pumped in z. If the machine M does not find such a decomposition it halts and rejects. Otherwise it continues with the next word in the list. In case there is no next word in the list, that is, by cycling trough all words in the list, M has found a p-length valid pumpable decomposition of each word, the Turing machine halts and accepts, since by Theorem 8 the problem instance requires a positive answer.   □

The proof of Theorem 9 relies heavily on Theorem 8, which does not obviously generalize to context-free pumping lemmata, because the simultaneous pumping of sub-words (context-free pumping) can occur at any position in a given word— and not necessarily only in a prefix of bounded length, as it does for the regular pumping lemma in question (Lemma 3). Thus, proving that the context-free pumping problem for regular languages is decidable requires a different proof strategy, which we develop in the next proof.

Proof (of Theorem 7). Recall, that A is a finite automaton with input alphabet Σ. Let p be natural number. We want to decide whether Lemma 1 holds for the regular language $L(A)$ with value p. First let us take a closer look at word decompositions used in Lemma 1 with value p when applied to a regular language. Let $L := L(A)$ and z be any word in L (not necessarily of length at least p). Any decomposition of z into $u, v, w, x, y \in \Sigma^*$ such that $z = uvwxy$, $|vx| \geq 1$, $|vwx| \leq p$, and $uv^t wx^t y \in L$ for $t \geq 0$, can be described by a five-tuple

$$(u', v, w, x, y'),$$

where u' (respectively y') is any representative for the Myhill-Nerode equivalence class $[u']_{\equiv_L}$ (respectively $[y']_{\equiv_L}$). For such a 5-tuple (induced by a word z and its decomposition), it is easy to see that any word z' with the property $z' \in [u']_{\equiv_L} \cdot vwx \cdot [y']_{\equiv_L}$, obeys a v-x-pumpable decomposition $z' = uvwxy$ with $u \in [u']_{\equiv_L}$ and $y \in [y']_{\equiv_L}$ satisfying $|vx| \geq 1$ and $|vwx| \leq p$ such that $uv^t wx^t y \in L$ for $t \geq 0$. Observe that the representatives of the equivalence classes in the first and last component can be chosen as a word of length at most 2^{n^2} each, because the syntactic monoid has at most 2^{n^2} elements. If this is the case, we call the corresponding five-tuple *valid*.

Let $D_{L,p}$ refer to the set of all valid five-tuples. It is easy to see that membership in $D_{L,p}$ is decidable. On input (u, v, w, x, y) a Turing machine first verifies the length requirements on these words, i.e., $|u|, |y| \leq 2^{n^2}$, $|vwx| \leq p$, and $|vx| \geq 1$. If the length requirements are fulfilled, then for the linear context-free grammar $G = (\{S, A\}, \Sigma, P, S)$ with the production set $P = \{ S \to uAy, A \to vAx \mid w \}$, which generates the language $\{ uv^t wx^t y \mid t \geq 0 \}$, it is verified whether $L(G) \subseteq L$ holds. This can be done by checking $L(G) \cap \overline{L}$ for being empty. If this is the case, the Turing machine halts and accepts the input (u, v, w, x, y). Otherwise, the Turing machine halts and rejects. Here $\overline{L}$ refers to the complement of L, i.e., $\overline{L} := \Sigma^* \setminus L$. Since all the necessary checks on the linear context-free grammar G described above can be decided, the whole algorithm decides membership in $D_{L,p}$.

Next we construct an NFA for the language

$$P := \bigcup_{(u,v,w,x,y) \in D_{L,p}} [u]_{\equiv_L} \cdot vwx \cdot [y]_{\equiv_L}.$$

This can be easily done by cycling through all elements of $D_{L,p}$. It is easy to see that P describes all words in L that can be pumped according to Lemma 1 w.r.t. the value p. Finally we build an automaton for the language $L \setminus P$. Then we consider two cases:

1. $L \setminus P$ contains a word w of length at least p. Then the pumping lemma for context-free languages (Lemma 1) w.r.t. the value p does not hold for L. This is witnessed by w. Hence, the input A and p has to be rejected.
2. $L \setminus P$ does not contain any word of length at least p. Then the pumping lemma for context-free languages w.r.t. the value p holds for L. Therefore the input A and p is accepted.

Since all constructions rely on basic operations on regular languages that can be done by a Turing machine, the problem in question is decidable. This proves the stated claim.   □

With the idea used in the previous proof one can also show the following decidability result for linear context-free pumping on regular languages.

Theorem 10. *Given a finite automaton A and a natural number p, it is decidable whether for the language $L(A)$ the statement of the pumping lemma for linear context-free languages holds for the value p.*   □

3.2 Undecidability of Context-Free Pumping for Context-Free Languages

In contrast to the previous section, where it was shown that the PUMPING-PROBLEM is decidable if we use the pumping lemmata under consideration on regular languages, here we show that the PUMPING-PROBLEM becomes undecidable if we consider context-free languages. In fact, the undecidability already holds for linear context-free languages. To this end, we exploit non-semi-decidable properties of Turing machines by encoding complex Turing machine computations into small grammars [9].

Basically, we consider *valid computations* of Turing machines. Roughly speaking, these are histories of accepting Turing machine computations. It suffices to consider deterministic Turing machines with one single tape and one single read-write head. Without loss of generality and for technical reasons, we assume that the Turing machine accepts by halting and cannot print blanks. A valid computation is a string built from a sequence of configurations passed through during an accepting computation. To be more precise, let Q be the state set of some Turing machine M, where q_0 is the initial state, T is the tape alphabet containing the blank symbol satisfying $T \cap Q = \emptyset$, and $\Sigma \subseteq T$ is the input alphabet. Then a *configuration* of M can be written as a word of the form T^*QT^* such that $a_1 a_2 \cdots a_i q a_{i+1} \cdots a_n$ is used to express that M is in state q, scanning tape symbol a_{i+1}, and a_1, a_2 to a_n is the support of the tape inscription.

Let $\mathsf{VAL}(M)$ be the set of all words of the form

$$ w_1 \$ w_2^R \$ w_3 \$ w_4^R \$ \cdots \$ w_{2k-1} \$ \quad \text{or} \quad w_1 \$ w_2^R \$ w_3 \$ w_4^R \$ \cdots \$ w_{2k}^R \$ $$

where $\$$ is a new symbol not contained in $T \cup Q$, sub-words $w_i \in T^* Q T^*$ are configurations of M, word w_1 is an *initial configuration* of the form $q_0 \Sigma^*$, word w_{2k-1} (w_{2k}, respectively) is a *halting configuration*, i.e., accepting configuration, and w_{i+1} is the *successor configuration* of w_i. The set of all *invalid computations* $\mathsf{INVAL}(M)$ is the complement of $\mathsf{VAL}(M)$ w.r.t. the alphabet $T \cup Q \cup \{\$\}$. From [9] it is known, that a linear context-free grammar generating the language $\mathsf{INVAL}(M)$ can be effectively constructed from a description of M.

It is worth mentioning that $\mathtt{mpcf}(\mathsf{INVAL}(M)) = 1$ holds. This is quite surprising, but due to the fact, that whenever the word under consideration contains the encoding of a state, then this state can be pumped without changing the membership of the pumped word. If there is no state in the considered word, then we pump any single letter such that the pumped string stays in the language $\mathsf{INVAL}(M)$. Nevertheless, we will use a particular encoding of $\mathsf{INVAL}(M)$ for our purpose to show that the PUMPING-PROBLEM becomes undecidable if one considers context-free languages and their pumping lemma.

Theorem 11. *Given a context-free grammar G and a natural number p, it is undecidable whether for the language $L(G)$ the statement of Lemma 1 for the value p holds. The statement remains valid if a linear context-free grammar and the pumping lemma for linear context-free languages is considered instead.*

Proof. We only prove the statement for the context-free pumping case (on a linear context-free language). The proof for the linear context-free case is similar and is left to the interested reader.

The emptiness problem for Turing machines, which is undecidable, is reduced to the problem in question. Let M be a Turing machine. Recall, that $L(M) = \emptyset$ if and only if $\mathsf{INVAL}(M) = \Sigma^*$, for some alphabet Σ that depends on M. Without loss of generality we assume that Σ contains the letters a and b.

Let $p \geq 2$ and $T_p \subseteq \{a, b\}^*$ be a regular language satisfying $\mathtt{mpcf}(T_p) = p$. Consider the language

$$ L_M = h(\Sigma^*) \cdot \# \cdot T_p \cup h(\mathsf{INVAL}(M)) \cdot \# \cdot \Sigma^+ \cup \Sigma^* \# \cup \#^*, $$

where $\#$ is a new symbol not contained in Σ and $h : \Sigma \to \Sigma^*$ is the homomorphism defined by $h(a) = a^2$, for $a \in \Sigma$. It is not hard to see that one can construct a linear context-free grammar for the language L_M—the details are omitted.

Next we show how to decide the emptiness problem for Turing machines using the linear context-free grammar G_M that generates L_M. First observe that whenever we have a word from the sub-languages $\Sigma^* \#$ or $\#^*$, respectively, then such a word can be pumped by using a single letter from Σ, or by using the symbol #, respectively. Thus for those words, the pumping constant can be chosen to be 1. It remains to consider the words from the remaining two sub-languages. To this end we consider two cases:

1. If $L(M) = \emptyset$, then $\mathsf{INVAL}(M) = \Sigma^*$. This implies that

$$L_M = h(\Sigma^*) \cdot \# \cdot \Sigma^+ \cup \Sigma^* \# \cup \#^*,$$

   and any word $u\#w$ from the sub-language $h(\Sigma^*) \cdot \# \cdot \Sigma^+$ can be pumped by any single letter from w, even if the word w consists of only a single letter. By the above argumentation we conclude that the pumping constant for the whole language L_M can be chosen to be $p = 1$.
2. Otherwise, let $L(M) \neq \emptyset$. Regardless of the assumption, note that any non-empty word in $h(\mathsf{INVAL}(M)) \cdot \# \cdot \Sigma^+$ can be pumped by any single letter that appears after the #-symbol. Thus, for these words we can choose the pumping constant $p = 1$.

   It remains to consider the words in $h(\Sigma^*) \cdot \# \cdot T_p$ that are not covered by the set $h(\mathsf{INVAL}(M)) \cdot \# \cdot \Sigma^+$. Since by assumption we have $L(M) \neq \emptyset$, there is at least one word $u = h(u')$ such that u' does not belong to $\mathsf{INVAL}(M)$. Next we consider the word $u\#w$, for a non-empty word $w \in T_p$ such that w is a witness for $\mathtt{mpcf}(T_p) = p$, i.e., any context-free pumping within w requires the length of the simultaneously pumped words (both together) are of length p. The word $u\#w$ can be properly pumped as follows:

   (a) If the pumping appears entirely in the sub-word u, then the total length of the simultaneously pumped words is even, since otherwise the length constraint being of even length is not satisfied for the prefix up to the #-symbol. Hence the minimal pumping constant is at least 2.
   (b) The pumping appears entirely in the sub-word w, which is a member of T_p and a witness for $\mathtt{mpcf}(T_p) = p$. Hence, the pumping constant for $u\#w$ must be chosen to be at least p. By assumption $p \geq 2$.
   (c) Finally, the pumping appears in both u and w—note that the #-symbol can not be part of any word for pumping. Here we exclude the minimal pumping constant 1, which implies that minimal pumping constant is at least 2. If the length of the simultaneously pumped word is 1, then either the pumped sub-word in u or w is empty. In both cases, pumping of a single letter is not possible, because either the length constraint being of even length is not satisfied for the prefix up to the #-symbol or the suffix doesn't belong to T_p anymore, since w was a witness for $\mathtt{mpcf}(T_p) = p$, for $p \geq 2$. Thus, in this case the minimal pumping constant is at least 2.

   Summarizing, in all sub-cases the minimal pumping constant is at least 2.

Thus, the case analysis shows we can decide whether $L(M)$ is empty or not by checking whether for the linear context-free grammar G_M that generates the language L_M the Bar-Hillel pumping lemma is satisfied w.r.t. the value $p = 1$. If this is the case, then $L(M) = \emptyset$. Otherwise, the Turing machine M accepts at least one word and thus its language is non-empty. This proves the stated claim on the undecidability of the Bar-Hillel pumping lemma applied to context-free languages.   □

We can reuse the proof for Theorem 11 to prove the following statement for regular pumping.

Theorem 12. *Given a context-free grammar G and a natural number $p \geq 3$, it is undecidable whether for the language $L(G)$ the statement of Lemma 3 for the value p holds. The statement remains valid if a linear context-free grammar is considered, or the constraint $|xy| \leq p$ is no longer a prerequisite.*

Proof. First we argue about the case where $|xy| \leq p$ is not required for a decomposition regarding Lemma 3. We observe that dropping the length constraint implies that we can pump a word $w \in L$ which fulfills $|w| \geq p$ if it can be decomposed into xyz for any words $x, y, z \in \Sigma$ such that $|y| \geq 1$ and $xy^t z \in L$ for $t \geq 0$. We refer to the minimal constant p fulfilling this statement for a language L by $\texttt{mpc}(L)$.

By inspecting the cases of the proof of Theorem 11 we obtain that in the case $L(M) = \emptyset$ we have $\texttt{mpc}(L) = 1$ and in the case $L(M) \neq \emptyset$ we get $\texttt{mpc}(L) \geq 2$. Therefore the statement of this theorem follows for the variant of Lemma 3. Unfortunately we obtain for the original version of the lemma that $\texttt{mpl}(L_M) = 2$ regardless of whether $L(M) \neq \emptyset$ or not. Therefore we use a variant of the language L_M instead by reversing the concatenations but not the individual languages. Let

$$L'_M = T_p \cdot \# \cdot h(\Sigma^*) \cup \Sigma^+ \cdot \# \cdot h(\mathsf{INVAL}(M)) \cup \# \cdot \Sigma^* \cup \#^*.$$

Then we consider two cases:

1. If $L(M) = \emptyset$, then $\mathsf{INVAL}(M) = \Sigma^*$ and we get that every word w in the language

   $$L'_M = \Sigma^+ \cdot \# \cdot h(\Sigma^*) \cup \# \cdot \Sigma^* \cup \#^*,$$

   can be pumped by its first letter except the words in $\# \cdot \Sigma^*$. All these words allow pumping by their second letter. Therefore $\texttt{mpl}(L'_M) = 2$ in this case.
2. For $L(M) \neq \emptyset$ there exists a word $w \# u \in T_p \cdot \# \cdot h(\Sigma^*)$ such that w is a witness for $\texttt{mpl}(T_p) = p$ and $u = h(u')$ with $u' \notin \mathsf{INVAL}(M)$. Without loss of generality we assume that $|w| \geq p - 1$ and $u \neq \lambda$. Pumping w.r.t. Lemma 3 inside the prefix of length $p - 1$ of w results in a word $x \# u$ such that $x \notin T_p$. This implies that $x \# u \notin L^R_M$ since $u \notin h(\mathsf{INVAL}(M))$. Hence, we obtain $\texttt{mpl}(L^R_M) \geq p$ in the case that $L(M) \neq \emptyset$.

Therefore we obtain that for a linear context-free grammar it is undecidable whether the statement of Lemma 3 holds for a given value p.   □

3.3   More on Context-Free Pumping for Context-Free Languages

As mentioned above, it is undecidable whether for a given context-free grammar G and a value p the Bar-Hillel's pumping w.r.t. the value p is satisfied. So, we are interested to explore how hard the problem is. Is it semi-decidable or is it placed at a higher degree of unsolvability? To this end, we consider the *arithmetic hierarchy*, which is defined as follows—see, e.g., [19]:

$$\Sigma_1^0 = \{\, L \mid L \text{ is recursively enumerable}\,\},$$
$$\Sigma_{n+1}^0 = \{\, L \mid L \text{ is recursive enumerable in some } \in \Sigma_n^0\,\},$$

for $n \geq 1$. Here a language L is said to be recursively enumerable in some B if there is a Turing machine with oracle B that semi-decides L. Here Π_n^0 is the complement of Σ_n^0, i.e., $\Pi_n^0 = \{\, L \mid \overline{L} \text{ is in } \Sigma_n^0\,\}$. Observe that the intersection $\Sigma_1^0 \cap \Pi_1^0$ is the class of all recursive sets. Completeness and hardness are always meant with respect to many-one reducibilities.

A well-known Π_1^0-complete problem, which we will refer to, is the emptiness problem for Turing machines [19], denoted by EMPTY. Here EMPTY $= \{\, \langle M \rangle \mid L(M) = \emptyset \,\}$, where $\langle M \rangle$ is the index, or Gödel number, of M. As usual, $\overline{\text{EMPTY}}$ denotes the complement of EMPTY, which is Σ_1^0-complete. In fact, both problems are related to the PUMPING-PROBLEM in question. Recall the proof of Theorem 11. There we reduced the problem EMPTY to the PUMPING-PROBLEM for context-free languages w.r.t. Bar Hillel's lemma by

$$\langle M \rangle \in \text{EMPTY} \quad \text{if and only if} \quad \langle G_M, 1 \rangle \text{ is a positive instance of the}$$
$$\text{PUMPING-PROBLEM w.r.t. Bar Hillel's lemma,}$$

where G_M generates the language L_M described in proof of Theorem 11. Thus, we can state the following result:

Lemma 13. *The EMPTY-problem reduces to the PUMPING-PROBLEM for context-free languages w.r.t. Bar Hillel's pumping lemma via a many-one reduction.*   □

In order to show Π_1^0-completeness it remains to show that the PUMPING-PROBLEM for context-free languages w.r.t. Lemma 1 is contained in Π_1^0. This is shown next—compare with [15]:

Lemma 14. *Given a context-free grammar G and a natural number p, it belongs to Π_1^0 to check whether for the language $L(G)$ the statement of Lemma 1 for the value p holds, i.e., the problem is co-recursively enumerable.*

Proof. The encoding $\langle G, p \rangle$ is a *negative* instance of the PUMPING-PROBLEM w.r.t. Bar Hillel's lemma if and only if there exists a word z of length at least p, with $z \in L(G)$, such that for each 5-tuple of words (u, v, w, x, y) with

1. $z = uvwxy$,
2. $|vwx| \leq p$, and $|vx| \geq 1$

there exists an integer i such that uv^iwx^iy is not in $L(G)$.

We find it easier to describe a nondeterministic Turing machine which accepts all negative instances. To begin the Turing machine nondeterministically guesses a word z of length at least p, and verifies that $z \in L(G)$ using the well-known

CYK's algorithm. If this is not the case, the Turing machine immediately rejects on this computation path. Then it generates a list of all 5-tuples (u, v, w, x, y) of words which constitute a decomposition $z = uvwxy$ and satisfy both $|vwx| \leq p$ and $|vx| \geq 1$. For each of these decompositions, the machine guesses a nonnegative integer i, and verifies that $uv^iwx^iy \notin L$. If this is not the case, the machine immediately rejects on this computation path. After all such decompositions are verified, the Turing machine accepts on this computation path.

It is clear from the above description that an accepting computation witnesses the existence of a word z of length at least p with $z \in L(G)$, such that each p-admissible decomposition of z can be pumped to a word not in $L(G)$. Conversely, from a given word z of length at least p in $L(G)$ and a list of exponents $i_1, i_2, \ldots, i_j, \ldots$, such that for each admissible decomposition $w = u_jv_jw_jx_jy_j$ the pumped word $u_jv_j{}^{i_j}w_jx_j{}^{i_j}y_j$ is not in $L(G)$, we can construct an accepting computation of the nondeterministic Turing machine described above.

Since there is a nondeterministic Turing machine which halts exactly on the negative instance of the PUMPING-PROBLEM w.r.t. Bar Hillel's lemma, we can see that the set of positive instances is co-semidecidable, and thus in Π_1^0.   □

In summary, we have obtained the following result:

Theorem 15. *Given a context-free grammar G and a natural number p to check whether for the language $L(G)$ the statement of Lemma 1 for the value p holds. is Π_1^0-complete.*   □

4   Conclusion

In this paper, we continued our research on the PUMPING-PROBLEM, recently started in [7]. Here we focused on Bar-Hillel's pumping lemma for context-free languages and its variants. It turned out that the PUMPING-PROBLEM for context-free languages w.r.t. Bar-Hillel's lemma is undecidable and complete for the level Π_1^0 of the arithmetical hierarchy. It remains undecidable even if regular pumping is considered. On the other hand, context-free pumping becomes decidable if the underlying language family is regular. This result relies heavily on the congruence representation of regular languages. Whether this result can be extended to other language families that allow for some congruence representation, such as the family of visibly pushdown languages [1,2], is subject of further research. Furthermore, for the decidable variants of the problem, it remains to consider their computational complexity, which would nicely complement our previous investigations mentioned above.

References

1. Alur, R., Kumar, V., Madhusudan, P., Viswanathan, M.: Congruences for visibly pushdown languages. In: Caires, L., Italiano, G.F., Monteiro, L., Palamidessi, C., Yung, M. (eds.) ICALP 2005. LNCS, vol. 3580, pp. 1102–1114. Springer, Heidelberg (2005). https://doi.org/10.1007/11523468_89

2. Alur, R., Madhusudan, P.: Adding nesting structure to words. J. ACM **56**(3), Art. 16 (2009)
3. Bar-Hillel, Y., Perles, M., Shamir, E.: On formal properties of simple phrase structure grammars. Zeitschrift für Phonetik, Sprachwissenschaft und Kommunikationsforschung **14**, 143–177 (1961)
4. Brauer, W.: Automatentheorie: Eine Einführung in die Theorie endlicher Automaten. Leitfäden und Monographien der Informatik, Teubner Stuttgart (1984). https://doi.org/10.1007/978-3-322-92151-2, (in German)
5. Dassow, J., Jecker, I.: Operational complexity and pumping lemmas. Acta Inform. **59**, 337–355 (2022). https://doi.org/10.1007/s00236-022-00431-3
6. Ginsburg, S.: The Mathematical Theory of Context-Free Languages. McGraw-Hill, New York (1966)
7. Gruber, H., Holzer, M., Rauch, C.: The pumping lemma for regular languages is hard. In: Nagy, B. (eds.) Implementation and Application of Automata. CIAA 2023. LNCS, vol. 14151, pp. 128–140. Springer, Cham (2023). https://doi.org/10.1007/978-3-031-40247-0_9
8. Harrison, M.A.: Introduction to Formal Language Theory. Addison-Wesley, Boston (1978)
9. Hartmanis, J.: Context-free languages and Turing machine computations. In: Proceedings of Symposia in Applied Mathematics, vol. 19, pp. 42–51. American Mathematical Society, Providence, Rhode Island (1967)
10. Holzer, M., Rauch, C.: On Jaffe's pumping lemma, revisited. In: Bordihn, H., Tran, N., Vaszil, G. (eds.) Descriptional Complexity of Formal Systems. DCFS 2023. LNCS, vol. 13918, pp. 65–78. Springer, Cham (2023). https://doi.org/10.1007/978-3-031-34326-1_5
11. Holzer, M., Rauch, C.: On minimal pumping constants for regular languages. In: Gazdag, Z., Iván, S., Kovásznai, G. (eds.) Proceedings of the 16th International Conference on Automata and Formal Languages, pp. 127–141. No. 386 in EPTCS, Eger, Hungary (2023). https://doi.org/10.4204/EPTCS.386.11
12. Hopcroft, J.E., Ullman, J.D.: Introduction to Automata Theory, Languages and Computation. Addison-Wesley, Boston (1979)
13. Horváth, S.: The family of languages satisfying Bar-Hillel's lemma. RAIRO-Informatique théorique et Applications/Theoretical Informatics and Applications **12**(3), 193–199 (1978)
14. Jaffe, J.: A necessary and sufficient pumping lemma for regular languages. SIGACT News **10**(2), 48–49 (1978). https://doi.org/10.1145/990524.990528
15. Kalociński, D.: On computability and learnability of the pumping lemma function. In: Dediu, A.-H., Martín-Vide, C., Sierra-Rodríguez, J.-L., Truthe, B. (eds.) LATA 2014. LNCS, vol. 8370, pp. 433–440. Springer, Cham (2014). https://doi.org/10.1007/978-3-319-04921-2_35
16. Kozen, D.C.: Automata and Computability. Undergraduate Texts in Computer Science, Springer, New York (1997). https://doi.org/10.1007/978-1-4612-1844-9
17. Lange, M., Leiß, H.: To CNF or not to CNF? An efficient yet presentable version of the CYK algorithm. Inform. Didact. **8** (2009)
18. Rabin, M.O., Scott, D.: Finite automata and their decision problems. IBM J. Res. Dev. **3**, 114–125 (1959). https://doi.org/10.1147/rd.32.0114
19. Rogers Jr, H.: Theory of Recursive Functions and Effective Computability. Higher Mathematics. McGraw-Hill, New York (1967)
20. Sommerhalder, R.: Classes of languages proof against regular pumping. RAIRO-Informatique théorique et Applications/Theoretical Informatics and Applications **14**(2), 169–18 (1980)

Techniques for Showing the Decidability of the Boundedness Problem of Language Acceptors

Oscar H. Ibarra[1] and Ian McQuillan[2(✉)]

[1] Department of Computer Science, University of California,
Santa Barbara, CA 93106, USA
ibarra@cs.ucsb.edu
[2] Department of Computer Science, University of Saskatchewan,
Saskatoon, SK S7N 5A9, Canada
mcquillan@cs.usask.ca

Abstract. There are many types of automata and grammar models that have been studied in the literature, and for these models, it is common to determine whether certain problems are decidable. One problem that has been difficult to answer throughout the history of automata and formal language theory is to decide whether a given system M accepts a bounded language (whether there exist words $w_1, \ldots, w_k$ such that $L(M) \subseteq w_1^* \cdots w_k^*$?). Boundedness was only known to be decidable for regular and context-free languages until recently when it was shown to also be decidable for finite automata and pushdown automata augmented with reversal-bounded counters, and for vector addition systems with states. However, decidability of this problem has still gone unanswered for the majority of automata/grammar models with a decidable emptiness problem that have been studied in the literature.

In this paper, we develop new techniques to show that the boundedness problem is decidable for larger classes of one-way nondeterministic automata and grammar models by reducing the problem to the decidability of boundedness for simpler classes of automata. One technique involves characterizing the models in terms of multi-tape automata. We give new characterizations of finite-turn Turing machines, finite-turn Turing machines augmented with various storage structures (like a pushdown, multiple reversal-bounded counters, partially-blind counters, etc.), and simple matrix grammars. The characterizations are then used to show that the boundedness problem for these models is decidable. Another technique uses the concept of the store language of an automaton. This is used to show that the boundedness problem is decidable for pushdown automata that can "flip" their pushdown a bounded number of times. Boundedness remains decidable even if we augment this device with additional stores.

The research of I. McQuillan was supported, in part, by Natural Sciences and Engineering Research Council of Canada Grant 2022-05092 (Ian McQuillan).

J. D. Day and F. Manea (Eds.): DLT 2024, LNCS 14791, pp. 156–172, 2024.
https://doi.org/10.1007/978-3-031-66159-4_12

1   Introduction

There are many well-studied models of automata/grammars that are more powerful than finite automata (denoted by NFA) but less powerful than Turing machines. Perhaps the most well-studied is the one-way nondeterministic pushdown automata (NPDA) which accept the context-free languages. This model is quite practical—for example, the *non-emptiness problem* ("given a machine M, is $L(M) \neq \emptyset$?"), as well as the *infiniteness problem* ("given a machine M, is $L(M)$ infinite?"), can both be determined in polynomial time for NPDA [20].

Authors have studied models that are more powerful than NPDA, such as t-flip NPDA (resp. finite-flip NPDA), which have the ability to flip (or reverse) their pushdown stack at most t (resp. a finite number of) times [18,19]. Non-emptiness and infiniteness are decidable for this model as well [19] (implied from their semilinear Parikh image). Another more powerful model is simple matrix grammars, which are a class of grammars that generates a family of languages properly between the context-free languages and the matrix languages [23].

Other well-studied models with power between that of finite automata and Turing machines is the one-way nondeterministic reversal-bounded multicounter machines [24] (NCM). This is an NFA with some number of counters, where each counter contains a non-negative integer, and transitions can detect if a counter is non-zero or not, and can either subtract one, leave it unchanged, or add one. The condition of being r-reversal-bounded (resp. reversal-bounded) enforces that in each accepting computation, the number of changes between sequences of non-decreasing transitions (adding one or zero at each step) and sequences of non-increasing (subtracting one or zero at each step) on each counter is at most r (resp. a finite number). It is also possible to combine different types of stores. For example, another class of automata is NPDA augmented by reversal-bounded counters, denoted by NPCM. This device, which is strictly more powerful than either NPDA or NCM, has an NP-complete non-emptiness problem [15].

We will also consider nondeterministic Turing machines with a one-way read-only input tape and a single two-way read/write worktape, denoted by NTM. All of the problems above are undecidable for NTM. A t-*turn* (resp. *finite-turn*) NTM are machines with at most t (resp. some number of) changes in direction on the worktape in every accepting computation (called reversal-bounded in [14], but we call it finite-turn here). Here, as a machine reads the worktape from either left-to-right or from right-to-left between two consecutive turns, we interchangeably refer to it as a *pass* or a *sweep* of the worktape. Again, the non-emptiness and infiniteness problems are decidable for this model.

Another important property beyond emptiness and infiniteness is boundedness. A language $L \subseteq \Sigma^*$ is *bounded* if there exist non-empty words $w_1, \ldots, w_k$ such that $L \subseteq w_1^* \cdots w_k^*$. Indeed, boundedness is intimately connected with decidability of the *containment problem* ("given M_1, M_2, is $L(M_1) \subseteq L(M_2)$?"). The containment problem is undecidable for even the simplest non-NFA class of machines: NFA with one 1-reversal-bounded counter [2]. Furthermore, Hopcroft showed that for a context-free language $L_0 \subseteq \Sigma^*$, the problem of deciding, "given a context-free language $L \subseteq \Sigma^*$, is $L_0 \subseteq L$?" is decidable if and only if L_0 is bounded [21].

Here we further explore the important decision problem called the *boundedness problem*: "given a machine M, is $L(M)$ a bounded language?". In the early years of the study of formal language theory, this property was shown to be decidable for NFA and NPDA by Ginsburg and Spanier using a rather complicated procedure [9,10]. In contrast, if a class of machines with an undecidable emptiness problem accepts languages that are closed under concatenation with the language $\$\Sigma^*$ (where $\$$ is a new symbol and Σ is an alphabet with at least two letters), then the boundedness problem is also undecidable for the class, because $\$\Sigma^*$ concatenated with anything non-empty is not bounded, and so $L\$\Sigma^*$ is bounded if and only if L is empty. Until recently, the status of the boundedness problem had been elusive (e.g. it was stated as an open problem for Parikh automata which are equivalent to NCM [4]) or unsolved for essentially all other machine/grammar models (besides NPDA) studied in the literature that have a decidable emptiness problem. Finally, Czerwinski, Hofman, and Zetzsche showed that the boundedness problem is decidable for vector addition systems with states [5] (equivalent to one-way partially blind multicounter machines [11], denoted by PBCM, that properly contain NCM). With PBCM, machines can add and subtract from counters but cannot detect whether counters are zero or not, and therefore no differences in the transitions to be executed are allowed regardless of the counter contents. However, a computation of a machine crashes if a counter goes below zero (i.e. a computation cannot continue with a counter going below zero), and a word is accepted if it reaches a final state after it has read the entire input word with all counters being zero. In [3], it was determined that the boundedness problem for NPCM (and NCM) is decidable, and also coNP-complete.

Here, we develop techniques for showing that the boundedness problem is decidable. One technique involves creating characterizations in terms of multi-tape versions of NFA, NCM, NPCM, and PBCM of the following:

1. finite-turn NTM in terms of multi-tape NFA,
2. finite-turn NTM augmented with reversal-bounded counters in terms of multi-tape NCM,
3. finite-turn NTM augmented with a pushdown and reversal-bounded counters where in each accepting computation, the pushdown can only be changed during one sweep of the worktape, in terms of multi-tape NPCM,
4. finite-turn NTM augmented with partially blind counters in terms of multi-tape PBCM.

These characterizations are used to show decidability of the boundedness (also emptiness/infiniteness) problem for each of the models. These are strong as machines consisting of two 1-turn stores have an undecidable emptiness and thus boundedness problem. In model (3) above, the restriction that the pushdown can only be used during one sweep of the read/write worktape cannot be dropped, as allowing one more sweep would make both emptiness and boundedness undecidable. Note that the model in (3) is more powerful than NPDA and finite-turn NTM, and can even accept non-indexed languages [1]. For model (4), this model is strictly more powerful than the family of PBCM languages. Hence, it is the most powerful model containing non-semilinear languages with a known decidable boundedness problem. Using a similar technique, we show that the boundedness problem is decidable for simple matrix grammars (even with reversal-bounded counters). We also use these characterizations to show that every finite-turn NTM language (even with reversal-bounded counters) is in NLOG, generalizing a result in [29].

Another technique to help with boundedness involves the *store language* of a machine, which is the set of strings that encode the contents of the internal stores that can appear in any accepting computation. There are some automata models in the literature where the family of store languages for that class can be accepted by a simpler type of automata. We use this to show that the boundedness problem is decidable for finite-flip NPDA. This is also true if finite-flip NPDA are augmented by reversal-bounded counters, and by a finite-turn worktape where the flip-pushdown is only used during one sweep of the worktape. Hence, this is the most powerful model properly containing the context-free languages with a known decidable boundedness problem. Although this result implies decidability of boundedness for models (1), (2), and (3) described above, we also present separate constructions for each of (1), (2), and (3), as the characterizations are useful on their own, they can have their own applications (e.g. in showing that every finite-turn NTM is in NLOG), and because the constructions build on top of each other.

There are necessarily many different automata models used in this paper as the two techniques to help with boundedness can be applied to many different models. However, many definitions and proofs are omitted due to space constraints, but more can be found in [22,27].

2 Preliminaries and Notation

We assume knowledge of introductory automata and formal language theory [20], including deterministic and nondeterministic finite automata, context-free grammars, pushdown automata, and Turing machines.

Let $\mathbb{N}$ be the set of positive integers and $\mathbb{N}_0$ be the set of non-negative integers. Given a set X and $t \in \mathbb{N}$, let $\langle X \rangle^t$ be the set of all t-tuples over X. Given a finite alphabet Σ, let Σ^* (resp. Σ^+) be the set of all words (resp. non-empty words) over Σ. The empty word is denoted by λ. A *language* L *over* Σ is any subset of Σ^*, and a t-tuple language L is any subset of $\langle \Sigma^* \rangle^t$. Given a word w, the *reverse* of w, denoted w^R is λ if $w = \lambda$, and $a_n a_{n-1} \cdots a_1$ if $w = a_1 a_2 \cdots a_n$, where $a_i \in \Sigma$ for $1 \leq i \leq n$, which is extended to languages L by $L^R = \{w^R \mid w \in L\}$. The *length* of w, denoted by $|w|$, is equal to the number of characters in w, and given $a \in \Sigma$, $|w|_a$ is the number of a's in w. Given the alphabet $\Sigma = \{a_1, \ldots, a_m\}$ and $w \in \Sigma^*$, the *Parikh image* of w is defined by $\psi(w) = (|w|_{a_1}, \ldots, |w|_{a_m})$; and the Parikh image of a language $L \subseteq \Sigma^*$ is $\psi(L) = \{\psi(w) \mid w \in L\}$. Although we will not provide the definition of semilinearity, a language is semilinear if and only if it has the same Parikh image as some regular language [11]. Similarly, the Parikh image of $(w_1, \ldots, w_t) \in \langle \Sigma^* \rangle^t$ is defined by $\psi(w_1, \ldots, w_t) = \psi(w_1 \cdots w_t)$, and for $L \subseteq \langle \Sigma^* \rangle^t$ it is defined by $\psi(L) = \{\psi(x) \mid x \in L\}$. A class of machines/grammars is said to be *effectively semilinear* if, given such a machine/grammar, a finite automaton that accepts a language with the same Parikh image can be effectively constructed.

We will informally describe a variety of different machine models. For many of our models, we will use multi-tape inputs. Multi-tape inputs have been studied for NPDA [17], NCM [24], and NPCM [26]. For $t \geq 1$, a *one-way t-tape nondeterministic finite automaton* (t-tape NFA) is a tuple $M = (Q, \Sigma, \delta, q_0, F)$ where Q is a finite set of states, Σ is the finite input alphabet, $q_0 \in Q$ is the initial state, $F \subseteq Q$ is the set of final states, and δ is a partial function from $Q \times (\Sigma \cup \{\lambda\}) \times \{i \mid 1 \leq i \leq t\}$ (for 1-tape machines, we unambiguously leave off the last component) to finite subsets of Q. We usually denote an element $q' \in \delta(q, a, i)$ by $\delta(q, a, i) \to q'$. A *configuration* of M is a tuple $(q, (w_1, \ldots, w_t))$ where $q \in Q$ is the current state, and $(w_1, \ldots, w_t), w_1, \ldots, w_t \in \Sigma^*$ is the remainder of the t-tape input. Two configurations change as follows:

$$(q, (w_1, \ldots, w_{i-1}, aw_i, w_{i+1}, \ldots, w_t)) \vdash (q', (w_1, \ldots, w_t)),$$

if there is a transition $\delta(q, a, i) \to q'$. We let $\vdash^*$ be the reflexive and transitive closure of $\vdash$. An *accepting computation* on $(w_1, \ldots, w_t) \in \langle \Sigma^* \rangle^t$ is a sequence $(q_0, (w_1, \ldots, w_t)) \vdash \cdots \vdash (q_n, (\lambda, \ldots, \lambda))$, where $q_n \in F$. The *language accepted* by M, $L(M) \subseteq \langle \Sigma^* \rangle^t$, is the set of all $(w_1, \ldots, w_t)$ for which there is an accepting computation.

We will augment t-tape NFAs with additional stores. A t-tape NPDA is a t-tape NFA with an additional pushdown alphabet Γ (and fixed bottom-of-stack marker Z_0), and δ becomes a partial function with rules of the form $\delta(q, a, i, x) \to (q', \gamma)$, where q, q', a, i are as with t-tape NFAs, $x \in \Gamma$ is the topmost symbol of the pushdown which gets replaced by the word $\gamma \in \Gamma^*$. Configurations include a third component in $Z_0(\Gamma - Z_0)^*$ which contains the current pushdown contents, as is standard for pushdown automata [20]. The exact definition appears in [27].

We can similarly define machines with multiple stores by defining the transitions to only read and change one store at a time. As an example, a t-tape k-counter machine has transitions $\delta(q, a, i, s, j) \to (q', x)$ where $q, q' \in Q, a \in \Sigma \cup \{\lambda\}, 1 \leq i \leq t, s \in \{0, 1\}, x \in \{-1, 0, 1\}, 1 \leq j \leq k$, which can only be applied if q is the current state, a is next on the input of tape i, and counter j is zero if $s = 0$ and counter j is not zero if $s = 1$; causing the state to switch to q' and counter j to be increased by x. A machine is reversal-bounded if there is an $r \geq 1$ such that in each accepting computation, the number of changes between sequences of non-decreasing transitions (adding one or zero at each step) and sequences of non-increasing (subtracting one or zero at each step) on each counter is at most r. A t-tape NCM is a t-tape reversal-bounded k-counter machines for some k. A machine is partially-blind (a t-tape PBCM) if $\delta(q, a, i, 0, j) = \delta(q, a, i, 1, j)$ for each $q \in Q, a \in \Sigma \cup \{\lambda\}, 1 \leq i \leq t, 1 \leq j \leq k$, and therefore there are no differences in transitions between a counter being zero or not. With this model, acceptance is by final state with all counters being zero.

Some other one-tape machine models described in Sect. 1 including finite-flip NPDA and finite-turn NTM will be used, and we refer to the extended version of the paper for their definitions [27] due to space constraints. Intuitively, a t-flip NPDA is an NPDA but has an additional 'flip' instruction that causes the pushdown to change from $Z_0 z$ to $Z_0 z^R$, and at most t such transitions can be applied in each accepting computation.

For any class of multi-tape machines, it is effectively semilinear if and only if the class of 1-tape machines of that type is effectively semilinear. This is true because given a t-tape M, we can construct a 1-tape machine M' that reads $a \in \Sigma$ from the tape whenever M can read a from any tape. That is, any accepting computation of M on input $(w_1, \ldots, w_t)$ has some order in which characters of the t tapes are read, and therefore there is a word w that is a permutation of $w_1 \cdots w_t$ that is accepted by M'. Conversely, in an accepting computation on input w by M', each letter read of w is simulating the reading of that letter by one of the t tapes. Thus, there is some t-tuple input $(w_1, \ldots, w_t)$ to M that accepts such that $w_1 \cdots w_t$ is a permutation of w. Hence, $\psi(L(M)) = \psi(L(M'))$.

It is known that for any type of nondeterministic machine model with reversal-bounded counters, one can equivalently use monotonic counters [25] instead of reversal-bounded counters. Such machines have an even number k of counters that we identify by $C_1, D_1, \ldots, C_{k/2}, D_{k/2}$ that can only be incremented but not decremented. Transitions do not detect the counter status, and acceptance occurs when the machine enters an final state with counters C_i and D_i having the same value for each i. Due to the equivalence, we will use the same notation as above (NPCM, etc.) to mean machines with monotonic counters. Monotonic counters are helpful in this paper because if we simulate an accepting computation of a machine with another machine that applies the same changes but in a different order, then the resulting simulation will still have matching monotonic counters.

3   Boundedness Using Multi-tape Characterizations

3.1   Characterizations of Finite-Turn Turing Machines

We first look at finite-turn NTM, and finite-turn NTM with reversal-bounded counters (denoted by finite-turn NTCM). These machines have previously been studied both without counters [14] and with counters [16]. We give characterizations of these machines in terms of multi-tape NFA and multi-tape NCM.

Example 1. Consider $L = \{w\#w\$v\#v \mid w, v \in \{a, b\}^*, |w|_a = |v|_a, |w|_b = |v|_b\}$. L can be accepted by a 4-turn NTCM M with four monotonic counters as follows: on input $w_1\#w_2\$v_1\#v_2$, M reads w_1 and writes it to the tape while in parallel recording $|w|_a$ and $|w|_b$ in two monotonic counters C_1 and C_2. When it hits $\#$, it turns and moves the read/write head to the left end of the worktape, and verifies $w_2 = w_1$. When it hits $\$$, it does the same procedure with the read/write tape to the right to verify $v_1 = v_2$, while in parallel putting $|v_1|_a$ and $|v_2|_b$ on two monotonic counters D_1 and D_2. It then accepts if the contents of C_1 equals D_1 and C_2 equals D_2.

We say M is in *state normal form* if: M makes exactly t turns on all inputs accepted; the read/write worktape head always moves left or right at every step that uses the worktape (no stay transitions); on every accepting computation, there is a worktape cell d, and M only turns left on cell d and right on cell 1; the worktape never moves left of cell 1 or right of cell d; the machine writes the current state in the first and last cell (1 and d) every time it reaches them; and M accepts only in cell 1 or d.

Lemma 2. *Let $t \geq 0$. Given a t-turn NTM (resp. t-turn NTCM) M, we can construct a t-turn NTM (resp. t-turn NTCM) M' in state normal form such that $L(M') = L(M)$.*

For such a t-turn NTM M in state normal form, define an alphabet Δ of $t+1$ tuples of the form $b = (b_1, \ldots, b_{t+1})$ where each b_i is in the worktape alphabet Γ. We can think of a word in Δ^* as representing a $t+1$ track worktape, where the ith component is the ith track. For $1 \leq i \leq t+1$, define a projection homomorphism h_i from Δ^* to Γ^* such that $h_i((b_1, \ldots, b_{t+1})) = b_i$. Given a t-turn NTM M in state normal form, define the *history language* $H(M)$ over Δ^* as follows: $H(M)$ contains all strings x where there is an accepting computation of M such that $h_i(x)$ is the string on the worktape after the ith sweep of the worktape and $h_{t+1}(x)$ is the string on the worktape at the end of the computation after it has made the final sweep after the last turn. This means if t is even (it is similar if odd)

$$h_1(x) = q_0 x_1 q_1, h_2(x) = q_2 x_2 q_1, \ldots, h_t(x) = q_t x_t q_{t-1}, h_{t+1}(x) = q_t x_{t+1} q_{t+1},$$

where q_0 is the initial state, M writes $q_0 x_1 q_1$ on the first sweep, etc. until $q_t x q_{t+1}$, which is the final worktape contents, and q_{t+1} is a final state.

Let $t \geq 1$. For a t-tuple $(w_1, \ldots, w_t)$, $w_i \in \Sigma^*$, let its *alternating pattern* be:

$$(w_1, \ldots, w_t)^A = \begin{cases} w_1 w_2^R \cdots w_{t-1} w_t^R & \text{if } t \text{ is even,} \\ w_1 w_2^R \cdots w_{t-1}^R w_t & \text{if } t \text{ is odd.} \end{cases}$$

If there is a $t \geq 1$ with $L \subseteq \langle \Sigma^* \rangle^t$, let $L^A = \{(w_1, \ldots, w_t)^A \mid (w_1, \ldots, w_t) \in L\}$. We now show that from every t-turn NTM M, we can construct a $(t+1)$-tape NFA M' such that $L(M')^A = L(M)$. Starting with M in state normal form, M' guesses a $(t+1)$-track string $x \in \Delta^*$ letter-by-letter from left-to-right while checking in parallel that the input on tape i would be read by the simulated moves on track i thereby verifying that $x \in H(M)$.

Lemma 3. *Let $t \geq 0$, and M be a t-turn NTM (resp. t-turn NTCM). We can construct a $(t+1)$-tape NFA (resp. $(t+1)$-tape NCM) M' such that $L(M')^A = L(M)$.*

For the opposite direction, on the first sweep of the worktape, M' guesses and writes a guessed sequence of transition labels of M, and then sweeps the worktape once for each tape i to make sure the next section of the input word of M' would be read by tape i in the simulation.

Lemma 4. *Let $t \geq 0$, and let M be a $(t + 1)$-tape NFA (resp. $(t + 1)$-tape NCM). Then we can construct a t-turn NTM (resp. t-turn NTCM) M' such that $L(M') = L(M)^A$.*

From the two lemmas above, we obtain:

Proposition 5. *Let $t \geq 0$. There is a $(t+1)$-tape NFA (resp. $(t+1)$-tape NCM) M if and only if there is a t-turn NTM (resp. t-turn NTCM) M' such that $L(M') = L(M)^A$.*

As a generalization of bounded languages, we say $L \subseteq \langle \Sigma^* \rangle^t$ is a *bounded t-tuple language* if $L \subseteq B_1 \times \cdots \times B_t$, where each B_i is of the form $w_1^* \cdots w_n^*$ for some $w_1, \ldots, w_n \in \Sigma^+$. Given $L \subseteq \langle \Sigma^* \rangle^t$, let $L_i = \{w_i \mid (w_1, \ldots, w_t) \in L\}$.

Let $\mathcal{M}$ be a class of multi-tape machines consisting of an NFA with zero or more stores. Given a t-tape $M \in \mathcal{M}$, for each i, $1 \leq i \leq t$, let M_i be the one tape machine in $\mathcal{M}$ that simulates moves that read from tape i by reading from the input tape, but replaces an input symbol a in a transition that reads from any tape other than tape i with λ. So, $L(M_i) = L_i$ for all i, $1 \leq i \leq t$. The following is easily verified:

Lemma 6. *A t-tape $M \in \mathcal{M}$ is a bounded (resp. non-empty, finite) t-tuple language if and only if $L(M_i)$ is a bounded (resp. non-empty, finite) language for each $1 \leq i \leq t$.*

Proof. The proofs for non-emptiness and finiteness are clear.

Assume $L(M)$ is a bounded t-tuple language, and therefore there exists $B_1, \ldots, B_t$, where each B_i is of the form $w_1^* \cdots w_n^*$ and $L(M) \subseteq B_1 \times \cdots \times B_t$. Thus, for each i, $L(M_i) \subseteq B_i$, and therefore $L(M_i)$ is a bounded language.

Assume each $L(M_i)$ is bounded, and let B_i be such that $L(M_{(i)}) \subseteq B_i$ and B_i is of the form $w_1^* \cdots w_n^*$. Then, $L(M) \subseteq B_1 \times \cdots \times B_t$. □

Using this characterization, we can show the following.

Proposition 7. *The boundedness, non-emptiness, and infiniteness problems for finite-turn NTM (resp. finite-turn NTCM) are decidable, and they are effectively semilinear.*

Proof. Due to Proposition 5, given a t-turn NTM M', there is a $t + 1$-tape NFA M with $L(M)^A = L(M')$. We will decide if $L(M)^A$ is bounded; indeed, we will show $L(M)^A$ is bounded if and only if, for each i, $1 \leq i \leq t + 1$, $L(M_i)$ is bounded.

Assume $L(M)^A = L(M')$ is bounded. Assume that t is odd (with the even case being similar). Then $L(M)^A = \{w_1 w_2^R \cdots w_t w_{t+1}^R \mid (w_1, \ldots, w_{t+1}) \in L(M)\}$ is bounded. It is known that given any bounded language L, the reverse of L is bounded, the set of subwords of L is bounded, and any subset of L is bounded [10]. Hence, for each i, $\{w_i \mid (w_1, \ldots, w_{t+1}) \in L(M')\}$ is bounded as, if i is odd, then this is a subset of the set of subwords of $L(M)^A$, and for i even, it is the reverse. This set is $L(M_i)$, and so each $L(M_i)$ is bounded.

Conversely, assume each $L(M_i)$ is bounded for $1 \le i \le t+1$, and hence $L(M_i)^R$ is also bounded for i even. By Lemma 6, $L(M)$ is a bounded $t+1$-tuple language. Since the finite concatenation of bounded languages is bounded [10], $L(M)^A$ is bounded.

Hence, $L(M')$ is bounded if and only if, for each i, $1 \le i \le t+1$, $L(M_i)$ is bounded. Since these are each regular, we can decide this property.

The proof is the same for finite-turn NTCM using decidability of boundedness for NCM [3,5].

Semilinearity follows from semilinearity of NFA and NCM [24], and since $\psi(L(M)) = \psi(L(M'))$. □

Although decidability for non-emptiness, finiteness, and semilinearity were previously known for both finite-turn NTM and finite-turn NTCM [16], decidability of boundedness for both finite-turn NTM and for finite-turn NTCM were not previously known.

We conclude this section with another application of this characterization. This generalizes the result that every finite-turn NPDA is in NLOG [29]. To note, finite-turn NPDA is a restriction of NPDA and is therefore less powerful than NPDA (and also less powerful than finite-turn NTM) as opposed to finite-flip NPDA which is a generalization of NPDA and is more powerful than NPDA.

Proposition 8. *Every finite-turn* NTCM *language is in* NLOG.

3.2 Finite-Turn NTM with Pushdown and Counters

In this section, we provide a further generalized model by augmenting finite-turn NTM with not only monotonic counters but also a pushdown where the pushdown can only be used in a restricted manner:

A t-turn NTM augmented with monotonic counters and a pushdown is called a t-turn NTPCM. Such a machine is called *i-pd-restricted* if, during every accepting computation, the pushdown is only used in the ith pass of the worktape (either left-to-right or right-to-left, after the $i - 1^{\text{st}}$ turn if $i > 1$ and the start of the computation otherwise, but before the ith turn if $i \le t$, and the end of the computation otherwise); and a machine is *pd-restricted* if in every accepting computation, the pushdown is only used in a single pass (it can be different passes depending on the computation).

Example 9. Let D_1 be the language over the alphabet $\{[,]\}$ generated by the context-free grammar with productions $S \to [S]S$ and $S \to \lambda$. This language is known as the Dyck language over one set of parentheses. Let $L = \{x\#x\#x \mid x \in D_1\}$. L can be accepted by a pd-restricted 4-turn NTPCM machine (even without counters). Indeed, on input $x_1\#x_2\#x_3$, a machine M can use the pushdown to verify $x_1 \in D_1$ while in parallel copying x_1 to the read/write tape. Since this is the only pass where the pushdown is used, the machine is 1-pd-restricted. Then it can match x_1 against the input to verify $x_1 = x_2 = x_3$. It follows from [8] and [30] that L is not even an indexed language, a family that strictly contains the context-free languages [1], and is equal to the family of languages accepted by automata with a "pushdown of pushdowns" [7]. Thus, pd-restricted NTPCM is quite a powerful model, containing all of NPDA, finite-turn NTM, and even some non-indexed languages.

The characterization will use a restriction of multi-tape NPCM as follows. Let i satisfy $1 \leq i \leq t$. A t-tape NPCM is i-pd-restricted if, for every accepting computation, the pushdown is only used when it reads from the ith input tape.

The following is easy to verify. Any transition α that reads $a \in \Sigma \cup \{\lambda\}$ from input tape $j \neq i$ and uses the pushdown can be simulated by first reading a from tape j but not changing another store, then in the next transition, reading and changing the pushdown as in α while reading λ from tape i.

Lemma 10. *Every t-tape* NPCM *can be converted to an equivalent i-pd-restricted t-tape* NPCM, *for any $1 \leq i \leq t$.*

The proof of the next proposition follows a proof similar to Lemma 3 for one direction, where the pushdown is only used in one track because it is only used on one sweep of the NTPCM worktape; and for the other direction it first uses Lemma 10 and then follows the proof of Lemma 4 where it is verified that the pushdown changes properly according to the guessed transition sequence by simulating the pushdown in the ith sweep of the worktape.

Proposition 11. *Let $t \geq 0$. There is a $(t+1)$-tape* NPCM *M if and only if there is an i-pd-restricted t-turn* NTPCM *M' such that $L(M') = L(M)^A$, for any $0 \leq i \leq t$.*

For the next proof, we are able to prove that boundedness, emptiness, and infiniteness are decidable not just for i-pd-restricted machines, but also for pd-restricted machines. This is true because given a pd-restricted machine M, we can construct $M_1, \ldots, M_{t+1}$, where each M_i accepts the strings accepted by M for which the pushdown is used in pass i. Thus, M_i is i-pd-restricted. Furthermore, $L(M)$ is bounded if and only if $L(M_i)$ is bounded for each i. The remaining proof is similar to that of Proposition 7 using the fact that emptiness, infiniteness, and boundedness for NPCM are decidable [3,24].

Proposition 12. *For every $t \geq 0$, the boundedness, emptiness, and infiniteness problems for pd-restricted t-turn* NTPCM *are decidable, and they are effectively semilinear.*

Briefly, the result above can be generalized by replacing the pushdown with other potential types of stores. Consider an NFA augmented with a storage structure S and the specification for updating S and possibly some necessary condition(s) on S for acceptance, in addition to the machine entering a final state. The storage structure S can include multiple storage structures. We do not define such a storage structure formally for simplicity, and the following result can be thought of as a template for other models where decidability of boundedness can be shown. However, definitions such as storage structures [6] and store types [22] work. Examples of S are: pushdown; reversal-bounded counters (or equivalent storage structures such as monotonic counters); partially blind counters; and combinations of the structures, e.g., a pushdown and reversal-bounded counters.

We can examine S-restricted t-turn NTM augmented by S (denoted NTM(S)) where in every accepting computation, S can only be changed within a single sweep of the worktape. We can show the following seeing that the pushdown in the proof of the previous proposition can be replaced with other storage types.

Proposition 13. *Let $\mathcal{M}$ be a class of NFA with storage structure S, whose languages are closed under reversal, where the boundedness (resp. emptiness, infiniteness) problem for $\mathcal{M}$ are decidable. Then for every $t \geq 0$, the boundedness (resp. emptiness, infiniteness) problem for S-restricted t-turn NTM(S) are decidable. Furthermore, if $\mathcal{M}$ is augmented with additional reversal-bounded counters (no restrictions on their use) has a decidable boundedness (resp. emptiness, infiniteness) problem, the corresponding problem for S-restricted t-turn machines with reversal-bounded counters are decidable.*

This provides new results for certain general types of automata with decidable properties. For example, checking stack automata (which we will only describe informally and refer to [13] for the formal definition) function like an NFA with an additional worktape that can be written to before the first turn, and thereafter can only be used in a two-way read-only fashion. They have a decidable emptiness and infiniteness problem. If we augment these with a finite-turn worktape where the checking stack could only be used in a single sweep, emptiness and infiniteness are decidable.

As with other results in this paper, a t-turn NTM (combined with other stores) can be replaced with a checking stack that is limited to make at most t-turns. The restriction on NTPCM to be S-restricted in Proposition 12 is needed, as the next proposition shows. Let DCSA be deterministic checking stack automata. The first point uses undecidability of non-emptiness of the intersection of two 1-turn deterministic pushdown automata [2], the second problem uses undecidability of the halting problem for Turing machines [20], the third point uses the undecidability of the halting problem for 2-counter machines [28], and the fourth point uses the third point.

Proposition 14. *The emptiness (boundedness, infiniteness) problems are undecidable for the following models:*

1. *1-turn NTM (or DCSA) with a 1-turn pushdown.*
2. *2-turn DCSA with a 1-turn pushdown, even when the pushdown is used only during the checking stack reading phase (i.e., after turn 1).*
3. *1-turn NTM with an unrestricted counter.*
4. *1-turn deterministic pushdown automata with an unrestricted counter.*

3.3   Finite-Turn NTM with Partially Blind Counters

Partially blind counter machines are multicounter machines (PBCM) where the counters can be incremented or decremented but not tested for zero, however the machine crashes (i.e. it cannot continue) if any of the counters becomes negative, and acceptance occurs when the machine enters an accepting state with all the counters being zero. The emptiness, finiteness, [11] and boundedness problems [3] have been shown decidable for vector addition systems with states, which are equivalent to partially blind multicounter machines [11]. The family of languages accepted by these machines is a recursive family, and does not contain all context-free languages (as in the example below), but contains non-semilinear languages [11] (unlike all the other models considered so far in this paper).

Here, we look at t-turn NTM augmented with partially blind counters, called t-turn NTPBCM. It is pointed out in [11] that $L = \{w\#w^R \mid w \in \{a, b\}^*\}$ is not accepted by any PBCM. However, it is easily accepted by a NTPBCM (or even a 2-turn NTM), which are therefore strictly more powerful.

We could augment a NTPBCM (or PBCM, t-tape PBCM) with monotonic counters, but it is straightforward to see that each pair of monotonic counters can be simulated by a pair of partially-blind counters, and we therefore do not consider these machines with additional reversal-bounded counters.

The results in Sect. 3.1 concerned finite-turn NTM, optionally augmented with monotonic counters. We will see next that these results hold if "monotonic counters" is replaced by "partially blind counters". However, monotonic counters are easy to handle, as we can permute the order that counter changes are applied in an accepting computation and the resulting computation does not change the counter values. But this is not so for partially blind counters as changing the orders can cause counters to go below zero, which is not allowed. But we can modify the proof as follows:

Proposition 15. *Let $t \geq 0$. There is a $(t + 1)$-tape PBCM M if and only if there is a t-turn NTPBCM M' such that $L(M') = L(M)^A$.*

Proof. One half of the proof follows an identical construction to that in the proof of Lemma 4 where all counters are simulated on the first sweep while guessing the transition sequence and the order of counter changes is the same.

For the reverse direction, the construction is a modification to that of Lemma 3. We describe the construction of M from M'. Note that M' makes $s = (t + 1)$ left-to-right and right-to-left sweeps on its worktape. If M' has k partially blind counters $C_1, \ldots, C_k$, M will have $s \cdot k$ partially blind counters called $C_{1,j}, \ldots, C_{k,j}$ for $1 \leq j \leq s$. The simulation of all sweeps of the computation of M' on its worktape are done in parallel. The counters in $C_{1,j}, \ldots, C_{k,j}$ are used to simulate the counters of M in sweep j. For odd j, the simulation is exactly as done with M', but for even j, the simulation is backwards. The counters in $C_{i,1}$ are initially zero, as are $C_{i,s}$ if s is even. For all j even, $C_{i,j}$ and $C_{i,j+1}$ are set nondeterministically to be the same guessed values. The simulation of the computation of M' on the even tracks of the worktape (using counters $C_{i,j}$, j even) is done in reverse and in parallel with the simulation of the odd tracks (using counters in $C_{i,j}$ j odd). When the simulation reaches the end of the worktape and M' enters an accepting state, for all odd j, $j < s$, the counters in $C_{i,j}$ and $C_{i,j+1}$ are decremented simultaneously a nondeterministically guessed number of times to verify that they are the same (and if $j = s$ is not changed thereby verifying

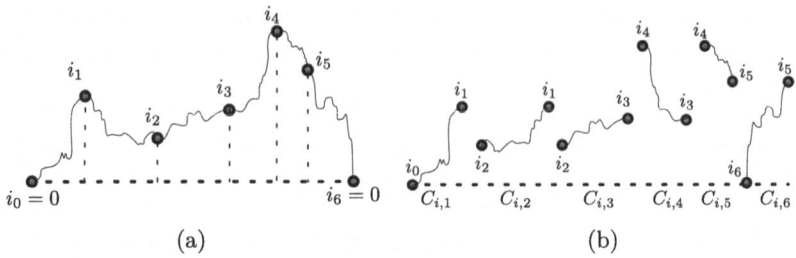

Fig. 1. In (a), it visually shows counter values in some counter i in an original accepting computation of M', where i_j is the value at turn j of the worktape. In (b), we see the modified computation of M with different counters (shown on x-axis) simulating the changes in counter i between each two turns, in parallel.

that it is zero). Then M enters a final state. Figure 1 demonstrates an example. This technique allows counters to be adjusted in a different order.   □

From Proposition 15 and decidability of boundedness for PBCM, we obtain:

Theorem 16. *The boundedness, emptiness, and finitenesss problems for finite-turn* NTPBCM *are decidable.*

This is a new result for all three decision problems, and in particular, decidability of boundedness is quite powerful.

3.4   Simple Matrix Grammars

A matrix grammar has a finite set of matrix rules of the form $[A_1 \to w_1, \ldots, A_k \to w_k]$, where each $A_i \to w_i$ is a context-free production. In the derivation, at each step a matrix is chosen nondeterministically, whereby the context-free rules of the matrix must be applied in order to the sentential form to produce the next sentential form. An n-simple matrix grammar (n-SMG) (from [23]), a restricted form of a matrix grammar, is a tuple $G = (V_1, \ldots, V_n, \Sigma, P, S)$, where $V_1, \ldots, V_n$ are disjoint sets of nonterminals, Σ is the terminal alphabet, S is a start nonterminal not in $(V_1 \cup \cdots \cup V_n)$, and P is a finite set of rules of the form:

1. $S \to A_1 \cdots A_n$, where each $A_i \in V_i$,
2. $[X_1 \to w_1, \ldots, X_n \to w_n]$, where $X_i \in V_i$ and $w_i \in (V_i \cup \Sigma)^*$, and the number of nonterminals in w_i is equal to the number of nonterminals in w_j for all $i \neq j$.

The derivation relation enforces that in each rule of type 2, always the leftmost nonterminal of V_i in the sentential form is rewritten (precise definition of the derivation relation is in [23]). The language $L(G)$ consists of all strings $w \in \Sigma^*$

that can be derived starting from S and applying the rules in such a leftmost derivation. Note that a 1-SMG is just a context-free grammar. It is known that without either the restriction of the number of nonterminals being the same, or not requiring leftmost derivation, additional languages could be generated by the grammars [23].

The following result follows from [23] and [17].

Proposition 17. *L is generated by an n-SMG G if and only if there is an n-tape NPDA M accepting $L' \subseteq \langle \Sigma^* \rangle^n$ such that $L = \{x_1 \cdots x_n \mid (x_1, \ldots, x_n) \in L'\}$.*

We can generalize the definition of a simple matrix grammar by augmenting it with an even number $2 \cdot k$ of monotonic counters. Then, in every matrix rule, each context-free production includes $2 \cdot k$ components that describe the increments in each counter, and for a string to be generated, the counter values in counter i and $i + 1$ have to be equal, for i odd. Using this, Proposition 17 can be generalized to include counters.

Proposition 18. *L is generated by an n-SMG with $2 \cdot k$ monotonic counters G if and only if there is an n-tape NPDA with $2 \cdot k$ monotonic counters (which is equivalent to an n-tape NPCM) M accepting $L' \subseteq \langle \Sigma^* \rangle^n$ such that $L = \{x_1 \cdots x_n \mid (x_1, \ldots, x_n) \in L'\}$.*

Using a proof similar to the decidability problems shown using other multi-tape characterizations in this paper, we obtain:

Theorem 19. *The emptiness, infiniteness, and boundedness problems for simple matrix grammars (resp. with monotonic counters) are decidable, and they are effectively semilinear.*

4   Store Languages for the Boundedness Problem

To summarize, so far we have determined several new classes of machines for which the boundedness problem is decidable. One of the most powerful in terms of languages accepted is the class of finite-turn NTM with reversal-bounded counters and a pushdown where, in each accepting computation, the pushdown can only be used within a single sweep of the Turing worktape. In this section, we determine one more class that is even more general than this one. The algorithm provides an entirely different technique than multi-tape characterizations that we have used thus far.

We focus on finite-flip NPDA [19]. A t-flip (resp. finite-flip) NPCM augments a t-flip NPDA with k-reversal-bounded counters. With this model, configurations are of the form $(q, w, Z_0\gamma, i_1, \ldots, i_k)$ where q is the current state, w is the remaining input, $Z_0\gamma$ is the current pushdown contents, and i_j is the current contents of counter j.

In [22,26], the authors study the concept of a *store language of a machine* M for arbitrary types of automata, which is essentially a language description of all the store contents that can appear in any accepting computation of the machine. So, for a t-flip NPCM $M = (Q, \Sigma, \delta, q_0, F)$, the store language of M,

$$S(M) = \{qZ_0\gamma c_1^{i_1} \cdots c_k^{i_k} \mid (q_0, w, Z_0, 0, \ldots, 0) \vdash_M^* (q, w', Z_0\gamma, i_1, \ldots, i_k) \vdash_M^*$$
$$(q_f, \lambda, Z_0\gamma', i'_1, \ldots, i'_k),$$
$$q_f \in F, w, w' \in \Sigma^*, \gamma' \in \Gamma^*, i'_1, \ldots, i'_k \geq 0\},$$

where $c_1, \ldots, c_k$ are new special symbols associated with the counters. In [26], the authors showed that the store language of every t-flip NPDA (resp. t-flip NPCM) is in fact a regular (resp. NCM) language. Therefore, the pushdown can be essentially eliminated. This is a generalization of the important result that the store language of any NPDA is regular [12].

The proof of the next proposition uses an inductive procedure (informally described without counters) where we know 0-flip NPDAs (equal to the context-free languages) have a decidable boundedness problem. And inductively, if we have an $r + 1$-flip NPDA, we can construct two machines, an r-flip machine that accepts the parts of the inputs of M read during the first r flips that eventually leads to acceptance, and a 0-flip machine that accepts the parts of the inputs of M from which, with no flips, it will eventually accept. These two languages use the store languages, which can be accepted by finite automata. The purpose of the store languages should be noted. Simply using the fact that $(r+1)$-flip NPDA are closed under gsm mappings (see definition in [20]), it is immediately evident that both of these languages can be accepted by $(r + 1)$-flip NPDA (just using closure properties). But, by using the store language, it is possible to accept the first with only an r-flip NPDA and the second with a 0-flip NPDA. This is needed to make the induction work, so that essentially we can decide boundedness up to any given r.

Proposition 20. *The boundedness, emptiness, and infiniteness problems are decidable for finite-flip NPDA (resp. finite-flip NPCM).*

Lastly, we consider machines with a finite-flip pushdown, reversal-bounded counters, and a finite-turn worktape. Such a machine is *pd-restricted* if, in every accepting computation, the finite-flip pushdown can only be used in one left-to-right sweep or right-to-left sweep of the worktape. Finally, by Proposition 13:

Theorem 21. *The class of pd-restricted finite-flip NPCM augmented with a finite-turn worktape has a decidable boundedness, emptiness, and infiniteness problem.*

5   Conclusions

In this paper, we study powerful one-way nondeterministic machine models, and find new models where the boundedness, emptiness, and infiniteness problems are decidable. The largest of these are finite-turn Turing machines augmented by partially blind counters, and finite-turn Turing machines augmented by a pushdown that can be flipped a finite number of times, and reversal-bounded counters, where the pushdown can only be used in one sweep of the Turing worktape. It also shows two new techniques to show these problems are decidable.

References

1. Aho, A.V.: Indexed grammars—an extension of context-free grammars. J. ACM **15**, 647–671 (1968)
2. Baker, B.S., Book, R.V.: Reversal-bounded multipushdown machines. J. Comput. Syst. Sci. **8**(3), 315–332 (1974)
3. Baumann, P., et al.: Unboundedness problems for machines with reversal-bounded counters. In: 25th International Conference on Foundations of Software Science and Computation Structures (FoSSaCS) (2023)
4. Cadilhac, M., Finkel, A., McKenzie, P.: Bounded Parikh automata. Int. J. Found. Comput. Sci. **23**(08), 1691–1709 (2012)
5. Czerwinski, W., Hofman, P., Zetzsche, G.: Unboundedness problems for languages of vector addition systems. In: Chatzigiannakis, I., Kaklamanis, C., Marx, D., Sannella, D. (eds.) 45th International Colloquium on Automata, Languages, and Programming (ICALP 2018). Leibniz International Proceedings in Informatics (LIPIcs), vol. 107, p. 119. Schloss Dagstuhl–Leibniz-Zentrum fuer Informatik, Dagstuhl (2018)
6. Engelfriet, J.: The power of two-way deterministic checking stack automata. Inf. Comput. **80**(2), 114–120 (1989)
7. Engelfriet, J.: Iterated stack automata and complexity classes. Inf. Comput. **95**(1), 21–75 (1991)
8. Engelfriet, J., Skyum, S.: Copying theorems. Inf. Process. Lett. **4**(6), 157–161 (1976)
9. Ginsburg, S., Spanier, E.: Bounded Algol-like languages. Trans. Am. Math. Soc. **113**(2), 333–368 (1964)
10. Ginsburg, S.: The Mathematical Theory of Context-Free Languages. McGraw-Hill Inc., New York (1966)
11. Greibach, S.: Remarks on blind and partially blind one-way multicounter machines. Theoret. Comput. Sci. **7**, 311–324 (1978)
12. Greibach, S.: A note on pushdown store automata and regular systems. Proc. Am. Math. Soc. **18**, 263–268 (1967)
13. Greibach, S.: Checking automata and one-way stack languages. J. Comput. Syst. Sci. **3**(2), 196–217 (1969)
14. Greibach, S.A.: One way finite visit automata. Theoret. Comput. Sci. **6**, 175–221 (1978)
15. Hague, M., Lin, A.: Model checking recursive programs with numeric data types. In: Gopalakrishnan, G., Qadeer, S. (eds.) CAV 2011. LNCS, vol. 6806, pp. 743–759. Springer, Heidelberg (2011). https://doi.org/10.1007/978-3-642-22110-1_60
16. Harju, T., Ibarra, O., Karhumäki, J., Salomaa, A.: Some decision problems concerning semilinearity and commutation. J. Comput. Syst. Sci. **65**(2), 278–294 (2002)
17. Harrison, M.A., Ibarra, O.H.: Multi-tape and multi-head pushdown automata. Inf. Control **13**(5), 433–470 (1968)
18. Holzer, M., Kutrib, M.: Flip-pushdown automata: nondeterminism is better than determinism. In: Ésik, Z., Fülöp, Z. (eds.) DLT 2003. LNCS, vol. 2710, pp. 361–372. Springer, Heidelberg (2003). https://doi.org/10.1007/3-540-45007-6_29

19. Holzer, M., Kutrib, M.: Flip-pushdown automata: $k + 1$ pushdown reversals are better than k. In: Baeten, J.C.M., Lenstra, J.K., Parrow, J., Woeginger, G.J. (eds.) ICALP 2003. LNCS, vol. 2719, pp. 490–501. Springer, Heidelberg (2003). https://doi.org/10.1007/3-540-45061-0_40

20. Hopcroft, J.E., Ullman, J.D.: Introduction to Automata Theory, Languages, and Computation. Addison-Wesley, Reading (1979)

21. Hopcroft, J.E.: On the equivalence and containment problems for context-free languages. Math. Syst. **3**, 119–124 (1969)

22. Ibarra, O., McQuillan, I.: On store languages of languages acceptors. Theoret. Comput. Sci. **745**, 114–132 (2018)

23. Ibarra, O.H.: Simple matrix languages. Inf. Control **17**(4), 359–394 (1970)

24. Ibarra, O.H.: Reversal-bounded multicounter machines and their decision problems. J. ACM **25**(1), 116–133 (1978)

25. Ibarra, O.H.: Grammatical characterizations of NPDAs and VPDAs with counters. Theoret. Comput. Sci. **746**, 136–150 (2018)

26. Ibarra, O.H., McQuillan, I.: On store languages and applications. Inf. Comput. **267**, 28–48 (2019)

27. Ibarra, O.H., McQuillan, I.: Techniques for showing the decidability of the boundedness problem of language acceptors (2024). https://arxiv.org/abs/2405.08988

28. Minsky, M.L.: Recursive unsolvability of Post's problem of "tag" and other topics in theory of Turing Machines. Ann. Math. **74**(3), 437–455 (1961)

29. Moriya, E., Tada, T.: On the space complexity of turn bounded pushdown automata. Int. J. Comput. Math. **80**(3), 295–304 (2003)

30. Rozoy, B.: The Dyck language $D_1'^*$ is not generated by any matrix grammar of finite index. Inf. Comput. **74**(1), 64–89 (1987)

Semidirect Product Decompositions for Periodic Regular Languages

Yusuke Inoue[1(✉)], Kenji Hashimoto[2], and Hiroyuki Seki[1]

[1] Nagoya University, Nagoya, Japan
{y-inoue,seki}@sqlab.jp
[2] Kagawa University, Takamatsu, Japan
hashimoto.kenji@kagawa-u.ac.jp

Abstract. The definition of period in finite-state Markov chains can be extended to regular languages by considering the transitions of DFAs accepting them. For example, the language $(\Sigma\Sigma)^*$ has period two because the length of a recursion (cycle) in its DFA must be even.. This paper shows that the period of a regular language appears as a cyclic group within its syntactic monoid. Specifically, we show that a regular language has period p if and only if its syntactic monoid is isomorphic to a submonoid of a semidirect product between a specific finite monoid and the cyclic group of order p. Moreover, we explore the relation between the structure of Markov chains and our result, and apply this relation to the theory of probabilities of languages. We also discuss the Krohn-Rhodes decomposition of finite semigroups, which is strongly linked to our methods.

1 Introduction

Numerous algebraic approaches to formal languages have been conducted. Most of algebraic approaches to regular languages over an alphabet Σ are based on the fact that a language $L \subseteq \Sigma^*$ is regular iff there exist a finite monoid M, a monoid homomorphism $\eta : \Sigma^* \to M$ and a subset $S \subseteq M$ such that $L = \eta^{-1}(S)$. The smallest monoid M that satisfies this condition is called the syntactic monoid of L, which is described by a Cayley graph as follows. The Cayley graph of the syntactic monoid M is the directed labeled graph such that the set of vertices is M and the set of edges consists of $m_1 \xrightarrow{a} m_2$ where $m_1 \cdot \eta(a) = m_2$ with $m_1, m_2 \in M$ and $a \in \Sigma$. Then, the Cayley graph can be regarded as the DFA $\mathcal{A}$ recognizing L defined as: $\mathcal{A}$ accepts $w \in \Sigma^*$ iff there is a path $e_M \xrightarrow{w} m$ where e_M is the identity element of M and m is contained in the image $\eta(L)$.

In this paper, we focus on periods of regular languages. We say that a regular language $L \subseteq \Sigma^*$ has a period $P \geq 1$ with respect to $\Gamma \subseteq \Sigma$ if the Cayley graph of the syntactic monoid of L satisfies: if $m \xrightarrow{w} m$ is a path with some $m \in M$, then the number of occurrences of letters in Γ in w is a multiple of P. For example, the syntactic monoid of $(\Sigma\Sigma)^*$ is the cyclic group $C_2 = \{0, 1\}$ and the set of edges of the Cayley graph is $\{0 \xrightarrow{a} 1, 1 \xrightarrow{a} 0 \mid a \in \Sigma\}$. Therefore, $(\Sigma\Sigma)^*$ has period 2 with respect to Σ because w must be of even length if $m \xrightarrow{w} m$

© The Author(s), under exclusive license to Springer Nature Switzerland AG 2024
J. D. Day and F. Manea (Eds.): DLT 2024, LNCS 14791, pp. 173–188, 2024.
https://doi.org/10.1007/978-3-031-66159-4_13

is a path for each $m \in \{0,1\}$ and $w \in \Sigma^*$. Note that a period of a language $L \subseteq \Sigma^*$ is defined for any given non-empty subset $\Gamma \subseteq \Sigma$. As another example, we consider the language $L_1 = \{w \mid$ the numbers of occurrences of a and b in w are both even$\} \subseteq \{a,b\}^*$. Then, the syntactic monoid of L_1 is the direct product $C_2 \times C_2$ of two cyclic groups C_2. One C_2 counts the occurrences of a and the other C_2 counts the occurrences of b. Therefore, L_1 has period 2 with respect to both of $\Gamma_1 = \{a\}$ and $\Gamma_2 = \{b\}$.

Now, how are these periods represented by syntactic monoids? In the examples mentioned above, each period is simply represented by the corresponding syntactic monoid. In the first example, the period of $(\Sigma\Sigma)^*$ with respect to Σ is 2, and it is represented by the cyclic group C_2. In the second example, the periods of L_1 with respect to Γ_1 and Γ_2 are represented by two cyclic groups C_2, respectively. However, a general case is not always as simple as these examples. As we will see in later sections, there is a language that has period 2, and its syntactic monoid is the symmetry group $\mathcal{S}_3$, which consists of all permutations on $\{0,1,2\}$. Of course, $\mathcal{S}_3$ is not isomorphic to cyclic groups or their direct products. How does the syntactic monoid explain the period in such a case?

One of our goals is to provide an algebraic decomposition of the syntactic monoid of a periodic language. As the main result of this paper, we show that every syntactic monoid of a regular language with periods $P_1, \ldots, P_n > 1$ with respect to $\Gamma_1, \ldots, \Gamma_n \subseteq \Sigma$ can be decomposed into a submonoid of the semidirect product $N^G \rtimes G$ where N is a specific finite monoid and G is the direct product of cyclic groups of orders $P_1, \ldots, P_n$. The second component G of the decomposition clearly contains the period P_i as the i-th cyclic group for each $1 \le i \le n$, and the first component N^G simulates the behavior of each periodic class of the syntactic monoid with states of which number is smaller than its order. The contributions of our results and related studies are as follows.

(i) Our decomposition explains an iteration property of regular languages. Our definition of period is inspired by period in the context of Markov chains, which plays a crucial role in the analysis of Markov chains (see e.g., [9]). For example, the existence of the limit distribution of a Markov chain depends on its period. We extend the concept of period in Markov chains to regular languages by considering the transitions of DFA accepting them. An essential idea of the extension is using syntactic monoids rather than minimal DFAs or other DFAs.

(ii) Periods of Markov chains are strongly related to the probability of regular languages, and therefore, our results have applications in the study of the probability of regular languages. The probability $\mu_L(\ell)$ of a language L for length ℓ is the probability that w belongs to L when a word $w \in \Sigma^\ell$ is randomly chosen. Probabilities of languages have been studied in many different contexts. The most classical results on probabilities were obtained as an application of formal power series [13]. In recent years, there have been several approaches to probabilities using syntactic monoids (e.g., [14,15]), and our study contributes to this body of work.

(iii) Our decomposition theorem provides a partial Krohn-Rhodes decomposition of a transformation semigroup (M_L, M_L) where M_L is the syntactic monoid of a periodic regular language L. A transformation semigroup (X, M) is a pair of a set X and a semigroup M that acts on X. The Krohn-Rhodes prime decomposition theorem [6] states that every finite transformation semigroup can be decomposed into a wreath product of finite aperiodic semigroups and finite groups. This is a famous result in semigroup theory, and several related studies have been conducted (e.g., [3,5,12]). Because the wreath product of two semigroups N and G is defined as the transformation semigroup $(N \times G, N^G \rtimes G)$, our decomposition of the form $N^G \rtimes G$ with a group G provides a partial Krohn-Rhodes decomposition of syntactic monoids. Note that DFAs can be regarded as transformation semigroups, and therefore, the Krohn-Rhodes decomposition has been studied in the context of formal language theory. In particular, the holonomy decomposition is known as a decomposition of DFAs [2,16]. However, the holonomy decomposition is a decomposition of automata, not a decomposition of syntactic monoids.

This paper is organized as follows. Section 2 provides basic definitions of monoids, languages, and periods. In addition, we provide several examples of periodic languages that will be discussed in the later sections. In Sect. 3, we first prove the main theorem (Theorem 1). In the latter part of the section, we focus on periods with respect to a given alphabet and explain that the syntactic monoid can be decomposed into monoids corresponding to each residual of the period (Theorem 2). In Sect. 4, we discuss the applications (i) and (ii) described above in detail. Specifically, we connect the periods of languages to periods of Markov chains (Theorem 3), and provide an application of our results (Theorem 4). In Sect. 5, we consider the wreath products described in the application (iii), and provide a partial Krohn-Rhodes decomposition of periodic regular languages (Theorem 5). Please refer to the full version [4] for the proofs omitted due to space limitation.

2   Preliminaries

Let $|X|$ denote the cardinality of a set X. For sets X and Y, let $X \sqcup Y$ and Y^X denote the disjoint union of X and Y, and the set of all functions from X to Y, respectively. A function from X to itself is called a *transformation* on X. For a positive integer K, K is sometimes regarded as the set $\{0, \ldots, K - 1\}$.

2.1   Monoids

A *monoid* is a set M equipped with an associative binary operation $\cdot : M \times M \to M$, and containing the identity element $e_M \in M$. For monoids M and N, we say that $h : M \to N$ is a monoid homomorphism if h satisfies: (i) $h(e_M) = e_N$, and (ii) $h(m_1 \cdot m_2) = h(m_1) \cdot h(m_2)$ for each $m_1, m_2 \in M$. We say that a subset $N' \subseteq N$ is a submonoid of N if N' also forms a monoid. Note that for every monoid homomorphism $h : M \to N$, the homomorphic image $h(M)$ is a submonoid of N. Moreover, M is isomorphic to the submonoid $h(M)$ if h is injective. Therefore, we also say that M is a submonoid of N if there exists an injective monoid homomorphism $h : M \to N$. In this case, we identify M with $h(M)$, and each $m \in M$ with $h(m) \in N$ if the embedding $h : M \to N$ is clear from the context. We say that N is a *quotient* of M if there exists a surjective monoid homomorphism $\psi : M \to N$.

Example 1. Let K be a positive integer. The followings are examples of monoids used in this paper.

- Let $U_K = \{e, \iota_1, \ldots, \iota_K\}$ be the right-zero monoid, i.e., $s \cdot \iota_i = \iota_i$ for each $1 \leq i \leq K$ and $s \in U_K$. Also, let $\overline{U}_K = \{e, \iota_1, \ldots, \iota_K\}$ be the left-zero monoid, i.e., $\iota_i \cdot s = \iota_i$ for each $1 \leq i \leq K$ and $s \in \overline{U}_K$.
- Let $C_K = \{0, \ldots, K-1\}$ be the cyclic group of order K. That is, $i \cdot j = i + j$ mod K for each $i, j \in C_K$. We write $+$ for the operation of C_K.
- Let $\mathcal{S}_K$ be the symmetric group of order $K!$. That is, $\mathcal{S}_K$ consists of all permutations on K, and the operation $\cdot$ is the composition of functions.
- Let $\mathcal{T}_K$ be the monoid such that the carrier set consists of all transformations on K, and the operation is the composition of functions.   ∎

Note that any finite monoid is a submonoid of $\mathcal{T}_K$ for some integer K. This is because every element m in a monoid M can be regarded as the transformation $\tau : M \to M$ such that $\tau(s) = s \cdot m$ for each $s \in M$.

Let M be a finite monoid and $S \subseteq M$ be a generator of M. We say that (V, E) is the *Cayley graph* of M where $V = M$ is the set of vertices and $E = \{(m_1, s, m_2) \in M \times S \times M \mid m_1 \cdot s = m_2\}$ is the set of edges labeled by S.

2.2 Semidirect Products

Let X be a set and M be a monoid. A *left action* on X from M is a function $* : M \times X \to X$ satisfying: (i) $e_M * x = x$, and (ii) $m_1 * (m_2 * x) = (m_1 \cdot m_2) * x$ for each $m_1, m_2 \in M$ and $x \in X$. A *right action* is defined as the dual of a left action. In particular, the operation of M is a left (or right) action on M from M. When considering an action on M from M, the action refers to the operation of M unless otherwise specified.

For monoids M and N, a left action $*$ on M from N is *distributive* if $n * (m_1 \cdot m_2) = (n * m_1) \cdot (n * m_2)$ for all $n \in N$ and $m_1, m_2 \in M$. In this paper, all actions on monoids are supposed to be distributive. A left action $*$ on M from N is *unitary* if $n * e_M = e_M$ for all $n \in N$.

Let $\otimes$ and $\oplus$ be the operations of monoids M and N, respectively. For a distributive left action $* : N \times M \to M$, the monoid with operation $\cdot$ on $M \times N$ defined as

$$(m_1, n_1) \cdot (m_2, n_2) = (m_1 \otimes (n_1 * m_2), n_1 \oplus n_2) \tag{1}$$

for each $(m_1, n_1), (m_2, n_2) \in M \times N$ is called the *semidirect product* (with respect to $*$) of M and N, and denoted by $M \rtimes_* N$.

Example 2. – The direct product $M \times N$ of any two monoids M and N is the semidirect product with respect to the *trivial action* $*$ where $*$ is defined as $n * m = m$ for each $n \in N, m \in M$.

- The symmetric group $\mathcal{S}_3$ is isomorphic to $C_3 \rtimes_* C_2$ where $*$ is the unitary action defined as $0 * m = m$ and $1 * m = -m$ for each $m \in C_3$. ∎

Let M be a monoid with an operation $\otimes$, and $* : Y \times N \to Y$ be a right action on a set Y from a monoid N. Let M^Y be the monoid defined as $(f \cdot g)(y) = f(y) \otimes g(y)$ for each $f, g \in M^Y$. Then, the left action $\circledast : N \times M^Y \to M^Y$ is induced as

$$(n \circledast f)(y) = f(y * n) \tag{2}$$

for each $n \in N, y \in Y$ and $f \in M^Y$. That is, $\circledast$ is a pointwise extension of $*$ from M to M^Y. We let $M^Y \rtimes N$ denote the semidirect product of M^Y and N with respect to this action $\circledast$. Note that when considering the case $Y = N$ (i.e., considering $M^N \rtimes N$), the action $* : Y \times N \to Y$ refers to the operation on N.

Proposition 1. *Let M and N be monoids. The semidirect product $M \rtimes_* N$ is a submonoid of $M^N \rtimes N$ for every unitary left action $* : N \times M \to M$. In particular, $M \times N$ is a submonoid of $M^N \rtimes N$.* ∎

2.3   Periods of Regular Languages

Let Σ be a (finite) alphabet. For $w \in \Sigma^*$, $|w|$ denotes the length of w. We let $|w|_a$ denote the number of occurrences of $a \in \Sigma$ in $w \in \Sigma^*$, and $|w|_\Gamma = \sum_{a \in \Gamma} |w|_a$ for $\Gamma \subseteq \Sigma$. For example, $|w|_a = 3, |w|_{\{a,b\}} = 4$ and $|w| = |w|_\Sigma = 5$ with $w = aabac \in \{a, b, c\}^*$.

For a regular language $L \subseteq \Sigma^*$, we say that a monoid M *fully recognizes* L if there is a surjective monoid homomorphism $\eta : \Sigma^* \to M$ such that $L = \eta^{-1}(S)$ with some $S \subseteq M$. The smallest monoid that fully recognizes L is called the *syntactic monoid* of L, and the corresponding homomorphism is called the *syntactic morphism* of L. The uniqueness of the syntactic monoid and the syntactic morphism is guaranteed by the following proposition (see [7] in detail).

Proposition 2. *Let $L \subseteq \Sigma^*$ be a regular language, and let M_L and η_L be the syntactic monoid and the syntactic morphism of L, respectively. For any monoid M, if M fully recognizes L with a homomorphism η, then there exists a surjective homomorphism $\psi : M \to M_L$ such that $\eta_L = \psi \circ \eta$.* ∎

Because Σ^* is generated by Σ, the syntactic monoid M_L is generated by $\eta_L(\Sigma)$. When we illustrate the Cayley graph of M_L, the label $\eta_L(a)$ is often abbreviated as a for each $a \in \Sigma$.

We define periods of regular languages, which is the key concept of this paper. As discussed in Sect. 1, the concept of periods is inspired by studies on finite state Markov chains. Therefore, it is natural to define periods based on the graph structure of a DFA. However, there are more than one DFAs that recognize the same regular language. In this paper, we opt for the Cayley graph of the syntactic monoid to define periods. This is because the syntactic monoid more appropriately represents the periodicity of a regular language than the minimal DFA and other DFAs. For example, Lemma 1, discussed later, states that the syntactic monoid can be represented as the disjoint union of periodic classes, but a similar statement does not hold for the minimal DFA. We will discuss this aspect in detail in Sect. 4.

Definition 1. *Let $L \subseteq \Sigma^*$ be a regular language, and let M_L and η_L be the syntactic monoid and the syntactic morphism of L. For a non-empty subset $\Gamma \subseteq \Sigma$, we say that L has a* period *P with respect to Γ if P satisfies: for every $w \in \Sigma^*$ such that $t \cdot \eta_L(w) = t$ for some $t \in M_L$, $|w|_\Gamma$ is a multiple of P.* ∎

By the definition of period, every language has period one with respect to each subset of Σ. We are mainly interested in the maximum number of all the periods for each subset of Σ. For example, all periods mentioned in Example 3 below are maximum periods. We often say that a language $L \subseteq \Sigma^*$ is *periodic* if there is a period greater than one with respect to some non-empty subset $\Gamma \subseteq \Sigma$.

Example 3. The followings are examples of periodic regular languages.

(1) Let $L_1 \subseteq \{a,b\}^*$ be the language defined by DFA $\mathcal{A}_1$ (see Fig. 1a). The syntactic monoid of L_1 is $M_{L_1} = C_2 \times C_2$, and the Cayley graph of M_{L_1} has the same shape as $\mathcal{A}_1$. Therefore, L_1 has period 2 with respect to $\{a\}, \{b\}$ and $\{a,b\}$.

(2) Let $L_2 \subseteq \{a,b\}^*$ be the language defined by DFA $\mathcal{A}_2$ (see Fig. 1b). The syntactic monoid of L_2 is $M_{L_2} = S_3$, and the Cayley graph of M_{L_2} is shown in Fig. 1b. (The identity element is pointed to by a dangling arrow.) It is easy to see that $\eta_{L_2}(a)$ is an odd permutation and $\eta_{L_2}(b)$ is an even permutation, and so L_2 has period 2 with respect to $\{a\}$.

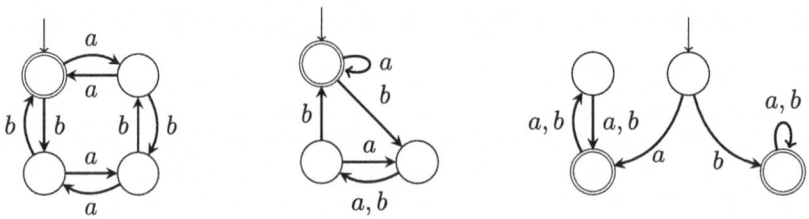

Fig. 1. (1a DFA $\mathcal{A}_1$) (1b DFA $\mathcal{A}_2$) (1c DFA $\mathcal{A}_3$)

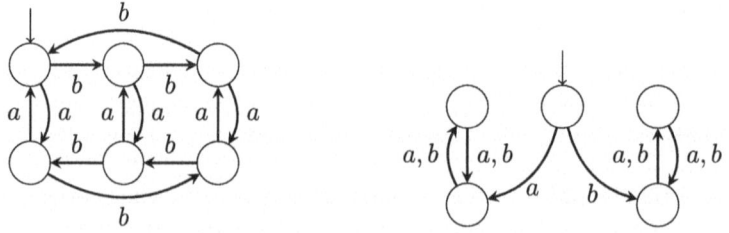

Fig. 2. (2a Cayley graph of M_{L_2}) (2b Cayley graph of M_{L_3})

(3) Let $L_3 \subseteq \{a, b\}^*$ be the language defined by DFA $\mathcal{A}_3$ (see Fig. 1c). By the shape of the Cayley graph of M_{L_3} shown in Fig. 1b, L_3 has period 2 with respect to $\{a, b\}$.   ∎

3   Decompositions of Periodic Regular Languages

Let $L \subseteq \Sigma^*$ be a regular language with periods $P_1, \ldots, P_n > 1$ with respect to $\Gamma_1, \ldots, \Gamma_n \subseteq \Sigma$, respectively. We define the monoid homomorphism $\rho : \Sigma^* \to (C_{P_1} \times \cdots \times C_{P_n})$ as $\rho(w) = (r_1, \ldots, r_n)$ where $r_i = |w|_{\Gamma_i} \bmod P_i$ for each $1 \leq i \leq n$. We call $\rho(w)$ the *residual* of w (modulo $P_1, \ldots, P_n$). For example, let $\Sigma = \{a, b\}$ and let L have periods $P_1 = 2$ and $P_2 = 2$ with respect to $\Gamma_1 = \{a\}$ and $\Gamma_2 = \{a, b\}$, respectively. Then, $\rho(aabac) = (1, 0)$ holds.

The next lemma states that the syntactic monoid is partitioned into $N_{\mathbf{r}} = \eta_L(\{w \in \Sigma^* \mid \rho(w) = \mathbf{r}\})$ for residuals $\mathbf{r} \in C_{P_1} \times \cdots \times C_{P_n}$. Note that this property does not hold for the minimal DFA. (Compare the DFAs in Figs. 1c and 2b in Example 3.)

Lemma 1. *Let $L \subseteq \Sigma^*$ be a regular language, and M_L and η_L be the syntactic monoid and the syntactic morphism of L. Let L have periods $P_1, \ldots, P_n > 1$ with respect to $\Gamma_1, \ldots, \Gamma_n \subseteq \Sigma$, respectively. Then, $M_L = \bigsqcup_{\mathbf{r} \in C_{P_1} \times \cdots \times C_{P_n}} N_{\mathbf{r}}$ holds for $N_{\mathbf{r}} = \eta_L(\{w \in \Sigma^* \mid \rho(w) = \mathbf{r}\}) \subseteq M_L$.*   ∎

Let $N_{\mathbf{r}}$ be the subset defined in Lemma 1 for each $\mathbf{r} \in C_{P_1} \times \cdots \times C_{P_n}$. We define $\overline{\rho} : M_L \to C_{P_1} \times \cdots \times C_{P_n}$ as $\overline{\rho}(t) = \rho(w)$ with any $w \in \Sigma^*$ such that $t = \eta_L(w)$. By Lemma 1, $\overline{\rho}$ is well-defined and is a monoid homomorphism. We call $\overline{\rho}(t)$ the *residual* of t (modulo $P_1, \ldots, P_n$).

3.1   Semidirect Product Decompositions with Cyclic Groups

Our goal is to show that M_L can be decomposed into a submonoid of $N^{C_{P_1} \times \cdots \times C_{P_n}} \rtimes (C_{P_1} \times \cdots \times C_{P_n})$ for some appropriate monoid N described below. This decomposition means that when a residual $\mathbf{r}$ (modulo $P_1, \ldots, P_n$) is given, we can obtain the fragment of M_L with respect to $\mathbf{r}$ ($N_{\mathbf{r}}$ in Fig. 3) as follows: First, retrieve information from N (by using $\mathbf{r}$ as a retrieval key) and then take the semidirect product of the retrieved fragment and $\mathbf{r}$. By Lemma 1, M_L can be represented as the disjoint union of $N_{\mathbf{r}}$ for every residual $\mathbf{r}$ modulo $P_1, \ldots, P_n$. That is, $N_{\mathbf{r}}$ is exactly the information on M_L with respect to $\mathbf{r}$ mentioned above. To embed $N_{\mathbf{r}}$ for every residual $\mathbf{r}$ into N, we set $N = T_K$ for an appropriate number $K \geq 1$. In fact, taking $K = \max_{\mathbf{r}} \{|N_{\mathbf{r}}|\}$ suffices. In Example 5 mentioned later, we demonstrate this decomposition using the periodic language L_2 defined in Example 3.

Theorem 1. *Let $L \subseteq \Sigma^*$ be a regular language, and M_L and η_L be the syntactic monoid and the syntactic morphism of L. For any non-empty subsets $\Gamma_1, \ldots, \Gamma_n \subseteq \Sigma$, the following conditions are equivalent:*

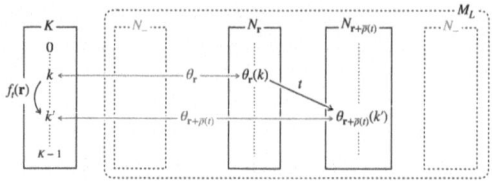

Fig. 3. The definition of f_t

(i) L has periods $P_1, \ldots, P_n$ with respect to $\Gamma_1, \ldots, \Gamma_n \subseteq \Sigma$, respectively.

(ii) M_L is a submonoid of $\mathcal{T}_K^{C_{P_1} \times \cdots \times C_{P_n}} \rtimes (C_{P_1} \times \cdots \times C_{P_n})$ where $K = \max\{|N_\mathbf{r}| \mid \mathbf{r} \in C_{P_1} \times \cdots \times C_{P_n}\}$. Furthermore, $\eta_L(w) \in \mathcal{T}_K^{C_{P_1} \times \cdots \times C_{P_n}} \times \{\rho(w)\}$ for each $w \in \Sigma^*$.

Proof Sketch. It is easy to show (ii)$\Rightarrow$(i) by the definition of semidirect product. To show (i)$\Rightarrow$(ii), it suffices to prove that if L has periods $P_1, \ldots, P_n$ with respect to $\Gamma_1, \ldots, \Gamma_n \subseteq \Sigma$, then there is an injective homomorphism $\mathbf{Can} : M_L \to \mathcal{T}_K^{C_{P_1} \times \cdots \times C_{P_n}} \rtimes (C_{P_1} \times \cdots \times C_{P_n})$ such that $\mathbf{Can}(t) \in \mathcal{T}_K^{C_{P_1} \times \cdots \times C_{P_n}} \times \{\overline{\rho}(t)\}$ for each $t \in M_L$.

Let $\theta_\mathbf{r} : |N_\mathbf{r}| \to N_\mathbf{r}$ be an arbitrary bijection for each $\mathbf{r} \in C_{P_1} \times \cdots \times C_{P_n}$. Intuitively, $\theta_\mathbf{r}^{-1}$ provides a total ordering of $N_\mathbf{r}$. For each $t \in M_L$, define $\mathbf{Can}(t) = (f_t, \overline{\rho}(t)) \in \mathcal{T}_K^{C_{P_1} \times \cdots \times C_{P_n}} \rtimes (C_{P_1} \times \cdots \times C_{P_n})$ where

$$f_t(\mathbf{r})(k) = \begin{cases} \theta_{\mathbf{r}+\overline{\rho}(t)}^{-1}(\theta_\mathbf{r}(k) \cdot t) & \text{if } k < |N_\mathbf{r}|, \\ k & \text{if } k \geq |N_\mathbf{r}| \end{cases}$$

for each $\mathbf{c} \in C_{P_1} \times \cdots \times C_{P_n}$ and $0 \leq k < K$ (see Fig. 3). Then, we can show that $\mathbf{Can}$ is a well-defined injective homomorphism. $\square$

Example 4. As mentioned in Example 3-(1), L_1 has period 2 with respect to both of $\Gamma_1 = \{a\}$ and $\Gamma_2 = \{b\}$. Therefore, M_{L_1} is a submonoid of $\mathcal{T}_1^{C_2 \times C_2} \rtimes (C_2 \times C_2)$ where $K = 1$ for $\mathcal{T}_K$ because $\max\{|N_{(r_1, r_2)}| \mid r_1, r_2 \in C_2\} = \max\{1\} = 1$. In fact, $M_{L_1} = C_2 \times C_2$ is isomorphic to $\mathcal{T}_1^{C_2 \times C_2} \rtimes (C_2 \times C_2)$. ∎

Example 5. As mentioned in Example 3-(2), L_2 has period 2 with respect to $\Gamma_1 = \{a\}$. We give the decomposition with this period. The syntactic monoid M_{L_2} consists of two fragments $N_0 = \{\eta(w) \mid |w|_a \text{ is even}\}$ and $N_1 = \{\eta(w) \mid |w|_a \text{ is odd}\}$, corresponding to the upper three states and the lower three states in Fig. 2a, respectively. Then, $f_b(0) \in \mathcal{T}_3$ is the transformation on the three states that moves one state to the right, and $f_b(1) \in \mathcal{T}_3$ is the transformation that moves one state to the left. Similarly, $f_{bb}(0)$ is the transformation that moves two states to the right, and $f_{bbb}(0) = f_a(0)$ is the identity transformation. Thus, each fragment can be described as a submonoid of $\mathcal{T}_3$. Because the transition between two fragments can be represented by C_2, the overall behavior of M_{L_2} can be represented as $\mathcal{T}_3^{C_2} \rtimes C_2$. In fact, M_{L_2} is isomorphic to $\mathcal{S}_3 = C_3 \rtimes_* C_2$ with a unitary action $*$ (see Example 2), and a submonoid of $\mathcal{T}_3^{C_2} \rtimes C_2$. ∎

Theorem 1 claims that the syntactic monoid can be decomposed into the first component $\mathcal{T}_K^{C_{P_1} \times \cdots \times C_{P_n}}$ and the second component $C_{P_1} \times \cdots \times C_{P_n}$ of the semidirect product stated in (ii). The significance of this decomposition is as follows. For the second component $C_{P_1} \times \cdots \times C_{P_n}$, the last condition $\eta_L(w) \in \mathcal{T}_K^{C_{P_1} \times \cdots \times C_{P_n}} \times \{\rho(w)\}$ in the statement of the theorem is crucial. Intuitively, this condition says that the periodicity of the syntactic monoid is explicitly extracted as $C_{P_1} \times \cdots \times C_{P_n}$. Next, let us consider the first component $\mathcal{T}_K^{C_{P_1} \times \cdots \times C_{P_n}}$. If we naively consider the behavior of each element in M_L as an action on the set M_L, the corresponding transformation monoid is $\mathcal{T}_{|M_L|}$. On the other hand, Theorem 1 states that by fixing a residual $\mathbf{r}$, each element can be described as an action on a set of size at most $K < |M_L|$, and the order of $\mathcal{T}_K$ is much smaller than that of $\mathcal{T}_{|M_L|}$. More intuitively, to describe a periodic regular language, it suffices to have at most K states for each residue $\mathbf{r}$. Nevertheless, $\mathcal{T}_K$ is still a large monoid. The possibility of replacing the first component $\mathcal{T}_K^{C_{P_1} \times \cdots \times C_{P_n}}$ with a simpler monoid needs further investigation.

The following is a corollary of Theorem 1 for $n = 1$ and $\Gamma_1 = \Sigma$.

Corollary 1. *Let $L \subseteq \Sigma^*$ be a regular language, and M_L and η_L be the syntactic monoid and the syntactic morphism of L. The following conditions are equivalent:*

(i) L has a period P with respect to Σ.
(ii) M_L is a submonoid of $\mathcal{T}_K^{C_P} \rtimes C_P$ where $K = \max\{|\eta_L(\Sigma^r(\Sigma^P)^*)| \mid r \in C_P\}$. Furthermore, $\eta_L(w) \in \mathcal{T}_K^{C_P} \times \{\rho(w)\}$ for each $w \in \Sigma^*$. ∎

Example 6. As mentioned in Example 3-(3), L_3 has period 2 with respect to $\Sigma = \{a, b\}$. Therefore, M_{L_3} is a submonoid of $\mathcal{T}_3^{C_2} \rtimes C_2$ and $\eta_{L_3}(\Sigma) \subseteq \mathcal{T}_3^{C_2} \times \{1\}$. We can show that M_{L_3} is a submonoid of $\overline{U}_2 \times C_2$ by the mapping $\eta_{L_3}(a) \mapsto (\iota_1, 1)$ and $\eta_{L_3}(b) \mapsto (\iota_2, 1)$. Furthermore, $\overline{U}_2 \times C_2$ is a submonoid of $\overline{U}_2^{C_2} \rtimes C_2$ by Proposition 1, and also a submonoid of $\mathcal{T}_3^{C_2} \rtimes C_2$. ∎

3.2 Residual Monoids

In this subsection, we focus on periods with respect to a fixed alphabet Σ. Let $L \subseteq \Sigma^*$ be a regular language that has a period P with respect to Σ. Our goal is to extract a monoid T_r that recognizes $L_w = \{u \in (\Sigma^P)^* \mid wu \in L\}$ with $w \in \Sigma^r$ for each $0 \le r < P$, where T_r is designed to treat a word of length P as a single letter. Intuitively, L_w is the 'periodic image' of L with residual $r = \rho(w)$. For each $0 \le r < P$, we define the subset $T_r \subseteq \mathcal{T}_K$ as $T_r = \{\tau \in \mathcal{T}_K \mid \tau = f(r)$ with $\mathbf{Can}^{-1}(f, 0) \in M_L\}$ where $\mathbf{Can}$ is the homomorphism defined in the proof of Theorem 1. We call T_r the *residual monoid* (with residual r), and this definition is justified by the following fact.

Lemma 2. *Let $L \subseteq \Sigma^*$ be a regular language that has a period P with respect to Σ. For each $0 \le r < P$, the residual monoid T_r is a submonoid of $\mathcal{T}_K$.* ∎

Note that $L = \bigsqcup_{w \in \Sigma^{<P}}(L \cap w(\Sigma^P)^*) = \bigsqcup_{w \in \Sigma^{<P}} wL_w$ where $\Sigma^{<P} = \bigcup_{0 \le i < P} \Sigma^i$. We can show that each language $L_w = \{u \in (\Sigma^P)^* \mid wu \in L\} \subseteq (\Sigma^P)^*$ is fully recognized by the residual monoid $T_{\rho(w)}$.

Theorem 2. *Let $L \subseteq \Sigma^*$ be a regular language that has a period P with respect to Σ. For each $w \in \Sigma^{<P}$, the residual monoid T_r fully recognizes $L_w = \{u \in (\Sigma^P)^* \mid wu \in L\} \subseteq (\Sigma^P)^*$ where $r = \rho(w)$.* $\blacksquare$

This theorem implies that there is a surjective homomorphism $\eta_w : (\Sigma^P)^* \rightarrow T_{\rho(w)}$ recognizing L_w for each $w \in \Sigma^{<P}$. Now, let M_{L_w} and η_{L_w} be the syntactic monoid and the syntactic morphism of $L_w \subseteq (\Sigma^P)^*$. Then, there is a surjective homomorphism $\psi : T_{\rho(w)} \rightarrow M_{L_w}$ by Theorem 2 and Proposition 2. The following commutative diagram illustrates the relationship among the above monoids:

$$
\begin{array}{ccc}
(\Sigma^P)^* & \xrightarrow{\eta_w} & T_{\rho(w)} \\
& \searrow^{\eta_{L_w}} & \downarrow^{\psi} \\
& & M_{L_w} \,.
\end{array}
\tag{3}
$$

4   Probabilities of Regular Languages

As stated in Sect. 1, our definition of period is inspired by the study of Markov chains. In this section, we discuss the probability of languages that directly connects periods of a regular language defined in this paper and those of Markov chains.

For a language $L \subseteq \Sigma^*$, the *probability* (or *density*) of L for length ℓ, denoted by $\mu_L(\ell)$, is defined as $|L \cap \Sigma^\ell|/|\Sigma^\ell|$. Note that $0 \le \mu_L(\ell) \le 1$ for every $L \subseteq \Sigma^*$ and $\ell \ge 0$. The probability of L, denoted by μ_L, is the behavior of $\lim_{\ell \to \infty} \mu_L(\ell)$. The probabilities of languages have been extensively studied. In particular, the following proposition was proved in the study of formal power series [13].

Proposition 3. *The probability μ_L of every regular language $L \subseteq \Sigma^*$ has only finitely many accumulation points. Precisely, there exists a sequence of numbers $(\mu_0, \ldots, \mu_{P-1})$ with some $P \ge 1$ such that $\lim_{\ell \to \infty} \mu_L(\ell \cdot P + r) = \mu_r$ for each $0 \le r < P$.* $\blacksquare$

For a regular language $L \subseteq \Sigma^*$ that has P accumulation points, we write $\mu_L = (\mu_0, \ldots, \mu_{P-1})$. If $P = 1$, we simply write $\mu_L = \mu_0$. For example, the probability of $L_3 = a(\Sigma\Sigma)^* \cup b\Sigma^*$ (see also Example 3-(3)) can be represented as $\mu_{L_3} = (\frac{1}{2}, 1)$ with two accumulation points.

4.1   Periods and Markov Chains

The probability of a regular language can be computed by a finite state Markov chain. For a regular language $L \subseteq \Sigma^*$, let $\mathcal{M}$ be the finite state Markov chain (with uniform transition probabilities) defined as:

- the state space is the set Q of states of a DFA $\mathcal{A}$ that recognizes L, and
- the transition matrix $\varPi \in [0,1]^{Q \times Q}$ is defined as $\varPi(q_i, q_j) = |\{a \in \varSigma \mid q_i \xrightarrow{a} q_j$ is a transition of $\mathcal{A}\}|/|\varSigma|$ for each $q_i, q_j \in Q$.

Then, it is clear that the probability $\mu_L(\ell)$ is equal to $\sum_{q \in F} \varPi^\ell(q_0, q)$ where q_0 and F are the initial state and the set of final states of $\mathcal{A}$, respectively.

Example 7. Let $\mathcal{A}_3$ be the DFA that recognizes $L_3 = a(\varSigma\varSigma)^* \cup b\varSigma^*$ defined in Example 3-(3). Then, the corresponding Markov chain $\mathcal{M}_3$ is as shown in Fig. 4 and the transition matrix is

$$\varPi = \begin{pmatrix} 0 & 1/2 & 0 & 1/2 \\ 0 & 0 & 1 & 0 \\ 0 & 1 & 0 & 0 \\ 0 & 0 & 0 & 1 \end{pmatrix} .$$

The probability of L_3 for length ℓ is equal to $\varPi^\ell(1,2) + \varPi^\ell(1,4)$ because q_1 and $\{q_2, q_4\}$ are the initial state and the set of final states of $\mathcal{A}_3$, respectively. For example, $\mu_{L_3}(2) = \varPi^2(1,2) + \varPi^2(1,4) = 1/2$. In fact, $|L_3 \cap \varSigma^2|/|\varSigma^2| = |\{ba, bb\}|/4 = 1/2$. ∎

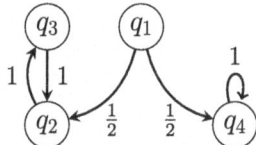

Fig. 4. Markov chain $\mathcal{M}_3$

A strongly connected component of a directed graph is said to be a *sink* if there are no outgoing edges from itself. A sink Q' of a Markov chain is said to be *P-periodic* if P is the maximum number satisfying: ℓ is a multiple of P if $\varPi^\ell(q,q) > 0$ with some $q \in Q'$. For example, Markov chain $\mathcal{M}_3$ (see Fig. 4) has two sinks. 2-periodic sink $\{q_2, q_3\}$ and 1-periodic sink $\{q_4\}$. Note that for a DFA $\mathcal{A}$ and the corresponding Markov chain $\mathcal{M}$, $\varPi^\ell(q,q) > 0$ iff there is a run $q \xrightarrow{w} q$ for some $w \in \varSigma^\ell$. Therefore, we also say that a sink Q' of a DFA is P-periodic if P is the maximum number satisfying: for each $w \in \varSigma^*$, $|w|$ is a multiple of P if $q \xrightarrow{w} q$ with some $q \in Q'$. P-periodicity affects the limit distribution of the Markov chain. In fact, we can show that for the transition matrix $\varPi$ of every finite state Markov chain, the limit distribution $\lim_{\ell \to \infty} \varPi^\ell$ oscillates around P accumulation points where P is a divisor of the least common multiple of the periods of all sinks. By the correspondence between the probability of a regular language and the limit distribution of the Markov chain, we have the following property.

Proposition 4. *Let $L \subseteq \varSigma^*$ be a regular language, and let $\mathcal{A}$ be a DFA that recognizes L. The probability of L can be represented as $\mu_L = (\mu_0, \ldots, \mu_{P-1})$ with $\mu_0, \ldots, \mu_{P-1} \in [0,1]$ where P is the least common multiple of $\{P' \in \mathbb{N} \mid Q'$ is P'-periodic where Q' is a sink of $\mathcal{A}\}$.* ∎

As can be seen from the definition, P-periodicities of sinks correspond to the periods of a language with respect to Σ. The next lemma describes this correspondence in detail.

Lemma 3. *Let $L \subseteq \Sigma^*$ be a regular language, and M_L be the syntactic monoid of L. If a P-periodic sink exists in the Cayley graph of M_L, then the maximum period of L with respect to Σ is a multiple of P.* ∎

This lemma implies that for any sinks $Q', Q'' \subseteq M_L$, if Q' is P'-periodic, then Q'' is P''-periodic where P'' is a multiple of P'. Therefore, the periodicity of each sink of the syntactic monoid is equal to each other. For example, the Cayley graph of M_{L_3} in Example 3 has two sinks, and both are 2-periodic. (Compare the minimal DFA $\mathcal{A}_3$. The right two states in the Cayley graph collapse in $\mathcal{A}_3$, and the periodicities of the two sinks of $\mathcal{A}_3$ are different.) Moreover, by Lemma 1, we see that the subsets $N_0, \ldots, N_{P-1} \subseteq M_L$ defined in Sect. 3 naturally correspond to the *periodic classes* in the context of Markov chains.

Note that the Cayley graph of the syntactic monoid M_L of a regular language L can be regarded as a DFA that recognizes L. By Proposition 4 and Lemma 3, we have the following fact.

Theorem 3. *Let $L \subseteq \Sigma^*$ be a regular language, and P be the maximum period of L with respect to Σ. Then, the probability of L can be represented as $\mu_L = (\mu_0, \ldots, \mu_{P-1})$ with $\mu_0, \ldots, \mu_{P-1} \in [0,1]$.* ∎

The number of accumulation points of the probability can be smaller than the maximum period P. For example, the maximum period of the language $L = a(\Sigma\Sigma)^* \cup b\Sigma(\Sigma\Sigma)^* \subseteq \{a,b\}^*$ with respect to Σ is 2, and $\mu_L = (\frac{1}{2}, \frac{1}{2})$ holds. However, $\frac{1}{2}$ is the only accumulation point, and therefore, μ_L can also be represented as $\mu_L = \frac{1}{2}$.

4.2 Applications

In this subsection, we give an application of the algebraic decomposition discussed in Sect. 3. A monoid M is said to be *zero* if there exists the zero element $\iota \in M$, i.e., $\iota \cdot m = \iota = m \cdot \iota$ for each $m \in M$. Note that every monoid M can have at most one zero element. As a relationship of regular languages and the zero monoids, the following interesting fact is known.

Proposition 5 ([14]). *Let $L \subseteq \Sigma^*$ be a regular language. Then, either $\mu_L = 0$ or $\mu_L = 1$ iff the syntactic monoid of L is zero.* ∎

This proposition implies that if the probability of a regular language L has only one accumulation point μ_0, then we can determine whether $\mu_0 \in \{0,1\}$ by examining the existence of the zero element in the syntactic monoid. We would like to generalize this result to determine whether $\mu_r \in \{0,1\}$ for arbitrary $0 \leq r < P$ when μ_L is represented as $(\mu_0, \ldots, \mu_{P-1})$. Remember that $L_w = \{u \in (\Sigma^P)^* \mid wu \in L\}$ is a language over Σ^P. Because $\mu_r = \sum_{w \in \Sigma^r} (|\Sigma|^{-r} \cdot \mu_{L_w})$ holds by a simple discussion on probabilities, we have $\mu_r = 1$ iff $\mu_{L_w} = 1$ for each $w \in$

Σ^r, and $\mu_r = 0$ iff $\mu_{L_w} = 0$ for each $w \in \Sigma^r$. Therefore, a naive way to determine whether $\mu_r \in \{0, 1\}$ is to construct the syntactic monoid of $L_w \subseteq (\Sigma^P)^*$ and examine the existence of the zero element for each $w \in \Sigma^r$. Nevertheless, we already obtain the diagram (3), and thus we can directly determine whether $\mu_r \in \{0, 1\}$ without constructing the syntactic monoid of each L_w. Intuitively, the r-th accumulation point of the probability is either 0 or 1 iff there exists a sink consisting only of either rejecting states or accepting states in the r-th periodic class of the syntactic monoid.

To state this fact precisely, we define an ideal of a monoid. For a subset I of a monoid M, I is an *ideal* of M if $IM = I = MI$. For example, the singleton of the zero element $\{\iota\} \subseteq M$ is an ideal of every zero monoid M. It is well known that every homomorphic image (or inverse image) of an ideal is also an ideal (e.g., see Chapter II-3 of [10]). In particular, any ideal can be collapsed to the zero element by an appropriate monoid homomorphism. (The quotient is called the *Rees factor monoid*, which originates from [11].)

Theorem 4. *Let $L \subseteq \Sigma^*$ be a regular language, and let P be the maximum period with respect to Σ. For each $w \in \Sigma^r$ with $0 \le r < P$, the limit $\mu_{L_w} = \lim_{\ell \to \infty} \mu_{L_w}(\ell)$ exists and the following conditions are equivalent:*

(i) $\mu_{L_w} = 0$ or $\mu_{L_w} = 1$.
(ii) There exists a non-empty ideal I of T_r such that either $I \cap \eta_w(L_w) = \emptyset$ or $I \subseteq \eta_w(L_w)$ holds where η_w is the homomorphism in the diagram (3). ∎

Example 8. Let $L_3 = a(\Sigma\Sigma)^* \cup b\Sigma^*$ (see also Example 3-(3)). Note that the maximum period with respect to Σ is 2, and the probability is $\mu_{L_3} = (\frac{1}{2}, 1)$. For any $w \in \Sigma^*$, we abbreviate $(L_3)_w$ as L_w in the following. We have $\mu_{L_\varepsilon} \notin \{0, 1\}$ where $L_\varepsilon = ba(\Sigma\Sigma)^* \cup bb(\Sigma\Sigma)^*$, and $\mu_{L_a} = \mu_{L_b} = 1$ where $L_a = L_b = (\Sigma\Sigma)^*$. In fact, T_1 is the trivial group $\{e\}$, and $\{e\}$ itself is an ideal satisfying $\{e\} = \eta_a(L_a) = \eta_b(L_b)$. On the other hand, $T_0 = \overline{U}_2 = \{e, \iota_1, \iota_2\}$ and $\eta_{L_\varepsilon}(L_\varepsilon)$ contains only one of either ι_1 or ι_2. Because the only non-empty ideals of $\overline{U}_2$ are $\overline{U}_2$ itself and $\{\iota_1, \iota_2\}$, the condition (ii) in Theorem 4 cannot be satisfied. ∎

5   Krohn-Rhodes Decompositions

We get back to general periods, including the cases that $\Gamma \subsetneq \Sigma$. In Sect. 3, we showed that every language with periods can be decomposed into a semidirect product with cyclic groups. In this section, we show that every language with periods also can be decomposed into another kind of product, called the wreath product, with cyclic groups.

For a right action $* : X \times M \to X$, we say that (X, M) is a *transformation monoid*[1]. For example, (M, M) is a transformation monoid for every monoid M with the monoid operation $*$ of M. A transformation monoid is said to be finite if

---

[1] This is a conventional term. A transformation monoid is the pair of an action and a monoid, but not a monoid itself. Nevertheless, similarly to the existence of a monoid simulating a DFA, there exists a monoid 'isomorphic' to each transformation monoid.

both of X and M are finite. For two transformation monoids (X, M) and (Y, N), a pair (ϕ, ψ) of $\phi : X \to Y$ and $\psi : M \to N$ is a *homomorphism* if (i) ψ is a monoid homomorphism, and (ii) $\phi(x * m) = \phi(x) * \psi(m)$ for each $x \in X, m \in M$. We say that (ϕ, ψ) is surjective if both of ϕ and ψ are surjective. If there are a subset X' of X and a submonoid M' of M with a surjective homomorphism (X', M') to (Y, N), then (Y, N) is called a *divisor* of (X, M).

For two transformation monoids (X, M) and (Y, N), the transformation $(X \times Y, M^Y \rtimes N)$ where the action $* : (X \times Y) \times (M^Y \rtimes N) \to (X \times Y)$ is defined as

$$(x, y) * (f, n) = (x * f(y), y * n) \tag{4}$$

is called the *wreath product* of (X, M) and (Y, N), and denoted by $(X, M) \wr (Y, N)$. The following property is well known.

Proposition 6. *Let $(X_1, M_1), (X_2, M_2), (X_3, M_3)$ be transformation monoids. Then, $((X_1, M_1) \wr (X_2, M_2)) \wr (X_3, M_3)$ is isomorphic to $(X_1, M_1) \wr ((X_2, M_2) \wr (X_3, M_3))$.* ∎

Krohn and Rhodes showed the *decomposition theorem* of finite monoids ([6]).

Proposition 7 (Krohn-Rhodes). *Every finite transformation monoid (X, M) is a divisor of a wreath product of the form*

$$(Y_1, N_1) \wr \cdots \wr (Y_k, N_k)$$

where each (Y_i, N_i) with $1 \leq i \leq k$ is (U_2, U_2) or (G, G) with a nontrivial group G dividing M. ∎

Because every DFA can be regarded as a transformation monoid, Krohn-Rhodes theorem is also referred to as the decomposition theorem of regular languages. Nevertheless, known proofs for Proposition 7 are purely semigroup-theoretic (e.g., [3,12]), and there is a gap between the proofs and formal language theory. Moreover, a finite monoid is not always the syntactic monoid of a language. For example, it is explained in [8] that U_K with $K \geq 3$ is not a syntactic monoid (see Example 1 for the definition of U_K). For these reasons, there is a need for studies on Krohn-Rhodes decompositions for syntactic monoids. As an application of the results of this paper, we give a partial Krohn-Rhodes decomposition of syntactic monoids of periodic regular languages. More precisely, for any periodic regular language L that has periods $P_1, \ldots, P_n$, (M_L, M_L) is a divisor of $(\mathcal{T}_K, \mathcal{T}_K) \wr (G, G)$ for some K where G is the direct product of the cyclic groups C_{P_i} for $1 \leq i \leq n$.

Theorem 5. *Let $L \subseteq \Sigma^*$ be a regular language and M_L be the syntactic monoid of L. Let L have periods $P_1, \ldots, P_n > 1$ with respect to $\Gamma_1, \ldots, \Gamma_n \subseteq \Sigma$, respectively. Then, (M_L, M_L) is a divisor of the wreath product of the form*

$$(\mathcal{T}_K, \mathcal{T}_K) \wr (G, G)$$

where $G = C_{P_1} \times \cdots \times C_{P_n}$ and $K = \max\{|N_{\mathbf{r}}| \mid \mathbf{r} \in G\}$. Furthermore, G divides M_L. ∎

Note that this theorem provides only a partial decomposition: the monoid $\mathcal{T}_K$ is not represented as a wreath product of monoids of the form (U_2, U_2) and groups in general. For this partial decomposition to be useful, the structure embedded in $\mathcal{T}_K$ needs to be simpler than M_L. This aspect is related to the group complexity, which was introduced in [5]. The smallest number of components (Y_i, N_i) of the form (G, G) in Proposition 7 over all possible decompositions is called the *group complexity* of (X, M). There are several studies for computing group complexity, but any complete algorithm has not yet been obtained (see e.g. [12] for detail). Therefore, if the decomposition given in Theorem 5 leads to the minimal decomposition, that is, if the minimal decomposition of (M_L, M_L) can be represented as

$$(Y_1, N_1) \wr \cdots \wr (Y_k, N_k) \wr (G, G)$$

with $G = C_{P_1} \times \cdots \times C_{P_n}$, our result is beneficial for computing the group complexity. Proving this is left for future work, but it appears to be challenging.

6   Conclusions

We have provided an algebraic decomposition of regular languages with periods using cyclic groups and semidirect products. Furthermore, we have discussed several applications of our results including probabilities, Markov chains, and Krohn-Rhodes decompositions. In Sects. 3 and 5, we have already outlined the following future work:

- In the decomposition given by Theorem 1, is it possible to restrict the first component $\mathcal{T}_K^{C_{P_1} \times \cdots \times C_{P_n}}$ to a simpler monoid?
- Does the partial decomposition given in Theorem 5 lead to the minimal Krohn-Rhodes decomposition of the syntactic monoid?

We believe that these two problems are inherently related.

In addition, further investigation is needed regarding the relation between our decomposition and the holonomy decomposition of automata [2,16]. As mentioned in Sect. 1, the holonomy decomposition is known as a decomposition of automata, and is closely related to the Krohn-Rhodes decomposition. Therefore, we believe that there exists a connection between the decomposition of syntactic monoids we provided and the holonomy decomposition.

Finally, extending the definition of periods is also left for future work. For example, each non-empty subset $\Gamma \subseteq \Sigma$ can be regarded as a *code* because every word in Γ^* has a unique factorization in *codewords* in Γ (see e.g., [1] in detail). Therefore, the definition of periods could be extended by the number of occurrences of codewords in Γ with another code $\Gamma \subseteq \Sigma^*$.

References

1. Berstel, J., Perrin, D., Reutenauer, C.: Codes and Automata, Encyclopedia of Mathematics and Its Applications, vol. 129. Cambridge University Press (2010)

2. Eilenberg, S.: Automata, Languages, and Machines. Academic Press (1976)
3. Diekert, V., Kufleitner, M., Steinberg, B.: The krohn-rhodes theorem and local divisors. Fund. Inform. **116**(1–4), 65–77 (2012)
4. Inoue, Y., Hashimoto, K., Seki, H.: Semidirect product decompositions for periodic regular languages. arXiv preprint arXiv:2403.05088 (2024)
5. Krohn, K., Rhodes, J.L.: Complexity of finite semigroups. Ann. Math. **88**(1), 128–160 (1968)
6. Krohn, K., Rhodes, J.L., Tilson, B.R.: Algebraic Theory of Machines, Languages, and Semigroups, vol. 1, pp. 81–125. Academic Press (1968)
7. Lallement, G.: Semigroups and Combinatorial Applications. John Wiley & Sons, Inc. (1979)
8. Lawson, M.V.: Finite Automata. Chapman and Hall/CRC (2003)
9. Norris, J.R.: Markov Chains. Cambridge University Press (1998)
10. Pin, J.É.: Mathematical Foundations of Automata Theory. Université Paris, Lecture notes LIAFA (2010)
11. Rees, D.,: On Semi-groups, Mathematical Proceedings of the Cambridge Philosophical Society, vol. 36, no. 4. Cambridge University Press (1940)
12. Rhodes, J.L., Tilson, B.: The q-theory of Finite Semigroups. Springer Science & Business Media (2009). https://doi.org/10.1007/b104443
13. Salomaa, A., Matti, S.: Automata-theoretic Aspects of Formal Power Series. Springer Science & Business Media (2012). https://doi.org/10.1007/978-1-4612-6264-0
14. Sin'ya, R.: An automata theoretic approach to the zero-one law for regular languages: algorithmic and logical aspects s. Electron. Proc. Theor. Comput. Sci. **193**, 172–185 (2015)
15. Sin'ya, R.: Measuring power of locally testable languages. In: International Conference on Developments in Language Theory, pp. 274–285 (2022)
16. Zeiger, H.P.: Cascade synthesis of finite-state machines. Inf. Control **10**, 419–433 (1967)

Approximate Cartesian Tree Pattern Matching

Sungmin Kim and Yo-Sub Han$^{(\boxtimes)}$

Department of Computer Science, Yonsei University, Seoul 03722, Republic of Korea
{rena_rio,emmous}@yonsei.ac.kr

Abstract. The Cartesian tree of a string is a binary tree, which is useful in capturing minimalities within strings. We study the approximate pattern matching problem for two Cartesian trees of two strings. We design a poly-time algorithm that computes the minimum edit cost when a given string is edited to match the Cartesian tree of the other string. Then, we adapt the algorithm to the approximate pattern matching problem, where we find all substrings of a given text that match a given Cartesian tree pattern within a given number of edit operations. We also consider variant problems such as the approximate Cartesian matching under Hamming distance, and present poly-time algorithms for the considered problems.

Keywords: Cartesian trees · Approximate pattern matching · Edit distance

1 Introduction

Many real-world problems can be formulated using discrete time-series data, or integer strings. For example, we can formalize vehicle routing through highly congested traffic using shortest-path problems where the edge cost changes depending on the current timestep. Since the problems are motivated by real-world instances, it is natural to consider settings where the changes according to time follows a certain pattern and is not entirely arbitrary. Therefore, many problems dealing with integer strings also try to capture various patterns in the data that heavily rely on the ordering between characters. For example, the longest increasing subsequence problem is to capture the longest subsequence with increasing characters of a given text [17]. The longest k-rollercoaster problem is to compute the longest subsequence of a given text that repeat increasing at least k times and decreasing at least k times [2,7]. These problems capture increases and decreases in the text.

Recently, Park et al. [15] defined the Cartesian tree pattern matching problem. Unlike rollercoasters or increasing substrings, the Cartesian tree pattern matching focuses on the minimum character of a given integer string. For a string w over the integer alphabet, the root of a Cartesian tree is the leftmost minimum character of w, and the left and right subtrees of the root are built

© The Author(s), under exclusive license to Springer Nature Switzerland AG 2024
J. D. Day and F. Manea (Eds.): DLT 2024, LNCS 14791, pp. 189–202, 2024.
https://doi.org/10.1007/978-3-031-66159-4_14

recursively with the substrings on the left and right of the leftmost minimum character of w, respectively [14,20]. For example, let $w = (7,8,9,7,2,7,6)$. Then, the root of w's Cartesian tree $\mathbb{CT}(w)$ is labeled 2. The left subtree of the root is the Cartesian tree of the prefix $(7,8,9,7)$ and the right subtree of the root is the Cartesian tree of the suffix $(7,6)$. See the leftmost box of Fig. 1 for an example. Originally, Vuillemin [20] defined the Cartesian tree for representing a set of Cartesian coordinates, which is constructed recursively by selecting the point with the least y-component and dividing the remaining set of points using the selected point's x-component. This definition was adapted to integer strings, where one can first sort the set of points using their x-components and then leave only their y-components to form an integer string.

We say that two strings w and u over an integer alphabet are a *match* if their Cartesian trees share the same structure. The Cartesian tree matching problem is to find all substrings of a text T that have the same Cartesian tree of a single pattern P or multiple patterns P_+, where $|\mathsf{P}_+|$ denotes the sum of the lengths of all patterns in P_+. Park et al. [14] solved the Cartesian tree matching for a single string in optimal $O(|\mathsf{T}| + |\mathsf{P}|)$ time, and gave an $O((|\mathsf{T}| + |\mathsf{P}_+|) \log k)$ algorithm for the multiple pattern matching problem with k patterns. Afterwards, Song et al. [19] gave alternative algorithms for the multiple pattern matching problem. Their first and third algorithms run in worst-case $O\left(|\mathsf{T}|(|\mathsf{P}_+| + b)\right)$ time and best-case $O\left(\frac{|\mathsf{T}|b}{m-b}\right)$ time, where m is the length of the shortest pattern in P_+ and b is a constant at most m. Their second algorithm runs in worst-case $O(|\mathsf{T}||\mathsf{P}|)$ time and best-case $O(|\mathsf{T}|)$ time. Kim and Cho [10] devised a data structure using $3|\mathsf{T}| + o(|\mathsf{T}|)$ bits that solves the counting version of the pattern matching problem for a pattern query P in $|\mathsf{P}|$ time. For the case where the text and pattern are strings over a finite ordered alphabet Σ, Nishimoto et al. [12] introduced Cartesian-tree position heaps, constructed in $O(|\mathsf{T}| \log |\Sigma|)$ time, which are capable of answering pattern matching queries in $O(|\mathsf{P}|(|\Sigma| + \log(\min(h, |\mathsf{P}|))) + x)$ time. Here, h is the height of the heap and x is the number of occurrences of P in T.

Researchers also considered variations of the Cartesian tree matching problem. Faro et al. [5] solved the longest common Cartesian substring problem in $O(|\mathsf{T}||\mathsf{P}|)$ time. They also presented a randomized $O(|\mathsf{P}| + |\mathsf{T}|)$ time algorithm that gives a correct answer with high probability. Oizumi et al. [13] solved the Cartesian tree subsequence matching problem. Auvray et al. [1] studied a version of the approximate Cartesian tree matching in $O(|\mathsf{P}||\mathsf{T}|)$ time, where the problem allowed matches up to a single transposition, i.e. exchanging a pair of adjacent characters. Since errors such as character omission often happen in practice, it is natural to study approximate pattern matching for Cartesian trees. However, to the best of our knowledge, there are no prior results on the approximate Cartesian tree matching problem for edit operations. Therefore, we define the approximate Cartesian tree matching problem in the context of edit distance and solve the problem in polynomial time.

In Sect. 2, we give definitions for the approximate Cartesian tree matching. Then, in Sect. 3, we formalize the approximate Cartesian tree matching problem and tackle the problem from the edit distance perspective. We design a recurrence relation-based algorithm that computes the Cartesian tree distance in polynomial time in Sect. 4, and modify the algorithm for the approximate Cartesian tree matching problem under insert, delete and substitute in Sect. 5. We also consider some variant problems with unit edit cost, or only allowing substitutions. Finally, we conclude our paper with a possible future direction in Sect. 6.

2 Preliminaries

Let $\mathbb{Z}$ denote the set of integers and $\mathbb{N}_0$ denote the set of non-negative integers. We consider strings over an alphabet consisting of integers from $\mathbb{Z}$. The length $|w|$ of a string w over the alphabet $\Sigma = \mathbb{Z}$ is the number of characters in w. We denote the empty string, the string with length 0, by λ. The ith character of w counted from the left is denoted as $w[i]$ for $0 \le i \le |w| - 1$. Conversely, the *index* of $w[i]$ is i. A function $f : \mathbb{N}_0 \to \mathbb{N}_0$ is an *embedding* if f is non-decreasing with respect to the input. We denote the *composite embedding* of embeddings f and g with $g \circ f$; namely, $(g \circ f)(i) = g(f(i))$ for all $i \in \mathbb{N}_0$. A string u is a *subsequence* of w if there exists a strictly increasing embedding f such that $f(|u| - 1) < |w|$ and $w[f(i)] = u[i]$ for $0 \le i \le |u| - 1$. If there exists such an embedding that satisfies $f(i+1) = f(i)+1$ for $0 \le i \le |u|-2$, then u is a *substring* of w. Then the substring u can be written as $w[f(0) : f(|u|-1)]$. We define $w[i : j] = \lambda$ whenever $i > j$. For an integer $i \in \{0, 1, \ldots, |w|\}$, the substring $w[0 : i]$ or $w[i : |w| - 1]$ is a *prefix* or *suffix* of w which we abbreviate as $w[: i]$ or $w[i :]$, respectively. For an embedding f, use $f[i : j]$ to denote the string satisfying $f[i : j][k - i] = f(k)$ for all $i \le k \le j$, by abuse of notation.

For a rooted tree T, let $\mathbf{r}(T)$ denote the root. Each node v is labeled with an integer $v.\texttt{label}$. We use $|T|$ to denote the number of nodes in T. For a node v in T, let $\texttt{st}(v)$ denote the *subtree* of T rooted at node v. The *height* of T is the maximum number of edges between the root and a leaf. The *Cartesian tree* $\mathbb{CT}(w)$ of an integer string w is a rooted binary tree that satisfies the followings.

- The tree consists of exactly $|w|$ nodes. If $w = \lambda$, the tree is empty.
- Let i be the index of the smallest character in w. If there are multiple smallest characters, let i be the least index. The label of the root is $w[i]$.
- The left subtree of $\mathbf{r}(\mathbb{CT}(w))$ is $\mathbb{CT}(w[: i - 1])$, and the right subtree of $\mathbf{r}(\mathbb{CT}(w))$ is $\mathbb{CT}(w[i + 1 :])$.

We define two Cartesian trees to be *equivalent* if they have the same structure. Here, two Cartesian trees with the same number of nodes have the same structure if both trees consist of a single node, or the left and right subtrees of the root of one Cartesian tree have the same number of nodes and the same structure as the left and right subtrees of the root of the other Cartesian tree, respectively. For example, two strings $(7, 5, 10)$ and $(2, 1, 2)$ yield equivalent Cartesian trees. The Cartesian tree of a string w can be constructed in $O(|w|)$ time [6]. The string obtained by reading node labels through the in-order traversal of a

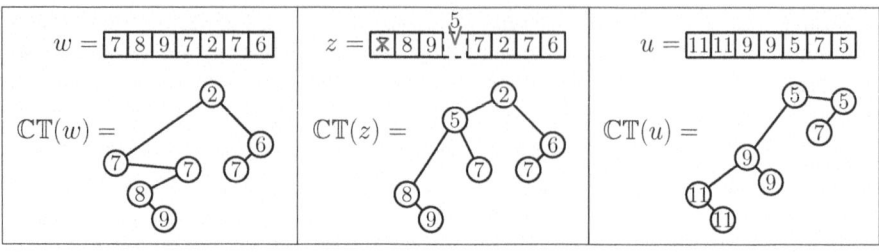

Fig. 1. An example of Cartesian trees for three strings $w = (7, 8, 9, 7, 2, 7, 6)$, $z = (8, 9, 5, 7, 2, 7, 6)$, and $u = (11, 11, 9, 9, 5, 7, 5)$. The Cartesian trees of z and u are equivalent. Since we can obtain z from w within two edit operations, the Cartesian tree distance from w to u is at most two when all edit costs are 1. Note that $\texttt{Cdist}(w \rightarrow u) = 1$ by substituting $w[2] = 9$ to any value between 6 and 3.

Cartesian tree T is denoted with $\texttt{treeLabel}(T)$. Then, it is easy to see that $\texttt{treeLabel}(\mathbb{CT}(u)) = u$ for any string u.

For strings w and u, a pair of embeddings (f_w, g_u) is an *alignment* of size k from w to u if $0 \le f_w(0) < f_w(1) < \cdots < f_w(k-1) < |w| \le f_w(k)$ and $0 \le g_u(f_w(0)) < g_u(f_w(1)) < \cdots < g_u(f_w(k-1)) < |u|$. We use this non-standard definition of alignments to distinguish between embedding a subsequence of w into u and embedding a subsequence of u into w. By abuse of notation, say that such an alignment is also an alignment from w to $\mathbb{CT}(u)$. Also we say that $w[f_w(i)]$ is *aligned* with $u[g_u(f_w(i))]$ or the node labeled $u[g_u(f_w(i))]$ in $\mathbb{CT}(u)$ for all $0 \le i < k$. An *edit* on string w is either an insertion, a deletion, or a substitution of a single character. Each edit is associated with a constant non-negative integral cost $\gamma_{\texttt{ins}}$, $\gamma_{\texttt{del}}$ and $\gamma_{\texttt{sub}}$. Since we consider integer alphabets for our problem, we assume that each edit of the same type costs the same, and that we do not substitute a character for the same character. Moreover, we assume that $\gamma_{\texttt{sub}} \le \gamma_{\texttt{del}} + \gamma_{\texttt{ins}}$, because otherwise we can use a pair of delete and insert operations instead of a single substitute operation while incurring a smaller cost. The *string edit distance* between two strings w and u is the minimum total cost of edits required to transform w to u. We define the *Cartesian tree distance* $\texttt{Cdist}(w \rightarrow u)$ from w to u to be the minimum total cost of edits required to transform w to a string yielding $\mathbb{CT}(u)$. Formally, for strings w, z and an alignment (f_w, g_z) of size k, we define

- the *substituted set* $\texttt{sub}(w, z, f_w, g_z) = \{i \mid 0 \le i < k, x[f_w(i)] \ne y[g_z(f_w(i))]\}$,
- the *deleted set* $\texttt{del}(w, z, f_w, g_z) = \{i \mid 0 \le i < |w|, \nexists j : f_w(j) = i\}$, and
- the *inserted set* $\texttt{ins}(w, z, f_w, g_z) = \{i \mid 0 \le i < |z|, \nexists j : g_z(f_w(j)) = i\}$.

Then, the cost $\texttt{cost}(w, z, f_w, g_z)$ of the alignment (f_w, g_z) is defined as

$$\gamma_{\texttt{sub}} \cdot |\texttt{sub}(w, z, f_w, g_z)| + \gamma_{\texttt{del}} \cdot |\texttt{del}(w, z, f_w, g_z)| + \gamma_{\texttt{ins}} \cdot |\texttt{ins}(w, z, f_w, g_z)|.$$

Finally, the Cartesian tree distance $\texttt{Cdist}(w \rightarrow u)$ from string w to string u is the minimum of $\texttt{cost}(w, z, f_w, g_z)$ over a string $z \in \mathbb{Z}^{|u|}$ with $\mathbb{CT}(z) = \mathbb{CT}(u)$

and an alignment (f_w, g_z) from w to z. An alignment (f, g) from w to u is *optimal* if there exists a string $z \in \mathbb{Z}^{|u|}$ such that $\mathtt{cost}(w, z, f, g) = \mathtt{Cdist}(w \to u)$.

3 The Cartesian Tree Distance

We give a formal definition of our problem and its motivation in Sect. 3.1. Then, we discuss properties of the Cartesian tree distance in Sect. 3.2.

3.1 Problem Definition and Motivation

Problem 1. [CARTEDITDIST] Given two strings w and u, compute $\mathtt{Cdist}(w \to u)$ the minimum total cost of edit operations required to transform w to a string yielding $\mathbb{CT}(u)$.

Here, we do not assume $\gamma_{\mathtt{ins}} = \gamma_{\mathtt{del}}$ because we want to tackle the most general version of the problem. Using the notion of Cartesian tree distance, we define the approximate Cartesian tree matching problem using edit operations. The problem is to compute all substrings of the text that, after some number of edit operations, yield a Cartesian tree equivalent to the Cartesian tree of the pattern.

Problem 2. [APPROXCARTMATCH] Given a pattern P, a text T and a non-negative integer t, compute all substrings w of T that satisfy $\mathtt{Cdist}(w \to \mathsf{P}) \leq t$.

Auvray et al. [1] studied a restricted version of the approximate Cartesian tree matching problem in which the problem allows a single character transposition. Moreover, Oizumi et al. [13] solved the Cartesian tree subsequence matching problem, which is a special case of Problem 2, where $t = 0$, $\gamma_{\mathtt{sub}} = \gamma_{\mathtt{ins}} = \infty$ and $\gamma_{\mathtt{del}} = 0$.

Recall that the motivation for the Cartesian tree matching is to capture patterns, where the position of the minimum element is crucial but requires more relaxed patterns rather than the order-preserving matching [4,9]. Specifically, an order-preserving match requires that if the ith character in w is larger than the jth character in w, then the ith character in u should also be larger than the jth character in u for all index pairs (i, j), $i < j$ of w. However, for the Cartesian tree matching, $w[i] < w[j]$ does not imply $u[i] < u[j]$ if there is a character c between $w[i]$ and $w[j]$ that satisfies $c \leq w[j]$ and $c < w[i]$. Now, if we want a more relaxed matching scheme than the Cartesian matching problem, it is sensible to consider the approximate Cartesian matching problem, which allows matching with more relaxed conditions. Considering that approximate matching for order-preserving matching has already been examined [8], our research inherits the line of reason behind both approximate matching and Cartesian tree matching. Furthermore, since the motivation behind Cartesian tree matching is to capture string patterns, we think that it is more appropriate to make edits on the string rather than the Cartesian tree.

3.2   An Edit Distance Perspective

Similar to the string edit distance problem, we have a trivial upper-bound on the Cartesian tree distance from w to u—linear in $|w| + |u|$. The bound comes from matching the length of w and u first via insertions or deletions, and then substituting as many characters as needed.

Proposition 1. *If $|w| \leq |u|$, then* $\mathtt{Cdist}(w \rightarrow u) \leq \gamma_{\mathtt{ins}} \cdot (|u| - |w|) + \gamma_{\mathtt{sub}} \cdot \max(|w|-1, 0)$. *Otherwise,* $\mathtt{Cdist}(w \rightarrow u) \leq \gamma_{\mathtt{del}} \cdot (|w|-|u|) + \gamma_{\mathtt{sub}} \cdot \max(|u|-1, 0)$.

Note that the trivial bound is indeed tight; if w is strictly increasing and u is strictly decreasing, then the optimal alignment must be of size $\min(|w|, |u|)$. Assuming $|w| \leq |u|$, we must first insert $|u| - |w|$ integers in decreasing order at the end of w to match the number of nodes, and then substitute the first $|w| - 1$ integers so that all the integers in w are decreasing. The same argument can hold when $|w| > |u|$, but we delete characters from w instead.

However, unlike the string edit distance problem, edits for Cartesian tree equivalence are asymmetric. Consider strings $w = (4, 4, 3, 2, 1, 1)$ and $u = (3, 3, 1, 1)$. The alignment (f, g), where $f[: 3] = (0, 1, 4, 5)$ and $(g \circ f)[: 3] = (0, 1, 2, 3)$, has a cost of two, but simply reversing the alignment from w to u gives a cost of three. Consider any alignment (f', g') from u to w, where $f'[: 3] = (0, 1, 2, 3)$ and $(g' \circ f')[: 3] = (0, 1, 4, 5)$. It is easy to verify that this alignment aligns the same characters in w and u compared to (f, g). However, we cannot insert two characters between $(3, 3)$ and $(1, 1)$ in u to yield a Cartesian tree same as $\mathbb{CT}(w)$. Lemma 1 proves the asymmetry of the Cartesian distance.

Lemma 1. *There exist strings w and u such that* $\mathtt{Cdist}(w \rightarrow u) \neq \mathtt{Cdist}(u \rightarrow w)$.

Proof. Consider strings $w = (3, 4, 5, 5, 4, 3)$ and $u = (4, 5, 5, 5, 4)$. Choose an alignment (f_w, g_u) of size five, where $f_w[0 : 4] = (0, 1, 2, 3, 5)$ and $(g_u \circ f_w)[0 : 4] = (0, 1, 2, 3, 4)$. The alignment gives a cost of one by deleting $w[4]$. On the other hand, no single edit on u yields a Cartesian tree equivalent to $\mathbb{CT}(w)$.   □

The asymmetry comes from the fact that $\mathbb{Z}$ is not dense. Indeed, we cannot arbitrarily scale up the integer characters without using the substitution operation. Therefore, we cannot directly operate on Cartesian trees, because Cartesian tree equivalence ignores labels. In other words, we cannot directly apply the general unordered tree edit distance algorithm, which takes $O(|T_1|^2 |T_2| \log |T_2|)$ time for two trees T_1 and T_2 [3,11]. Moreover, we point out that the Cartesian tree distance is not a distance metric, although we use the term "edit *distance*" to emphasize the similarity to the string edit distance. We show how to compute the edit distance in the next section.

4   A Dynamic Programming Approach for CartEditDist

Before we present our algorithm, let us consider a naive approach for the CARTE-DITDIST problem. If we try all possible combinations of edits up to the cost bound D, then the number of cases for choosing characters to delete in w is already exponential; in other words, assuming $\gamma_{\mathtt{del}} = 1$, we have $\sum_{i=0}^{\min(|w|,D)} \binom{|w|}{i}$ cases. Thus, we study to design an efficient algorithm for the CARTEDITDIST problem.

We establish a recurrence relation for the CARTEDITDIST problem. Since the target Cartesian tree $\mathbb{CT}(u)$ is fixed, we can break CARTEDITDIST from w to u into *subproblems* induced by subtrees of $\mathbb{CT}(u)$. Specifically, we aim to compute $\mathtt{Cdist}(w \to u)$ from the costs of converting prefixes of w to a string yielding the left subtree of $\mathbf{r}(\mathbb{CT}(u))$, and the costs of converting suffixes of w to a string yielding the right subtree of $\mathbf{r}(\mathbb{CT}(u))$. Formally, a subproblem of CARTEDITDIST from w to u is a 3-tuple (i, j, n), where i, j are integers satisfying $0 \le i \le j \le |w|$ and n is a node in $\mathbb{CT}(u)$. Each subproblem (i, j, n) corresponds to the case where string $w[i : j - 1]$ should be edited to yield the Cartesian tree rooted at n. A subproblem (i, k, l) is a *left subproblem* of (i, j, n) if l is the left child of n and $k \le j$. Symmetrically, a subproblem (k, j, r) is a *right subproblem* of (i, j, n) if r is the right child of n and $k \ge i$. The recurrence scheme is inspired from the CYK algorithm for context-free grammars [16, 18] or the basic Cartesian tree subsequence matching algorithm [13].

However, it is insufficient to compute only the minimum total edit cost for each subproblem. Figure 2 illustrates the need for a more complex approach.

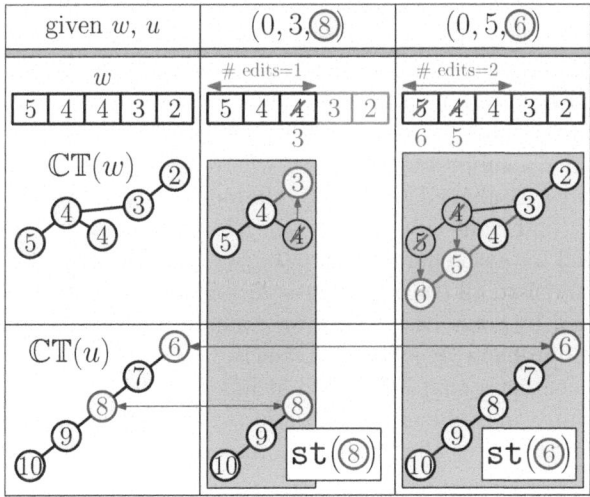

Fig. 2. Let $w = (5, 4, 4, 3, 2)$ and $u = (10, 9, 8, 7, 6)$. The subproblem $(0, 5, ⑥)$ depicted in the right column shows an optimal solution of two with the subtree rooted at ⑧ of $\mathbb{CT}(u)$ matching $w[0 : 2]$ under 2 edits. However, the subproblem $(0, 3, ⑧)$ depicted in the middle column has an optimal solution of one.

We solve the problem at the cost of one additional parameter in the recurrence relation. Recall that the Cartesian tree distance from w to u is bound linearly by $|w| + |u|$. Therefore, for all non-negative integers d no greater than the upper bound from Proposition 1, we compute whether or not there exists an alignment for each subproblem with cost d. Specifically, we define the optimal value $\mathsf{opt}(i, j, n, d)$ of each subproblem (i, j, n) with cost d by the following: consider the set of Cartesian trees of all strings z where there exists an alignment (f, g) from $w[i : j - 1]$ to z with $\mathsf{cost}(w[i : j - 1], z, f, g) = d$. Then, for a node n in $\mathbb{CT}(u)$, choose all the trees that are equivalent to $\mathsf{st}(n)$. Finally, $\mathsf{opt}(i, j, n, d)$ is the maximum label value of the root node among the trees that are chosen this way.

We first establish the base cases for the recurrence relation. For leaf nodes n in $\mathbb{CT}(u)$ and all integers i, j satisfying $0 \leq i < j \leq |w|$, we have

1. $\mathsf{opt}(i, i, n, \gamma_{\mathsf{ins}}) = \infty$,
2. $\mathsf{opt}(i, j, n, (j - i - 1)\gamma_{\mathsf{del}}) = \max\{w[k] \mid k = i, i + 1, \ldots, j - 1\}$ and
3. $\mathsf{opt}(i, j, n, (j - i - 1)\gamma_{\mathsf{del}} + \gamma_{\mathsf{sub}}) = \infty$.

Here, if $\gamma_{\mathsf{ins}} = (j - i - 1)\gamma_{\mathsf{del}}$ or $\gamma_{\mathsf{ins}} = (j - i - 1)\gamma_{\mathsf{del}} + \gamma_{\mathsf{sub}}$, then the maximum value is chosen between the two coinciding cases. The first case solves subproblems with the empty string, where a single character needs to be inserted to yield n. The second and third case solves subproblems with substring length at least one. The second case covers alignments in which all but a single character are deleted. The third case covers alignments in which all but one character are deleted and the remaining character is substituted. Recall that the total edit cost of deleting and inserting once each is no less than that of substituting once. Therefore, we do not need to consider cases where all characters are deleted. Finally, for all integers i, j, x and a leaf node n, where $\mathsf{opt}(i, j, n, x)$ is undefined yet, we define $\mathsf{opt}(i, j, n, x) = -\infty$. Notice that only this case allows $i > j$.

Let us consider a subproblem (i, j, n), where n has both left and right children l and r and n is aligned to the kth character in w. If we fix the left subproblem (i, k_1, l) and right subproblem (k_2, j, r) with $k_1 \leq k < k_2$, then a cost of γ_{del} is incurred for each character in $w[k_1 : k_2 - 1]$ not aligned to n. However, the case that we delete all characters in $w[k_1 : k_2 - 1]$ except the kth character is already covered by some case where we examine the left subproblem (i, k, l) and the right subproblem $(k + 1, j, r)$. Likewise, for a subproblem (i, j, n), let n be aligned to a character that is inserted at index k. By the same logic, we only need to investigate subproblems (i, k, l) and (k, j, r) when merging the results. Therefore, these base cases allow us to ignore deletion operations when merging the results from two subtrees.

Note that we still need to consider deletion if an internal node n has a single child. Indeed, for a subproblem (i, j, n) and an alignment that aligns n to $w[a]$, we need to delete $w[a + 1 : j - 1]$ if n does not have a right child. The same argument holds for when n does not have a left child. Thus, we want to resolve the issue by treating all internal nodes as if they have two children by making a dummy $\mathsf{nullNode}$ child. Then, we let $\mathsf{opt}(i, j, \mathsf{nullNode}, (j - i)\gamma_{\mathsf{del}}) = \infty$ and $\mathsf{opt}(i, j, \mathsf{nullNode}, x) = -\infty$ for $x \neq (j - i)\gamma_{\mathsf{del}}$. This helps us handle all deletion operations at the leaf node and null node level (Fig. 3).

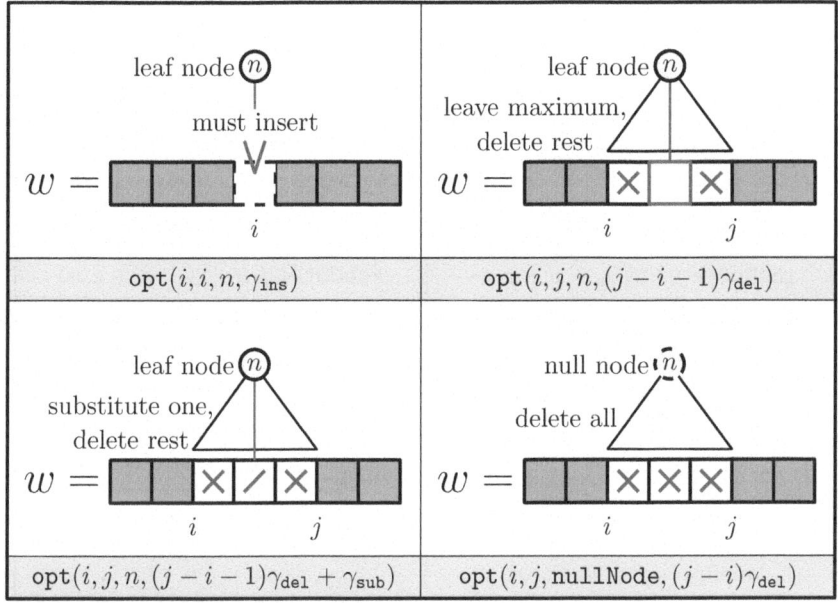

Fig. 3. An illustration of the base cases of opt.

Now, we move on to considering internal nodes. If the root n of the Cartesian tree is aligned to a character i in w, then the left child $left$ of n and the right child $right$ of n should satisfy $left.\texttt{label} > i$ and $right.\texttt{label} \geq i$ if $left$ and $right$ exist, respectively. In other words, if the label of either subtree does not satisfy $left.\texttt{label} > i$ or $right.\texttt{label} \geq i$, then an additional cost of $\gamma_{\texttt{sub}}$ should be incurred to substitute i into an integer no greater than $\max(left.\texttt{label} - 1, right.\texttt{label})$. Therefore, the recurrence relation should compute the maximum possible label value of n under the edit cost d for each subproblem (i, j, n).

Next, we consider substituting a single character into another character. In this case, for a fixed subproblem pair, the substituted character's new value can be up to the minimum between the maximum value that the root of the left subtree can take minus one, and the maximum value that the root of the right subtree can take. The total edit cost of the two subproblems should be exactly $d - \gamma_{\texttt{sub}}$, and thus the final maximum value that the root can take becomes the maximum of such values over all subproblem pairs with the total cost $d - \gamma_{\texttt{sub}}$.

Finally, we consider insertion costs between the left and right subproblems. In this case, we investigate all pairs $(i, k, l), (k, j, r)$ of left and right subproblems for $k = i, i + 1, \ldots, j$. Similar to the substitution case, for a fixed subproblem pair, we can only insert integers up to the minimum between the maximum value that the root of the left subtree can take minus one, and the maximum value that the root of the right subtree can take. Moreover, the two subproblems should have the total edit cost exactly $d - \gamma_{\texttt{ins}}$ (Fig. 4).

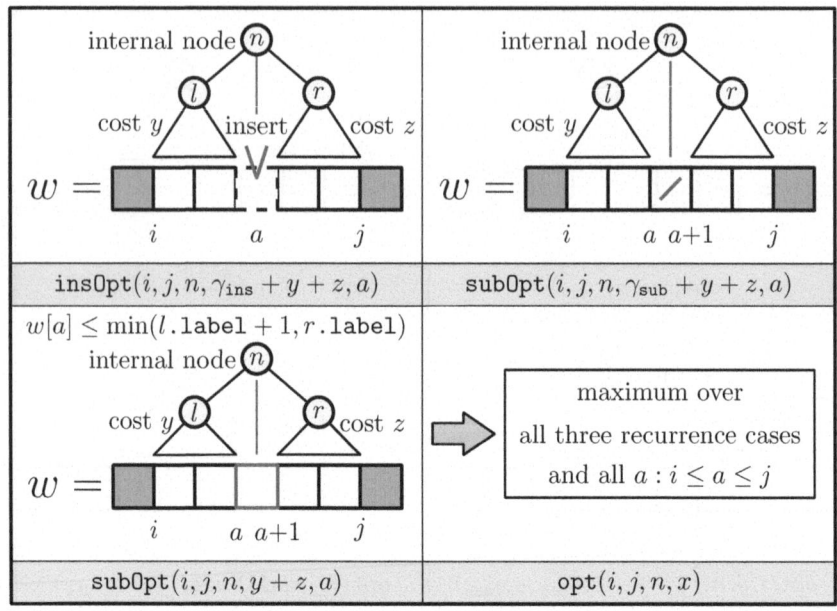

Fig. 4. An illustration of the recurrence relation of opt.

Summing up, the recurrence relation for subproblem (i, j, n) and cost x is as follows. Here, n is an internal node, and l and r are the left and right child of n, respectively. Moreover, x is less than the upper bound from Proposition 1.

$$\text{opt}(i, j, n, x) = \max_{a \in [i,j]} (\max(\text{insOpt}(i, j, n, x, a), \text{subOpt}(i, j, n, x, a))),$$

$$\text{insOpt}(i, j, n, x, a) = \max_{\substack{y, z : y, z \geq 0, \\ y + z = x - \gamma_{\text{ins}}}} (\min(\text{opt}(i, a, l, y) - 1, \text{opt}(a, j, r, z))),$$

$$\text{subOpt}(i, j, n, x, a) =$$
$$\max \begin{cases} \max_{\substack{y, z : y, z \geq 0, \\ y + z = x - \gamma_{\text{sub}}}} (\min(\text{opt}(i, a, l, y) - 1, \text{opt}(a + 1, j, r, z))) \\ w[a] \text{ if } w[a] \leq \max_{\substack{y, z : y, z \geq 0, \\ y + z = x}} (\min(\text{opt}(i, a, l, y) - 1, \text{opt}(a + 1, j, r, z))) \end{cases}.$$

Here, $\text{insOpt}(i, j, n, x, a)$ refers to the optimal value from the subproblems (i, a, l) and (a, j, r); that is, a character is inserted at a position a in w which n aligns to. Similarly, $\text{subOpt}(i, j, n, x, a)$ refers to the optimal value from the subproblems (i, a, l) and $(a + 1, j, r)$; then n aligns to $w[a]$ or a value substituted from $w[a]$. Finally, we let $\text{opt}(i, j, n, x) = -\infty$ for all cases left undefined.

After we compute $\text{opt}(0, |w|, \text{r}(u), x)$ for all possible x values, we compute the minimum x value for which $\text{opt}(0, |w|, \text{r}(u), x) \neq -\infty$. Lemma 2 proves that this value is the solution to CARTEDITDIST. Theorem 1 follows from Lemma 2.

Lemma 2. *We have* $\text{Cdist}(w \to u) = \min\{x \mid \text{opt}(0, |w|, \text{r}(\mathbb{CT}(u)), x) \neq -\infty\}$.

Algorithm 1. Cartesian tree distance from string w to string u

1: Compute the Cartesian tree $\mathbb{CT}(u)$ of u
2: Initialize DP table DP with dimension $|w| \times |w| \times |\mathbb{CT}(u)| \times D$, where D is the upper bound on $\texttt{Cdist}(w \to u)$ from Proposition 1
3: Initialize $\texttt{DP}[i][i][n][\gamma_{\texttt{ins}}] \leftarrow \infty$, $\texttt{DP}[i][j][n][(j-i-1)\gamma_{\texttt{del}}] \leftarrow \max\{w[k] \mid k = i, i+1, \ldots, j-1\}$, $\texttt{DP}[i][j][n][(j-i)\gamma_{\texttt{del}}+\gamma_{\texttt{ins}}] \leftarrow \infty$, and $\texttt{DP}[i][j][n][(j-i-1)\gamma_{\texttt{del}}+\gamma_{\texttt{sub}}] \leftarrow \infty$ for integers i, j satisfying $0 \le i \le j \le |w|$ and leaf nodes n of $\mathbb{CT}(u)$
4: Initialize all other cells with $-\infty$
5: **for** $n \leftarrow$ internal nodes in $\mathbb{CT}(u)$ in bottom-up order **do**
6:     $l \leftarrow$ left child of n, $r \leftarrow$ right child of n
7:     **for** $i \leftarrow 0, 1, \ldots, |w|$ and $j \leftarrow i, i+1, \ldots, |w|$ **do**
8:         **for** $x \leftarrow 0, 1, \ldots, D$ and $a \leftarrow i, i+1, \ldots, j$ **do**
9:             **for** $y \leftarrow 0, 1, \ldots, x - \gamma_{\texttt{ins}}$ **do**
10:                 $insCost \leftarrow \min(\texttt{DP}[i][a][l][y] - 1, \texttt{DP}[a][j][r][x - \gamma_{\texttt{ins}} - y])$
11:                 $\texttt{DP}[i][j][n][x] \leftarrow \max(\texttt{DP}[i][j][n][x], insCost)$
12:             **end for**
13:             **for** $y \leftarrow 0, 1, \ldots, x - \gamma_{\texttt{sub}}$ **do**
14:                 $subCost \leftarrow \min(\texttt{DP}[i][a][l][y] - 1, \texttt{DP}[a+1][j][r][x - \gamma_{\texttt{sub}} - y])$
15:                 $\texttt{DP}[i][j][n][x] \leftarrow \max(\texttt{DP}[i][j][n][x], subCost)$
16:             **end for**
17:             **for** $y \leftarrow 0, 1, \ldots, x$ **do**
18:                 **if** $w[a] \le \min(\texttt{DP}[i][a][l][y] - 1, \texttt{DP}[a+1][j][r][x - y])$ **then**
19:                     $\texttt{DP}[i][j][n][x] \leftarrow \max(\texttt{DP}[i][j][n][x], w[a])$
20:                 **end if**
21:             **end for**
22:         **end for**
23:     **end for**
24: **end for**
25: **return** $\min\{x \le D \mid \texttt{DP}[0][|w|][\texttt{r}(\mathbb{CT}(u))][x] \ne -\infty\}$

Theorem 1. *Given two strings w and u over the integer alphabet $\mathbb{Z}$, we can solve* CartEditDist *from w to u in $O(|w|^3 |u| D^2)$ time, where $D = \gamma_{\texttt{ins}} \cdot (|u| - |w|) + \gamma_{\texttt{sub}} \cdot \max(|w| - 1, 0)$ if $|w| \le |u|$, and $D = \gamma_{\texttt{del}} \cdot (|w| - |u|) + \gamma_{\texttt{sub}} \cdot \max(|u| - 1, 0)$ otherwise.*

5   Solutions for Problem Variations

We first solve the ApproxCartMatch problem using the same recurrence relation as CartEditDist. The problem is to return all nonempty substrings of T that can yield a Cartesian tree equivalent to $\mathbb{CT}(P)$ within the total edit cost t.

We let $D = t$ instead of the upper-bound from Proposition 1. Furthermore, after we compute $\texttt{opt}(0, |T|, \texttt{r}(\mathbb{CT}(P)), x)$'s for all $x \le t$, we return

$$\{(i, j) \mid 0 \le i < j \le |T|, \exists x \le t : \texttt{opt}(i, j, \texttt{r}(\mathbb{CT}(P)), x) \ne -\infty\}$$

instead of the original return value. This returns all integer pairs (i, j), where $\mathrm{T}[i : j - 1]$ can yield $\mathbb{CT}(\mathrm{P})$ within t. Note that computing the set takes $O(|\mathrm{T}|^2 t)$ time, and the computation of the outermost for loop takes $O(|\mathrm{T}|^3 |\mathrm{P}| t^2)$ time. Usually, t is a constant far smaller than the input size $|\mathrm{P}|$ or $|\mathrm{T}|$. Therefore, the overall time complexity of the modified algorithm is $O(|\mathrm{T}|^3 |\mathrm{P}| t^2) = O(|\mathrm{T}|^3 |\mathrm{P}|)$.

Theorem 2. *Given a pattern P, a text T and a constant approximate threshold t, we can compute all substrings w of T that satisfy* $\mathtt{Cdist}(w \to P) \leq t$ *in* $O(|T|^3 |P|)$ *time.*

Next, we consider a variation of the APPROXCARTMATCH problem, where all edit costs are 1. The setting is analogous to the Levenshtein distance in the context of the string edit distance—thus, we call this problem APPROX-CARTLEVENSHTEINMATCH. Since the deletion cost is nonzero, a match of P should be of length at most $|\mathrm{P}| + t$. Therefore, applying the same recurrence relation, we can solve APPROXCARTLEVENSHTEINMATCH in $O(|\mathrm{P}|^3 |\mathrm{T}|)$ time if we ignore subproblems (i, j, n) with $j - i \geq |\mathrm{P}| + t$. This is advantageous compared to the general setting because the length of the text is often far longer than the length of the pattern in common pattern matching settings.

Finally, we consider another variation of APPROXCARTMATCH, where we only allow substitutions of unit cost. Since the problem uses the same set of edit operations and costs as the Hamming distance—thus we call this problem APPROXCARTHAMMINGMATCH. In this setting, the length of a match of P is always fixed as $|\mathrm{P}|$. Therefore, we only need to consider subproblems (i, j, n), where $j - i = |\mathtt{st}(n)|$. Specifically, let n be a node in $\mathbb{CT}(\mathrm{P})$. Consider the Cartesian trees of strings obtained from $w[i : i + |\mathtt{st}(n)| - 1]$ with exactly x substitutions. Choose all the trees that are equivalent to $\mathtt{st}(n)$. We denote with $\mathtt{optH}(i, n, d)$ the maximum label value of the root node that is chosen this way. Note that $\mathtt{optH}(i, n, d)$ is analogous to $\mathtt{opt}(i, i + |\mathtt{st}(n)|, n, d)$. Now, we redefine the recurrence relation. For leaf nodes n of $\mathbb{CT}(\mathrm{P})$, the new base cases disallowing insertion and deletion are as such: $\mathtt{optH}(i, n, 0) = w[i]$, $\mathtt{optH}(i, n, 1) = \infty$ and $\mathtt{optH}(i, n, x) = -\infty$ for all $x : 2 \leq x \leq |\mathbb{CT}(\mathrm{P})|$. Furthermore, we let $\mathtt{optH}(i, \mathtt{nullNode}, 0) = \infty$ and $\mathtt{optH}(i, \mathtt{nullNode}, x) = -\infty$ for all $x : 1 \leq x \leq |\mathbb{CT}(\mathrm{P})|$. Finally, we define $\mathtt{st}(\mathtt{nullNode}) = 0$. This lets us treat all internal nodes as if they have two children, just as $\mathtt{opt}$.

Moving on to the internal nodes n, let l and r be the left and right child of n, respectively. If $w[i + |\mathtt{st}(l)|] \leq \max_{\substack{y, z : y, z \geq 0, \\ y + z = x}} (\min(\mathtt{optH}(i, l, y)) + 1, \mathtt{optH}(i + |\mathtt{st}(l)| + 1, r, z))$, then $\mathtt{optH}(i, n, x)$ is the maximum between $w[i + |\mathtt{st}(l)|]$ and

$$\max_{\substack{y, z : y, z \geq 0, \\ y + z = x - \gamma_{\mathrm{sub}}}} (\min(\mathtt{optH}(i, l, y)) + 1, \mathtt{optH}(i + |\mathtt{st}(l)| + 1, r, z)).$$

Otherwise, $\mathtt{optH}(i, n, x)$ is the latter. Using this recurrence relation, we find all the positions $i \leq |\mathrm{T}|$ in which there exists an integer $x \leq t$ satisfying $\mathtt{optH}(i, \mathtt{r}(\mathbb{CT}(\mathrm{P})), x) \neq -\infty$. This helps us to solve APPROXCARTHAMMING-MATCH more efficiently than the general case.

Theorem 3. *Given a pattern P, a text T and a constant approximate threshold t, we can compute all substrings w of T that yield $\mathbb{CT}(P)$ within t substitutions in $O(|T||P|)$ time.*

6   Conclusions

Cartesian trees are data structures representing patterns focusing on minimalities of integer strings. Recently, researchers studied Cartesian trees from the string pattern matching perspective [1,5,12,15]. We have considered Cartesian trees from the edit-distance viewpoint; given two strings over the integer alphabet, the CARTEDITDIST problem is to compute the minimum edit cost required to transform the first string into a string that has the same Cartesian tree as the second string. Then, the APPROXCARTMATCH problem is to compute all substrings of the first string which has Cartesian tree distance no greater than a given threshold to the second string.

We have devised a poly-time algorithm that solves the APPROXCARTMATCH problem based on our algorithm that solves the CARTEDITDIST problem. We have proposed more efficient solutions for special cases of APPROXCARTMATCH, which specify the cost of each edit operation. We think that the running time of our algorithm for APPROXCARTMATCH can be further improved, which is one of future work. Other open problems include extending the Cartesian tree distance measure to work between languages. It would be also interesting to consider neighborhoods and interiors with respect to the Cartesian tree distance.

Acknowledgements. We wish to thank the referees for the care they put into reading the previous version of this manuscript.

This work was supported by the NRF grant (RS-2023-00208094) and the AI Graduate School Program (No. RS-2020-II201361) funded by the Korean government (MSIT).

References

1. Auvray, B., David, J., Groult, R., Lecroq, T.: Approximate Cartesian tree matching: an approach using swaps. In: Proceedings of the 30th International Symposium on String Processing and Information Retrieval, pp. 49–61 (2023)
2. Biedl, T., et al.: Rollercoasters and caterpillars. In: Proceedings of the 45th International Colloquium on Automata, Languages, and Programming, pp. 18:1–18:15 (2018)
3. Bille, P.: A survey on tree edit distance and related problems. Theoret. Comput. Sci. **337**(1–3), 217–239 (2005)
4. Cho, S., Na, J.C., Park, K., Sim, J.S.: A fast algorithm for order-preserving pattern matching. Inf. Process. Lett. **115**, 397–402 (2015)
5. Faro, S., Lecroq, T., Park, K., Scafiti, S.: On the longest common Cartesian substring problem. Comput. J. **66**(4), 907–923 (2023)
6. Gabow, H.N., Bentley, J.L., Tarjan, R.E.: Scaling and related techniques for geometry problems. In: Proceedings of the 16th Annual ACM Symposium on Theory of Computing, pp. 135–143 (1984)

7. Gawrychowski, P., Manea, F., Serafin, R.: Fast and longest rollercoasters. Algorithmica **84**(4), 1081–1106 (2022)
8. Gawrychowski, P., Uznański, P.: Order-preserving pattern matching with k mismatches. Theoret. Comput. Sci. **638**, 136–144 (2016)
9. Kim, J., et al.: Order-preserving matching. Theoret. Comput. Sci. **525**, 68–79 (2014)
10. Kim, S., Cho, H.: A compact index for Cartesian tree matching. In: Proceedings of the 32nd Annual Symposium on Combinatorial Pattern Matching, pp. 18:1–18:19 (2021)
11. Klein, P.N.: Computing the edit-distance between unrooted ordered trees. In: Proceedings of the 6th Annual European Symposium on Algorithms, pp. 91–102 (1998)
12. Nishimoto, A., Fujisato, N., Nakashima, Y., Inenaga, S.: Position heaps for Cartesian-tree matching on strings and tries. In: Proceedings of the 28th International Symposium on String Processing and Information Retrieval, pp. 241–254 (2021)
13. Oizumi, T., Kai, T., Mieno, T., Inenaga, S., Arimura, H.: Cartesian tree subsequence matching. In: Proceedings of the 33rd Annual Symposium on Combinatorial Pattern Matching, pp. 14:1–14:18 (2022)
14. Park, S.G., Amir, A., Landau, G.M., Park, K.: Cartesian tree matching and indexing. In: Proceedings of the 30th Annual Symposium on Combinatorial Pattern Matching, pp. 16:1–16:14 (2019)
15. Park, S.G., Bataa, M., Amir, A., Landau, G.M., Park, K.: Finding patterns and periods in Cartesian tree matching. Theoret. Comput. Sci. **845**, 181–197 (2020)
16. Sakai, I.: Syntax in universal translation. In: Proceedings of the International Conference on Machine Translation and Applied Language Analysis, pp. 594–608 (1961)
17. Schensted, C.: Longest increasing and decreasing subsequences. Can. J. Math. **13**, 179–191 (1961)
18. Sipser, M.: Introduction to the Theory of Computation, 3rd edition. Cengage Learning (2012)
19. Song, S., Gu, G., Ryu, C., Faro, F., Lecroq, T., Park, K.: Fast algorithms for single and multiple pattern Cartesian tree matching. Theoret. Comput. Sci. **849**, 47–63 (2021)
20. Vuillemin, J.: A unifying look at data structures. Commun. ACM **23**(4), 229–239 (1980)

Deterministic Pushdown Automata with Translucent Input Letters

Martin Kutrib[1], Andreas Malcher[1], Carlo Mereghetti[2]([✉]),
Beatrice Palano[2], Priscilla Raucci[1,2], and Matthias Wendlandt[1]

[1] Institut für Informatik, Universität Giessen, Arndtstr. 2, 35392 Giessen, Germany
{kutrib,andreas.malcher,matthias.wendlandt}@informatik.uni-giessen.de
[2] Dipartimento di Informatica "G. Degli Antoni", Università degli Studi di Milano,
via Celoria 18, 20133 Milan, Italy
{carlo.mereghetti,beatrice.palano,priscilla.raucci}@unimi.it

Abstract. The use of translucent input letters is a variant of a discontinuous input processing in automata. In detail, the automaton performs several sweeps from left to right on the input. Depending on the current state of the automaton, some symbols are visible and can be processed, whereas some other symbols are invisible, and may be processed in another sweep. We also distinguish between the returning and non-returning mode, which differ from the fact that a new sweep starts or not, respectively, immediately after processing a visible input symbol. Here, we investigate deterministic pushdown automata with translucent letters both in the returning and non-returning mode. It turns out that the families of the languages accepted by these two types of devices can properly be ranked between the deterministic context-free languages and the deterministic context-sensitive languages. Moreover, both families are incomparable with the families of Church-Rosser languages, context-free languages, and growing context-sensitive languages. Finally, we study the closure of both language families under the Boolean operations and we obtain for both families the closure under complementation, and the non-closure under union and intersection.

Keywords: Translucent input letters · Deterministic pushdown automata · Returning and non-returning computations · Computational Capacity · Closure Properties

1 Introduction

Automata models with unidirectional motion of the reading head usually process their input in a continuous way. Namely, they read their input strictly from left to right and accept by entering an accepting state usually after having read the input completely. Typical examples are: (non)deterministic, probabilistic, and quantum finite state automata, pushdown automata, queue automata, transducers, one-way multihead automata (see, e.g., [1,4,5,10,11,13]). It is a recent trend to consider also automata models that process their input in a

© The Author(s), under exclusive license to Springer Nature Switzerland AG 2024
J. D. Day and F. Manea (Eds.): DLT 2024, LNCS 14791, pp. 203–217, 2024.
https://doi.org/10.1007/978-3-031-66159-4_15

discontinuous way. An example are *jumping finite automata* introduced in [15], which are classical finite automata with the difference that, after reading an input symbol, the input head can jump to any position inside the remaining input. Thus, the way the input is processed depends on the jumps of the reading head. From [15], it is known that such automata can accept in their deterministic variant even non-context-free languages, such as the language $L = \{\, w \in \{a, b, c\}^* \mid |w|_a = |w|_b = |w|_c \,\}$. The restricted variant of "right one-way jumping" has been considered in [2,7]. Another example of discontinuous computation devices are *input-revolving finite automata*, introduced in [3]. Such automata are able to move or interchange input symbols to the left end or right end of the input. It is shown in [3] that this particular way of discontinuously processing the input is sufficient to accept, for instance, the above-mentioned non-context-free language L. *Restarting automata* have been introduced in [12] as a restricted variant of linear bounded automata. These automata process their input in several sweeps from left to right by shifting a read/write window of size $k \geq 1$ along the input whose contents are rewritten exactly once by a shorter string at some time in every sweep. After rewriting, the automaton restarts in its initial state and the read/write window is reset to start with the leftmost symbol of the remaining input. Thus, the way the input is processed depends on the rewriting steps. There is a large literature on restarting automata that has grown up starting with [12] and many variants, generalizations, and aspects of the original model have been studied. A summary of some of the results may be found in the survey paper [23].

In this paper, we study deterministic pushdown automata *with translucent letters* (DPDAwtl). The key feature of this model is the existence of a translucency function which determines in which states which letters of the input are translucent. Now, a DPDAwtl is a classical deterministic pushdown automaton (DPDA), but it processes its input with respect to the translucency function. This means that the automaton skips the translucent part of the input and processes the first visible input symbol. After processing, in the *returning mode* the input head returns to the left end of the input while, in the *non-returning mode*, the input head continues to process the input according to its updated current state and the corresponding translucent symbols. In both modes, the input head returns to the left end when the right end of the input is reached. The concept of translucent letters has been introduced by Nagy and Otto in [19] for deterministic and nondeterministic finite automata. Such automata have connections with restarting automata, since they are, in the returning mode, equivalent to certain cooperating distributed systems of deterministic restarting automata of window size 1. Deterministic and nondeterministic finite automata with translucent letters are deeply investigated in the literature (see, e.g., [17,20,21,24]). However, many questions are still open. We point out that in a recent contribution [16], the translucency and jumping paradigms have been put together in the model of a deterministic or nondeterministic jumping finite automaton with translucent letters: in particular, an interesting investigation has been pursued, on the minimal amount of jumps for these devices to accept nonregular languages.

Returning pushdown automata with translucent letters have been considered by Nagy and Otto in [18,21] in terms of certain cooperating distributed systems of restarting automata with additional pushdown store. Most likely due to this connection, the pushdown automata are defined by Nagy and Otto in the deterministic and nondeterministic case to work in real time, i.e., without λ-transitions, and to accept by empty pushdown store. Both conditions are essential restrictions for DPDA, since DPDA with λ-transitions and accepting by accepting states are more powerful. Hence, to be closer to the classical definition of DPDA we define here DPDAwtl that accept by accepting states and that may perform λ-transitions. This enables to show that every deterministic context-free language can be accepted by a returning DPDAwtl and that the family of languages accepted by returning DPDAwtl is closed under complementation. We emphasize that both these results do not hold for the DPDAwtl in the definition of Nagy and Otto. We also investigate DPDAwtl working in the non-returning mode, analyzing their computational capabilities and closure properties. This computational model has not been studied in the literature yet.

The paper is organized as follows. In the next section, we give basic definitions and results. In Sect. 3, we study the computational capacity of returning DPDAwtl and non-returning DPDAwtl (nrDPDAwtl). We obtain that the family of the languages accepted by the latter automata model strictly contains the language family characterized by the former. Moreover, both families properly include the deterministic context-free languages and are properly included in the deterministic context-sensitive languages. We also present a non-semi-linear language that is accepted by an nrDPDAwtl. This result could shed some light on an open question in [17], concerning the possibility of accepting non-semilinear languages by non-returning deterministic finite automata with translucent letters. We also investigate the relations among the language families accepted by DPDAwtl and other well-known language families such as Church-Rosser languages, context-free languages, and growing context-sensitive languages. In all these three cases, we obtain incomparability results. Finally, in Sect. 4, we study closure properties under the Boolean operations. It turns out that our language families are closed under complementation, but are neither closed under union nor under intersection.

2   Definitions and Preliminaries

We write Σ^* for the *set of all words* over the finite alphabet Σ. The *empty word* is denoted by λ, and we set $\Sigma^+ = \Sigma^* \setminus \{\lambda\}$. The *reversal* of a word w is denoted by w^R, and for the *length* of w we write $|w|$. The number of *occurrences* of a symbol $a \in \Sigma$ in $w \in \Sigma^*$ is written as $|w|_a$. We use $\subseteq$ for *inclusions* and $\subset$ for *strict inclusions*. We write 2^S for the *power set* of the set S, and $|S|$ for its cardinality. For convenience, we use S_x to denote $S \cup \{x\}$, where S is a set and x is an element not belonging to S. The *complement* of a language $L \subseteq \Sigma^*$ is defined as $\overline{L} = \Sigma^* \setminus L$. We say that two language families $\mathscr{L}_1$ and $\mathscr{L}_2$ are *incomparable* if $\mathscr{L}_1$ is not a subset of $\mathscr{L}_2$ and vice versa.

Pushdown automata with translucent letters are extensions of classical push-down automata that do not have to read their inputs from left to right. Instead, depending on the current state of such devices, some of the input letters may be translucent (invisible). Accordingly, a pushdown automaton with translucent letters either performs a λ-transition without reading an input symbol or reads and processes the first visible input letter from left. Here we are particularly interested in deterministic computations.

Formally, a *deterministic pushdown automaton with translucent letters* (DPDAwtl) is a system $M = \langle Q, \Sigma, \Gamma, q_0, \triangleleft, \bot, \tau, \delta \rangle$, where Q is the finite set of *internal states*, Σ is the finite alphabet of *input symbols*, with $\Sigma \cap Q = \emptyset$, Γ is the finite set of *pushdown symbols*, $q_0 \in Q$ is the *initial state*, $\triangleleft \notin \Sigma$ is the *endmarker*, $\bot \notin \Gamma$ is the *bottom-of-pushdown symbol*, $\tau \colon Q \to 2^\Sigma$ is the *translucency mapping*, and $\delta \colon Q \times (\Sigma \cup \{\lambda, \triangleleft\}) \times \Gamma_\bot \to (Q \times \Gamma_\bot^*) \cup \{\texttt{accept}\}$ is the *partial transition function*. There must never be a choice between using an input symbol and using λ input. So, it is required that for all q in Q and z in $\Gamma_\bot$: if $\delta(q, \lambda, z)$ is defined, then $\delta(q, a, z)$ is undefined for all a in $\Sigma_\triangleleft$. For each state $q \in Q$, the letters from the set $\tau(q)$ are *translucent for q*, that is, in state q the automaton M does not see these letters.

To simplify matters, we require that in any configuration the bottom-of-pushdown symbol appears exactly once at the bottom of the pushdown store, that is, it cannot be deleted. Formally, we require that if $\delta(q, a, z) = (p, \beta)$ then either $z \neq \bot$ and $\beta \in \Gamma^*$, or $z = \bot$ and $\beta = \beta' \bot$, where $\beta' \in \Gamma^*$.

A *configuration* of M is a pair $(qw\triangleleft, \gamma)$ or $\texttt{accept}$, where $q \in Q$ is the *current state*, $w \in \Sigma^*$ is the *remaining part of the input*, and $\gamma \in \Gamma^* \bot$ denotes the current pushdown content, where the leftmost symbol is at the top of the pushdown store. On input w the *initial configuration* is defined to be $(q_0 w \triangleleft, \bot)$.

During the course of its computation, M runs through a sequence of configurations. Being in some configuration $(qw\triangleleft, z\gamma)$, a step of M occurs as follows. First, M first checks whether $\delta(q, \lambda, z)$ is defined. If it is (p, β), then the successor configuration is $(pw\triangleleft, \beta\gamma)$. Otherwise, $\delta(q, \lambda, z)$ is undefined and M determines an input letter by taking the first letter from left that is visible in state q, i.e., if $w = uav$ with $u \in \tau(q)^*$ and $a \notin \tau(q)$, then M takes a. Now, M halts and rejects whenever $\delta(q, a, z)$ is undefined, otherwise it computes $\delta(q, a, z) = (p, \beta)$ and the successor configuration is $(puv\triangleleft, \beta\gamma)$. In the case $w \in \tau(q)^*$, automaton M sees the endmarker $\triangleleft$ and halts and accepts if and only if $\delta(q, \triangleleft, z) = \texttt{accept}$.

One step from a configuration to its *successor configuration* is denoted by $\vdash$ which is specified as follows. Let $p, q \in Q$, $a \in \Sigma$, $u, v, w \in \Sigma^*$, $z \in \Gamma_\bot$ and $\beta\gamma, z\gamma \in \Gamma^* \bot$. We set

1. $(qw\triangleleft, z\gamma) \vdash (pw\triangleleft, \beta\gamma)$, if $(p, \beta) = \delta(q, \lambda, z)$,
2. $(quav\triangleleft, z\gamma) \vdash (puv\triangleleft, \beta\gamma)$, if $u \in \tau(q)^*$, $a \notin \tau(q)$ and $(p, \beta) = \delta(q, a, z)$,
3. $(qw\triangleleft, z\gamma) \vdash (pw\triangleleft, \beta\gamma)$, if $w \in \tau(q)^*$ and $(p, \beta) = \delta(q, \triangleleft, z)$,
4. $(qw\triangleleft, z\gamma) \vdash \texttt{accept}$, if $w \in \tau(q)^*$ and $\texttt{accept} = \delta(q, \triangleleft, z)$.

We denote the reflexive and transitive (resp., transitive) closure of $\vdash$ by $\vdash^*$ (resp., $\vdash^+$). The language accepted by the DPDAwtl M is the set $L(M)$ of those words in Σ^* for which the computation beginning in the initial configuration eventually halts accepting, i.e., $L(M) = \{\, w \in \Sigma^* \mid (q_0 w \triangleleft, \bot) \vdash^+ \texttt{accept}\,\}$.

In general, the family of all languages that are accepted by some device X is denoted by $\mathscr{L}(X)$. The following example clarifies the behavior of DPDAwtl.

Example 1. Let $A = \langle Q, \Sigma, \Gamma, q_0, \triangleleft, \bot, \tau, \delta \rangle$ be a DPDAwtl with $Q = \{q_0, q_1, q_2\}$, $\Sigma = \{a, b, c\}$, $\Gamma = \{c\}$, and the functions τ and δ defined as follows. Let $z \in \Gamma_\bot$.

$$
\begin{aligned}
\tau(q_0) &= \{b\}, & \delta(q_0, a, z) &= (q_1, z), \\
& & \delta(q_0, c, c) &= (q_2, \lambda), \\
& & \delta(q_0, \triangleleft, \bot) &= \texttt{accept}, \\
\tau(q_1) &= \{a\}, & \delta(q_1, b, z) &= (q_0, cz), \\
\tau(q_2) &= \emptyset, & \delta(q_2, c, c) &= (q_2, \lambda), \\
& & \delta(q_2, \triangleleft, \bot) &= \texttt{accept},
\end{aligned}
$$

and where $\delta(q_1, c, z)$, $\delta(q_2, a, z)$ and $\delta(q_2, b, z)$ are not defined. E.g., on the input word $w = abbacc$, A will perform the following computation:

$$
\begin{aligned}
(q_0 abbacc \triangleleft, \bot) &\vdash (q_1 bbacc \triangleleft, \bot) \vdash (q_0 bacc \triangleleft, c\bot) \vdash (q_1 bcc \triangleleft, c\bot) \\
&\vdash (q_0 cc \triangleleft, cc\bot) \vdash (q_2 c \triangleleft, c\bot) \vdash (q_2 \triangleleft, \bot) \vdash \texttt{accept}.
\end{aligned}
$$

It is not hard to see that the language accepted by A is the set $L(A) = \{\, xy \mid x \in \{a, b\}^* \wedge n = |x|_a = |x|_b \wedge y = c^n \,\}$, which cannot be accepted by any DFAwtl, since it does not contain any letter-equivalent regular sublanguage [20]. ∎

In [17], a variant of automata with translucent letters has been introduced. In terms of finite automata, the so-called non-returning mode has been considered. According to this paradigm, after processing a visible letter, the head of the automaton does not return to the left end of the input but it continues from the position of the letter just processed. When the endmarker is reached and the transition on the endmarker yields a new state, then the computation is continued in this state with the head placed at the left of the remaining input. So, here we also consider *deterministic pushdown automata with translucent letters working in the non-returning mode* (nrDPDAwtl). Basically, the definitions and notations of DPDAwtl apply to nrDPDAwtl as well. Let $M = \langle Q, \Sigma, \Gamma, q_0, \triangleleft, \bot, \tau, \delta \rangle$ be an nrDPDAwtl. Now a configuration of M is a pair $(xqw\triangleleft, \gamma)$ or $\texttt{accept}$, where $q \in Q$ is *the current state*, $xw \in \Sigma^*$ is the *remaining part of the input* with w to the right and x to the left of the input head, and $\gamma \in \Gamma^*\bot$ denotes the current pushdown content. The successor configuration given by $\vdash$ is now specified as follows. Let $p, q \in Q$, $a \in \Sigma$, $x, u, v, w \in \Sigma^*$, $z \in \Gamma_\bot$ and $\beta\gamma, z\gamma \in \Gamma^*\bot$. Then:

1. $(xqw\triangleleft, z\gamma) \vdash (xpw\triangleleft, \beta\gamma)$, if $(p, \beta) = \delta(q, \lambda, z)$,
2. $(xquav\triangleleft, z\gamma) \vdash (xupv\triangleleft, \beta\gamma)$, if $u \in \tau(q)^*$, $a \notin \tau(q)$ and $(p, \beta) = \delta(q, a, z)$,
3. $(xqw\triangleleft, z\gamma) \vdash (pxw\triangleleft, \beta\gamma)$, if $w \in \tau(q)^*$ and $(p, \beta) = \delta(q, \triangleleft, z)$,
4. $(xqw\triangleleft, z\gamma) \vdash \texttt{accept}$, if $w \in \tau(q)^*$ and $\texttt{accept} = \delta(q, \triangleleft, z)$.

Sometimes we say that an nrDPDAwtl performs *sweeps*, where a sweep is a sequence of transitions that starts with the input head at the left end of the (remaining) input and ends after the next return move on the endmarker (if it takes place).

Example 2. To exemplify the behavior of nrDPDAwtl, we consider the non-context-free language $L = \{\, x\#y\#z \mid x, y, z \in \{a, c\}^* \wedge |x|_a = |y|_a = |z|_c \,\}$. We are going to show that L can be accepted by the nrDPDAwtl defined as $M = \langle Q, \Sigma, \Gamma, q_0, \lhd, \bot, \tau, \delta \rangle$, where we let $Q = \{q_0, q_1, q_2, q_3, q_4\}$, $\Sigma = \{a, c, \#\}$, $\Gamma = \{a\}$. We explain the translucent function τ and the transition function δ while describing the computation of M. Let $z \in \Gamma_\bot$.

At the beginning of the computation, M starts reading the first block of the input string, i.e. x, and it stores in the pushdown as many symbols as the number of a's. In this phase, M reads (and erases) all characters in x, until the first $\#$ is reached. We set

$$\tau(q_0) = \emptyset, \quad \begin{aligned} \delta(q_0, a, z) &= (q_0, az), \\ \delta(q_0, c, z) &= (q_0, z), \\ \delta(q_0, \#, z) &= (q_1, z). \end{aligned}$$

The automaton then continues the computation by reading the second and the third block of the string, only deleting c's and a's respectively. It reads the right endmarker and it moves the head back to the beginning of the tape. We set

$$\begin{aligned} \tau(q_1) &= \{a\}, \quad & \delta(q_1, c, z) &= (q_1, z), \\ & & \delta(q_1, \#, z) &= (q_2, z), \\ \tau(q_2) &= \{c\}, \quad & \delta(q_2, a, z) &= (q_2, z), \\ & & \delta(q_2, \lhd, z) &= (q_3, z). \end{aligned}$$

In an accepting computation, at this point the configuration of the machine should be of the form $(q_3\, a^n c^n \lhd, a^n)$. It is now easy to check whether the number of a's and c's coincides with the number of symbols stored in the pushdown: for each a in the first block, M deletes one c in the second block, and pops one symbol of the pushdown. The automaton is accepting whenever it reaches the endmarker with empty pushdown and no symbol left on the tape. We set

$$\begin{aligned} \tau(q_3) &= \emptyset, \quad & \delta(q_3, a, z) &= (q_4, z), \\ & & \delta(q_3, \lhd, \bot) &= \texttt{accept}, \\ \tau(q_4) &= \{a\}, \quad & \delta(q_4, c, z) &= (q_5, z), \\ \tau(q_5) &= \{c\}, \quad & \delta(q_5, \lhd, a) &= (q_3, \lambda). \end{aligned}$$

Therefore, the whole computation of M can be considered as being divided into two parts. The first one consists of the first sweep, where $|x|_a$ symbols are stored in the pushdown and not relevant characters in the rest of the string are deleted. On a word $w = x\#y\#z \in L$, this first phase simply writes as $(q_0\, x\#y\#z\lhd, \bot) \vdash^* (q_3\, a^i\, c^j \lhd, a^h \bot)$, where $i = |y|_a$, $j = |z|_c$ and $h = |x|_a$.

In the second part of the computation, the head of M moves back and forth on the tape while deleting symbols in the string and from the pushdown. Note that the block structure of the string $a^n c^n \lhd$ at the beginning of this stage is guaranteed by the earlier computation phase. Also, the configuration $(q_3 \, a^n c^n \lhd, a^n \bot)$ is repeated at the beginning of each sweep in this second part. Beginning with this phase, M repeats the following cycle until the end of the computation:

$$(q_3 \, a^i c^j \lhd, a^h \bot) \vdash (q_4 \, a^{i-1} c^j \lhd, a^h \bot) \quad \vdash (a^{i-1} \, q_5 \, c^{j-1} \lhd, a^h \bot)$$
$$\vdash (q_3 \, a^{i-1} c^{j-1} \lhd, a^{h-1} \bot) \vdash (q_4 \, a^{i-2} c^{j-1} \lhd, a^{h-1} \bot)$$
$$\vdash (a^{i-2} \, q_5 \, c^{j-2} \lhd, a^{h-1} \bot) \vdash \cdots.$$

It is easy to see that the input string is accepted only when $|x|_a = |y|_a = |z|_c$ and the last reached configuration is $(q_3 \lhd, \bot)$. In any other case, M rejects. ∎

3 Computational Capacity

In this section, we are going to consider the computational capacity of DPDAwtl and nrDPDAwtl. To this end, we compare the accepted language families with other well-known language families. The results are summarized in Fig. 1 at the end of this section, page 12.

We first compare the two modes of translucent computation here considered, and show some crucial technical results for further considerations. The first one deals with halting computations. Clearly, a DPDAwtl or an nrDPDAwtl can run into an infinite loop while performing only λ-transitions. The following lemmas show how to overcome this problem. A similar result for classical deterministic pushdown automata is mentioned in [10, Exercise 10.2].

Lemma 3. *For a given DPDAwtl (resp., nrDPDAwtl), an equivalent DPDAwtl (resp., nrDPDAwtl) can effectively be constructed, in which acceptance only takes place after the whole input has been processed.*

Lemma 4. *For a given DPDAwtl (resp., nrDPDAwtl), an equivalent DPDAwtl (resp., nrDPDAwtl) can effectively be constructed, that halts on any input.*

Now, we turn to the relationship between DPDAwtl and nrDPDAwtl. We first provide the following preliminary

Lemma 5. *Let M be a DPDAwtl. Then an equivalent nrDPDAwtl can effectively be constructed.*

Next, we show that the inclusion induced by Lemma 5 is proper. The following result has been shown in [19] for finite automata with translucent letters, and can be directly adapted to translucent deterministic pushdown automata:

Proposition 6. *Let M be a DPDAwtl. Then a DPDA M' can effectively be constructed such that $L(M') \subseteq L(M)$ and $L(M')$ is letter-equivalent to $L(M)$.*

Example 7. An immediate corollary to Proposition 6 is that all languages in $\mathscr{L}$(DPDAwtl) are semi-linear. The language $L_{abc} = \{\, a^n b^n c^n \mid n \geq 0 \,\}$ does not contain a deterministic context-free sub-language that is letter-equivalent to L_{abc} itself. So, L_{abc} cannot be accepted by any DPDAwtl. ∎

Proposition 6 does not hold for the non-returning mode. Moreover, from [17], we have that L_{abc} is accepted even by a deterministic finite automaton with translucent letters working in the non-returning mode. So, trivially, it is accepted by some nrDPDAwtl as well. This, together with Lemma 5 and Example 7, yields

Theorem 8. *The family $\mathscr{L}(\mathsf{DPDAwtl})$ is properly included in $\mathscr{L}(\mathsf{nrDPDAwtl})$.*

The proper inclusion $\mathscr{L}(\mathsf{DPDAwtl}) \subset \mathscr{L}(\mathsf{nrDPDAwtl})$ is witnessed by the semi-linear language L_{abc}. So, an immediate question is whether all languages in $\mathscr{L}(\mathsf{nrDPDAwtl})$ are still semi-linear. The following theorem negatively answers this question by focusing on the language

$$L_{sq} = \{\, a\#a^3\#^2a^5\#^3\cdots\#^k a^{2k+1} \mid k \geq 0 \,\}.$$

Theorem 9. *The family $\mathscr{L}(\mathsf{nrDPDAwtl})$ includes a non-semi-linear language.*

Proof. The language L_{sq} is not semi-linear, the number of a's in its words being a square number. Thus, it remains for us to show that L_{sq} is accepted by some nrDPDAwtl. We define the nrDPDAwtl $M = \langle Q, \{a, \#\}, \Gamma, p_0, \vartriangleleft, \bot, \tau, \delta \rangle$, where $Q = \{p_i \mid 0 \leq i \leq 3\} \cup \{q_i \mid 0 \leq i \leq 5\}$, $\Gamma = \{A\}$, $\tau(p_0) = \tau(p_1) = \emptyset$, $\tau(p_2) = \{\#\}$, $\tau(p_3) = \{a\}$, $\tau(q_0) = \tau(q_1) = \tau(q_2) = \tau(q_4) = \emptyset$, $\tau(q_3) = \{\#\}$, $\tau(q_5) = \{a\}$. Let us now specify the transition function δ. Along the first sweep, M first checks whether the input string begins with an a by deleting it, and then it counts the number of blocks of the form $\#^i a^j$, with $i, j > 0$. To this aim, a symbol A is pushed per each of such blocks encountered. While doing this, each block $\#^i a^j$ is shrunk to $\#^{i-1} a^{j-1}$. To implement this first sweep, we formally define δ as:

(1) $\delta(p_0, a, \bot) = (p_1, \bot)$	(4) $\delta(p_2, a, A) = (p_3, A)$
(2) $\delta(p_1, \vartriangleleft, \bot) = \text{accept}$	(5) $\delta(p_3, \vartriangleleft, A) = (q_0, A)$
(3) $\delta(p_1, \#, \bot) = (p_2, A\bot)$	(6) $\delta(p_3, \#, A) = (p_2, AA)$

After the first sweep, an input string $a\#a^3\#^2a^5\#^3a^7\cdots\#^{k+1}a^{2(k+1)+1} \in L_{sq}$ reduces to $a^2\#a^4\#^2a^6\cdots\#^k a^{2(k+1)}$. In the next sweep, M begins by deleting the first two occurrences of a from the input, while popping A from the pushdown. Then, each block of the form $\#^i a^j$, with $i, j > 0$, shrinks to $\#^{i-1} a^{j-2}$. So, at the end of this sweep, the input string turns out to be reduced to $a^2\#a^4\#^2a^6\cdots\#^{k-1}a^{2k}$. Notice that this last string has the same form of the initial string processed along this sweep, but without the last block $\#^k a^{2(k+1)}$. So, the input form being maintained, this sweep can be iterated, completely consuming the input string and emptying the pushdown if and only if the input string belongs to L_{sq}. To implement these sweeps, we formally define δ as:

(7) $\delta(q_0, a, A) = (q_1, A)$	(11) $\delta(q_3, a, A) = (q_4, A)$
(8) $\delta(q_1, a, A) = (q_2, \lambda)$	(12) $\delta(q_4, a, A) = (q_5, A)$
(9) $\delta(q_2, \vartriangleleft, \bot) = \text{accept}$	(13) $\delta(q_5, \#, A) = (q_3, A)$
(10) $\delta(q_2, \#, A) = (q_3, A)$	(14) $\delta(q_5, \vartriangleleft, A) = (q_0, A)$

As a final observation, we remark that the block-counting performed in the first sweep and memorized in the pushdown ensures that each block $\#^i a^j$ in the input, with $i, j > 0$, does not disappear in the middle of the input string. In fact, if this happened, a block $\#^{i'} a^{j'}$, with $i' > i, j' > j$, occurring before $\#^i a^j$ would exist, thus implying that the input string is not in the language L_{sq}.   □

Remark 10. We quickly notice that the result in Theorem 9 may be seen as a first step towards approaching an open question posed by Mráz and Otto in [17], on the possibility of accepting non-semilinear languages on non-returning translucent deterministic finite automata.

Since our devices are equipped with pushdown stores and are deterministic, it is natural to study relationships with context-free languages. An immediate observation is that the family $\mathscr{L}(\mathsf{DPDAwtl})$ is a proper superset of the family of deterministic context-free languages (DCFL). In fact, up to the endmarker, a DPDAwtl without translucency relation is a deterministic pushdown automaton with endmarker. Since the family DCFL is closed under right quotient by a singleton, the inclusion follows straightforwardly. In addition, Example 1 shows that such an inclusion is proper. This "recognition power lower bound" does not seem far from being tight, in the sense that there are very simple, namely linear context-free languages, that do not even belong to $\mathscr{L}(\mathsf{nrDPDAwtl})$.

Lemma 11. *Let $L \subseteq \{\, b^m \#b^n \mid m, n \geq 0 \,\}$ be a language where, for all $m, n \geq 0$, there exist $x \geq m$ and $y \geq n$ satisfying $b^x \#b^y \in L$. If L is accepted by some nrDPDAwtl, then at least one of the two languages L and L^R is a deterministic context-free language.*

Proof. Let $M = \langle Q, \{b, \#\}, \Gamma, q_0, \lhd, \bot, \tau, \delta \rangle$ be an nrDPDAwtl that accepts L. By Lemma 3, we may assume that M accepts only after having processed the input entirely. We consider accepting computations on inputs $b^m \#b^n$ with m, n large enough. Basically, M has to start the computation by reading some constant number of b's from the prefix b^m of the input. If, in this phase, M reaches a state for which b's and $\#$'s are translucent, then possible λ-transitions can be directly simulated and the transition on the endmarker can be simulated in the finite control of M. If this transition is not accepting, the input head returns to the left of the input, where it currently is, its position does not change. So, we can modify M such that it does not enter any state for which b's and $\#$'s are translucent as long as the input head is at the left of the input. Now, we distinguish among three cases.

CASE 1: M does not reach a state for which b's are translucent while processing the prefix b^m, but reaches a state for which $\#$'s are translucent exactly when the $\#$ appears in the input. In this case, the computation continues by processing the suffix b^n. When M reaches the endmarker, only the $\#$ is left in the input. So, one can effectively construct a classical DPDA M' from M that just simulates M. When the $\#$ appears in the input, M' just reads and memorizes it. When the endmarker appears in the input, M' can simulate the remaining computation of M in its states.

CASE 2: M does not reach a state for which b's are translucent while processing the prefix b^m, and sees the # when it appears. Also in this case one can effectively construct a classical DPDA M' from M that simulates M until the # is processed. Subsequently, the simulation continues. Since the remaining input is b^n, then M' can still continue with the simulation. Even if the b's become translucent, M' knows what to do, since then the endmarker is seen by M and the input head (if it returns at all) returns to its current position at the left end of the input.

By inspecting the nrDPDAwtl M, we can check whether or not conditions for CASE 1 or CASE 2 are fulfilled. In the affirmative, we can effectively construct from M a classical DPDA M' that accepts if and only if M accepts. Thus, we conclude that in these cases L is a deterministic context-free language.

CASE 3: M reaches a state for which b's are translucent after processing some constant number k_1 of b's of the prefix b^m. We know from the preliminary considerations that # is not translucent for this state. So, M reads and processes the #. SUB-CASE 3.1: assume that M continues the computation by processing some constant number k_2 of b's and reaches again a state for which b's are translucent. Then, after reading the endmarker and returning to the left of the input, M sees only b's in the input. Again, from the modifications addressed at the beginning of this proof, we get that from now on the input head remains at the left of the input. So, the initial part of this computation phase looks like

$$(q_0 b^m \# b^n \triangleleft, \bot) \vdash^* (q_1 b^{m-k_1} \# b^n \triangleleft, \gamma_1) \quad \vdash^* (b^{m-k_1} q_2 b^n \triangleleft, \gamma_2)$$
$$\vdash^* (b^{m-k_1} q_3 b^{n-k_2} \triangleleft, \gamma_3) \vdash^* (q_4 b^{m-k_1+n-k_2} \triangleleft, \gamma_4)$$
$$\vdash^* (q_5 b^{m-k_1} \triangleleft, \gamma_5).$$

We design a classical DPDA M' accepting L^R to behave as follows. It starts with the state and pushdown content that M has reached after the constant number of steps that process k_1 symbols b, plus processing the # (this configuration can initially be created by M' with λ-transitions). Then, it directly simulates M until k_2 further b's have been processed. Now, M sees the endmarker and M' memorizes this fact as well as the return of M's input head. After having processed the next $n - k_2$ symbols b, that is, the number of symbols that M' sees until it reaches the #, it simulates the behavior of M on the endmarker. Next, M' processes the #, reads k_1 symbols b, and enters the state in which M would be in the same situation:

$$(q_2 b^n \# b^m \triangleleft, \gamma_2) \vdash^* (q_3 b^{n-k_2} \# b^m \triangleleft, \gamma_3) \vdash^* (q_4' b^{n-k_2} \# b^m \triangleleft, \gamma_4)$$
$$\vdash^* (q_5' \# b^m \triangleleft, \gamma_5) \vdash^* \quad \vdash^* (q_5 b^{m-k_1} \triangleleft, \gamma_5).$$

Therefore, in this case, M' accepts a word if and only if it belongs to L^R. SUB-CASE 3.2: assume that M processes the # and then does not reach a state for which b's are translucent. This sub-case is treated similarly as the SUB-CASE 3.1. The initial part of an accepting computation looks like

$$(q_0 b^m \# b^n \triangleleft, \bot) \vdash^* (q_1 b^{m-k_1} \# b^n \triangleleft, \gamma_1) \vdash^* (b^{m-k_1} q_2 b^n \triangleleft, \gamma_2)$$
$$\vdash^* (b^{m-k_1} q_3 \triangleleft, \gamma_3) \vdash^* \vdash^* (q_4 b^{m-k_1} \triangleleft, \gamma_4).$$

We design a classical DPDA M' for L^R to behave as follows. On the #, M' behaves as M on the endmarker, removing k_1 symbols from the input. The remaining computation is that of M where we know that the input head stays at the left end of the input:

$$(q_2 b^n \# b^m \vartriangleleft, \gamma_2) \vdash^* (q_3 \# b^m \vartriangleleft, \gamma_3) \quad \vdash^* (q_4' b^m \vartriangleleft, \gamma_4)$$
$$\vdash^* (q_4'' b^{m-k_1} \vartriangleleft, \gamma_4) \vdash^* (q_4 b^{m-k_1} \vartriangleleft, \gamma_4).$$

In conclusion, in CASE 3, a classical DPDA M' for L^R can effectively be constructed, thus showing that L^R is a deterministic context-free language. $\square$

Remark 12. A precondition in Lemma 11 states that m and n have to be unbounded at the same time. However, if both are bounded, we have a finite language which is certainly regular. If only one of them is bounded, then we have a unary language concatenated with a finite language. Lemma 16 below shows that each unary language in the family $\mathscr{L}(\mathsf{nrDPDAwtl})$ is semi-linear. This result can straightforwardly be generalized to hold also for the concatenation of a unary language and a finite language. So, the precondition can be relaxed.

Now, we apply Lemma 11 to witness the existence of a context-free language that does not belong to $\mathscr{L}(\mathsf{nrDPDAwtl})$:

Theorem 13. *The families $\mathscr{L}(\mathsf{DPDAwtl})$ and $\mathscr{L}(\mathsf{nrDPDAwtl})$ are both incomparable with the family of context-free languages.*

Proof. Example 1 provides a non-context-free language accepted by a DPDAwtl. So, it remains to show the existence of a context-free language not accepted by any nrDPDAwtl. To this end, we consider the linear context-free language $L = \{ b^n \# b^n \mid n \geq 1 \} \cup \{ b^n \# b^{2n} \mid n \geq 1 \}$. In [9], it is proved that L is not a deterministic context-free language. Since the reversal L^R can be written as $L^R = \{ b^n \# b^n \mid n \geq 1 \} \cup \{ b^n \# b^{\frac{n}{2}} \mid n \geq 2 \text{ even} \}$, the same argument in [9] yields that L^R does not belong to DCFL as well. Therefore, by Lemma 11, we conclude that L is not accepted by any nrDPDAwtl. $\square$

As a "recognition power upper bound" for our devices, we obtain the family of deterministic context-sensitive languages:

Proposition 14. *The family $\mathscr{L}(\mathsf{nrDPDAwtl})$ is properly included in the family of deterministic context-sensitive languages.*

Proof. By Lemma 4, we can assume that the computations of an nrDPDAwtl take a number of steps which is linearly bounded by the input length. This, in turn, implies that the pushdown store cannot grow beyond linear space. Therefore, an nrDPDAwtl can be simulated in linear space by a deterministic one-tape one-head Turing machine, i.e., by a deterministic linear bounded automaton. So, the family $\mathscr{L}(\mathsf{nrDPDAwtl})$ is included in the family of deterministic context-sensitive languages. Since $\mathscr{L}(\mathsf{nrDPDAwtl})$ is incomparable with the family of context-free languages, this inclusion turns out to be proper. $\square$

Let us now turn to another well-known language family. The so-called *growing context-sensitive grammars*, that is, context-sensitive grammars for which each production rule is strictly length-increasing, have been introduced in [8]. The induced language family GCSL lies strictly between CFL and DCSL. This family is interesting in our context, since each growing context-sensitive language is accepted in polynomial time by some one-way auxiliary pushdown automaton with a logarithmic space bound [6].

Theorem 15. *The families $\mathscr{L}(DPDAwtl)$ and $\mathscr{L}(nrDPDAwtl)$ are both incomparable with the family of growing context-sensitive languages.*

Another interesting language family to compare with is the family of *Church-Rosser languages* (CRL), which has been introduced in [14]. It lies strictly between the deterministic context-free and the growing context-sensitive languages. Church-Rosser languages are defined *via* finite, confluent, and length-reducing Thue systems. The family is incomparable with context-free languages [6] and has some neat properties. E.g., Church-Rosser languages parse rapidly in linear time, contain non-semi-linear as well as inherently ambiguous languages [14], are characterized by deterministic automata models [6,22], and contain the deterministic context-free languages as well as their reversals properly [14].

Though, according to Theorem 9, the family $\mathscr{L}(nrDPDAwtl)$ contains non-semi-linear languages, the situation changes when the special case of unary languages is considered.

Lemma 16. *Each unary language in the family $\mathscr{L}(nrDPDAwtl)$ is semi-linear and, thus, regular.*

Theorem 17. *The families $\mathscr{L}(DPDAwtl)$ and $\mathscr{L}(nrDPDAwtl)$ are both incomparable with the family of Church-Rosser languages.*

Proof. According to [14], the family CRL is known to include non-semi-linear unary languages, such as $\{\, a^{2^n} \mid n \geq 0 \,\}$, which do not belong to $\mathscr{L}(nrDPDAwtl)$ by Lemma 16. Conversely, by Theorem 15, there is a language in $\mathscr{L}(DPDAwtl)$ which is not even growing context-sensitive. $\square$

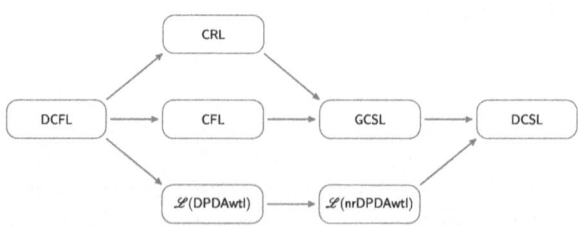

Fig. 1. Relationships between language families. An arrow between families indicates a strict inclusion. Any pair of families not connected by a path is incomparable.

4   Closure Properties

In the following, we will refer to Example 7, according to which the language $L_{abc} = \{\, a^n b^n c^n \mid n \geq 0 \,\}$ is not accepted by any DPDAwtl. Moreover, we will use the fact that language $L = \{\, w \in \{a, b, c\}^* \mid |w|_a = |w|_b = |w|_c \,\}$ can be accepted by a DPDAwtl, since in [20] it is shown that L is accepted by a deterministic finite automaton with translucent letters.

Proposition 18. *The language families* $\mathscr{L}(\mathsf{DPDAwtl})$ *and* $\mathscr{L}(\mathsf{nrDPDAwtl})$ *are closed under complementation, but they are neither closed under union nor under intersection. The language family* $\mathscr{L}(\mathsf{DPDAwtl})$ *is, moreover, neither closed under union with regular languages nor under intersection with regular languages.*

Proof. For complementation, let M be a DPDAwtl which, by Lemma 4, can be assumed to halt on every input. We build the DPDAwtl M' for $\overline{L(M)}$ by obtaining the transition function δ' of M' from δ of M as follows: for any instruction of the form $\delta(q, \lhd, z) = \texttt{accept}$, we let $\delta'(q, \lhd, z)$ undefined; instead, for any undefined instruction $\delta(q, \sigma, z)$, we let $\delta'(q, \sigma, z) = (s, z)$ and $\delta'(s, \lhd, z) = \texttt{accept}$, for a new state s with $\tau'(s) = \Sigma$. Apart from this modification, δ' and τ' coincide with δ and τ, respectively.

For closure under complementation for nrDPDAwtl, we can argue analogously since also an nrDPDAwtl can be assumed to halt on every input by Lemma 4.

As observed, the language $L = \{\, w \in \{a, b, c\}^* \mid |w|_a = |w|_b = |w|_c \,\}$ is accepted by a DPDAwtl. Moreover, the regular language $a^* b^* c^*$ can be accepted by a DPDAwtl as well. We assume that $\mathscr{L}(\mathsf{DPDAwtl})$ is closed under intersection with regular languages. Then, $L \cap a^* b^* c^* = L_{abc}$ belongs to $\mathscr{L}(\mathsf{DPDAwtl})$, and this contradicts Example 7. Since $\mathscr{L}(\mathsf{DPDAwtl})$ and REG are closed under complementation, it follows from the non-closure under intersection with regular languages that $\mathscr{L}(\mathsf{DPDAwtl})$ is not closed under union with regular languages as well. Since $\mathsf{REG} \subseteq \mathscr{L}(\mathsf{DPDAwtl})$, we obtain that $\mathscr{L}(\mathsf{DPDAwtl})$ is neither closed under union nor under intersection.

Finally, we use the language $\{\, b^n \# b^n \mid n \geq 1 \,\} \cup \{\, b^n \# b^{2n} \mid n \geq 1 \,\}$, which does not belong to $\mathscr{L}(\mathsf{nrDPDAwtl})$ by Theorem 13. Since both the languages $\{\, b^n \# b^n \mid n \geq 1 \,\}$ and $\{\, b^n \# b^{2n} \mid n \geq 1 \,\}$ are deterministic context-free, and hence they are both accepted by an nrDPDAwtl, we get the non-closure under union for $\mathscr{L}(\mathsf{nrDPDAwtl})$. Since $\mathscr{L}(\mathsf{nrDPDAwtl})$ is closed under complementation, then $\mathscr{L}(\mathsf{nrDPDAwtl})$ is not closed under intersection, either. □

Acknowledgements. The authors wish to thank the anonymous referees for their valuable and helpful comments.

References

1. Bednárová, Z., Geffert, V., Mereghetti, C., Palano, B.: Boolean language operations on nondeterministic automata with a pushdown of constant height. J. Comput. Sys. Sci. **90**, 99–114 (2017). https://doi.org/10.1016/j.jcss.2017.06.007

2. Beier, S., Holzer, M.: Nondeterministic right one-way jumping finite automata. Inf. Comput. **284**, 104687 (2022). https://doi.org/10.1016/J.IC.2021.104687
3. Bensch, S., Bordihn, H., Holzer, M., Kutrib, M.: On input-revolving deterministic and nondeterministic finite automata. Inf. Comput. **207**, 1140–1155 (2009). https://doi.org/10.1016/j.ic.2009.03.002
4. Bianchi, M.P., Mereghetti, C., Palano, B.: Size lower bounds for quantum automata. Theor. Comput. Sci. **551**, 102–115 (2014). https://doi.org/10.1016/j.tcs.2014.07.004
5. Bianchi, M.P., Mereghetti, C., Palano, B., Pighizzini, G.: On the size of unary probabilistic and nondeterministic automata. Fund. Inform. **112**, 119–135 (2011). https://doi.org/10.3233/FI-2011-583
6. Buntrock, G., Otto, F.: Growing context-sensitive languages and Church-Rosser languages. Inf. Comput. **141**, 1–36 (1998). https://doi.org/10.1006/inco.1997.2681
7. Chigahara, H., Fazekas, S.Z., Yamamura, A.: One-way jumping finite automata. Int. J. Found. Comput. Sci. **27**, 391–405 (2016). https://doi.org/10.1142/S0129054116400165
8. Dahlhaus, E., Warmuth, M.K.: Membership for growing context-sensitive grammars is polynomial. J. Comput. Syst. Sci. **33**, 456–472 (1986). https://doi.org/10.1016/0022-0000(86)90062-0
9. Ginsburg, S., Greibach, S.A.: Deterministic context-free languages. Inf. Control **9**, 620–648 (1966). https://doi.org/10.1016/S0019-9958(66)80019-0
10. Hopcroft, J.E., Ullman, J.D.: Introduction to Automata Theory, Languages, and Computation. Addison-Wesley, Reading (1979)
11. Jakobi, S., Meckel, K., Mereghetti, C., Palano, B.: Queue automata of constant length. In: Jurgensen, H., Reis, R. (eds.) DCFS 2013. LNCS, vol. 8031, pp. 124–135. Springer, Heidelberg (2013). https://doi.org/10.1007/978-3-642-39310-5_13
12. Jančar, P., Mráz, F., Plátek, M., Vogel, J.: Restarting automata. In: Reichel, H. (ed.) FCT 1995. LNCS, vol. 965, pp. 283–292. Springer, Heidelberg (1995). https://doi.org/10.1007/3-540-60249-6_60
13. Kutrib, M., Malcher, A., Mereghetti, C., Palano, B.: Deterministic and nondeterministic iterated uniform finite-state transducers: computational and descriptional power. In: Anselmo, M., Della Vedova, G., Manea, F., Pauly, A. (eds.) CiE 2020. LNCS, vol. 12098, pp. 87–99. Springer, Cham (2020). https://doi.org/10.1007/978-3-030-51466-2_8
14. McNaughton, R., Narendran, P., Otto, F.: Church-Rosser Thue systems and formal languages. J. ACM **35**, 324–344 (1988). https://doi.org/10.1145/42282.42284
15. Meduna, A., Zemek, P.: Jumping finite automata. Int. J. Found. Comput. Sci. **23**, 1555–1578 (2012). https://doi.org/10.1142/S0129054112500244
16. Mitrana, V., Păun, A., Păun, M., Couso, J.R.S.: Jump complexity of finite automata with translucent letters. Theor. Comput. Sci. **992**, 114450 (2024). https://doi.org/10.1016/j.tcs.2024.114450
17. Mráz, F., Otto, F.: Non-returning deterministic and nondeterministic finite automata with translucent letters. RAIRO Theor. Inform. Appl. **57**, 8 (2023). https://doi.org/10.1051/ita/2023009
18. Nagy, B., Otto, F.: CD-systems of stateless deterministic R(1)-automata governed by an external pushdown store. RAIRO Theor. Inform. Appl. **45**, 413–448 (2011). https://doi.org/10.1051/ITA/2011123
19. Nagy, B., Otto, F.: Finite-state acceptors with translucent letters. In: Bel-Enguix, G., Dahl, V., De La Puente, A. (eds.) International Workshop on AI Methods for Interdisciplinary Research in Language and Biology (BILC 2011), pp. 3–13. SciTePress (2011). https://doi.org/10.5220/0003272500030013

20. Nagy, B., Otto, F.: On CD-systems of stateless deterministic R-automata with window size one. J. Comput. Syst. Sci. **78**, 780–806 (2012). https://doi.org/10.1016/J.JCSS.2011.12.009

21. Nagy, B., Otto, F.: Deterministic pushdown-CD-systems of stateless deterministic R(1)-automata. Acta Inform. **50**, 229–255 (2013). https://doi.org/10.1007/S00236-012-0175-X

22. Niemann, G., Otto, F.: The Church-Rosser languages are the deterministic variants of the growing context-sensitive languages. Inf. Comput. **197**, 1–21 (2005). https://doi.org/10.1016/j.ic.2004.09.003

23. Otto, F.: Restarting automata and their relations to the Chomsky hierarchy. In: Ésik, Z., Fülöp, Z. (eds.) DLT 2003. LNCS, vol. 2710, pp. 55–74. Springer, Heidelberg (2003). https://doi.org/10.1007/3-540-45007-6_5

24. Otto, F.: A survey on automata with translucent letters. In: Nagy, B. (ed.) CIAA 2023. LNCS, vol. 14151, pp. 21–50. Springer, Cham (2023). https://doi.org/10.1007/978-3-031-40247-0_2

Network Topologies for Parallel Communicating Finite Automata: Token-Ring and Token-Bus

Philomena Moek$^{(\boxtimes)}$ (iD)

University of Potsdam, Am Neuen Palais 10, 14469 Potsdam, Germany
philomena.moek@uni-potsdam.de
https://www.uni-potsdam.de

Abstract. In this paper, parallel communicating finite automata (abbreviated as PCFA) that communicate by states will be investigated. We define new topologies for them, the Token-Ring and Token-Bus topology. The new topologies have two working modes and will be illustrated with an example. We study the computational power of deterministic Token-Ring and Token-Bus PCFA and prove that they are equal to deterministic one-way multi-head finite automata. The same is proven for their non-deterministic versions. We compare the topologies to each other in their different working modes. It will be derived that non-deterministic versions of Token-Ring and Token-Bus PCFA are strictly more powerful than deterministic ones. In addition it is shown that the language classes of Token-Ring and Token-Bus PCFA are strictly included in the complexity class NL and that all language classes accepted by them are incomparable to the class of (deterministic) (linear) context-free languages and the class of Church-Rosser languages. (The material published in this paper is based on the results presented in the Bachelor's thesis under the same name by Philomena Moek. The thesis was submitted in June of 2023.)

Keywords: parallel communicating finite automata · topologies of automata systems · Token-Ring and Token-Bus PCFA

1 Introduction

Traditionally, computer code is written for sequential machines. However, parallelism and cooperative devices have gained importance. That resulted in the development of new theoretic models to satisfy the new requirements, beginning in the late sixties. A simple way to develop models of cooperating devices is to generalise a single finite automaton to a system of automata.

This idea has been implemented differently in a variety of models: Multi-head finite automata [14] consist of a single finite automaton with multiple reading heads. In [9], two-way multi-head finite automata are introduced, where the reading heads can move in both directions. Multiple reading heads transform pushdown automata into multi-head pushdown automata [8].

© The Author(s), under exclusive license to Springer Nature Switzerland AG 2024
J. D. Day and F. Manea (Eds.): DLT 2024, LNCS 14791, pp. 218–235, 2024.
https://doi.org/10.1007/978-3-031-66159-4_16

Multiprocessor automata [7] showcase a different approach: They consist of a system of finite automata that cooperate in terms of a switching function. This function determines which processor is active in the next step.

Other models introduce explicit communication. This includes communicating finite-state machines [6] where automata are used to model communicating processes. Each pair of communicating processes is connected by a channel, through which they can send and receive messages. Parallel communicating finite automata (PCFA) systems also fall into this category of models. A PCFA(k) consists of a fixed number of k automata that are called *components* which can communicate by requesting the states of other components.

This paper is concerned with parallel communicating finite automata systems that were introduced by Martín-Vide et al. [12] in 2002. Instead of using their definition, we use the adjusted definition by Bordihn et al. [1] from 2008. The results of the 2008 paper were further investigated by looking into decidability problems for PCFA and the infinite hierarchies that are brought about by the number of components [2]. Another field of research is how the amount of communication in parallel communicating automata systems is connected with their other properties, like their computational power [3,4]. Additionally, it has been examined how reversibility of finite automata expands to PCFA systems [5].

In the definitions of Martín-Vide et al. [12] and Bordihn et al. [1], there are four working modes described for PCFA: In the *centralized* mode, only one component can query all other components for their states. In the *non-centralized* mode, every component is allowed to query all of the others. Additionally, it is differentiated between what happens once a component has sent its state. In the *returning* mode, they are set back to their initial state. Whereas in the *non-returning* mode, they remain in the same state.

The centralized and non-centralized working modes intuitively correspond to network topologies. Since the definition of PCFA already yields different network topologies, the aim of this paper is to expand on that.

We first give the definition of parallel communicating finite automata and multi-head automata. Then, the PCFA versions of some common network topologies – the Token-Ring [11] and the Token-Bus [10] topology – are defined.

In Scct. 3, the equivalence of deterministic returning Token-Ring PCFA(k) to deterministic non-returning Token-Ring PCFA(k) and deterministic one-way multi-head automata is proven. The same is done for Token-Bus PCFA. For that sake, we introduce a *cycling-token method* for each topology. This method allows us to transfer a finite data record – the *token* – to each automaton in a set order. We use it for storing states that would otherwise get lost once a component is reset to its initial state in the returning mode. Section 4 introduces some other properties that apply to PCFA and are transferable to Token-Ring/Token-Bus PCFA.

2   Definitions

A PCFA(k), as defined by Bordihn et al. in [1], is a network of $k \in \mathbb{N}$ parallel communicating finite automata, working on a common input tape. We use the notation 2^X to refer to the power set of the set X. The abbreviation "iff" stands for "if and only if".

A PCFA of degree k (PCFA(k) for short), for some positive integer k, is formally defined as a tuple $A = \langle \Sigma, A_1, A_2, ..., A_k, Q, \triangleleft \rangle$, consisting of the set of input symbols Σ, k finite state automata A_1 to A_k, the set of query states ($Q = \{q_1, q_2, ..., q_k\}$), and the end-of-input symbol $\triangleleft$ ($\triangleleft \notin \Sigma$).

Each automaton is a tuple $A_i = \langle S_i, \Sigma, \delta_i, s_{0,i}, F_i \rangle$ for $1 \leq i \leq k$ where S_i is a finite set of states, so that $Q \subseteq \bigcup_{1 \leq i \leq k} S_i$, $s_{0,i}$ is the initial state of the automaton, $F_i \subseteq S_i$ is the set of accepting states and $\delta_i : S_i \times (\Sigma \cup \{\lambda, \triangleleft\}) \rightarrow 2^{S_i}$ is the transition function with λ being the empty word.

The automata A_i are NFAs with a right delimiter where λ-transitions are allowed. They are also referred to as the *components* of A ($1 \leq i \leq k$).

A tuple $(s_1, x_1, s_2, x_2, ..., s_k, x_k)$ is referred to as a *configuration* of the PCFA(k) A. It represents the current state $s_i \in S_i$ of each component A_i of A and the parts of the input that A_i has not read yet. Hence the initial configuration for A and an input word $w \in \Sigma^*$ is $(s_{0,1}, w\triangleleft, s_{0,2}, w\triangleleft, ..., s_{0,k}, w\triangleleft)$. To show the transition from one configuration to another, the successor configuration relation with the symbol $\vdash$ is used. $\vdash^*$ refers to the transitive and reflexive closure of $\vdash$.

To complete a computation step, A undergoes two phases:

The *non-communicating step* results in the configuration change $(s_1, a_1y_1, s_2, a_2y_2, ..., s_k, a_ky_k) \vdash (p_1, z_1, p_2, z_2, ..., p_k, z_k)$ with the condition that $Q \cap \{s_1, s_2, ..., s_k\} = \emptyset$, $a_i \in \Sigma \cup \{\lambda, \triangleleft\}$, $p_i \in \delta_i(s_i, a_i)$ and $z_i = \triangleleft$ for $a_i = \triangleleft$ and $z_i = y_i$ otherwise.

The *communication phase* is entered iff at least one component is in a query state after the end of the last phase. Components communicate by requesting the state of other components. At this point, we have to further distinguish between two communication modes for a PCFA: the returning and non-returning modes.

The returning PCFA is abbreviated as RPCFA. In this mode, components get reset to their initial state after sending their state. In the case of an RPCFA(k), the communication phase is formally defined as:

$(s_1, x_1, s_2, x_2, ..., s_k, x_k) \vdash (p_1, x_1, p_2, x_2, ..., p_k, x_k)$ if for all $1 \leq i \leq k$ where $s_i = q_j$ and $s_j \notin Q$ we have $p_i = s_j$, $p_j = s_{0,j}$ and $p_r = s_r$ for all the other r ($1 \leq j \leq k, 1 \leq r \leq k$).

For non-returning PCFAs (abbreviated as PCFA) a component that sends its state does not get reset. In this case, the following configuration change is obtained: $(s_1, x_1, s_2, x_2, ..., s_k, x_k) \vdash (p_1, x_1, p_2, x_2, ..., p_k, x_k)$ if for all $1 \leq i \leq k$ where $s_i = q_j$ and $s_j \notin Q$ we have $p_i = s_j$ and $p_r = s_r$ for all the other r ($1 \leq j \leq k, 1 \leq r \leq k$).

The steps of the communication phase are repeated until all query states are resolved. If that is not possible, for example because all components are in a query state, the PCFA is in a deadlock. A PCFA halts when the successor configuration is undefined. That can – for example – happen when the PCFA is in a deadlock. The successor configuration is also undefined if the transition function for at least one component is undefined.

Formally, the accepted language of a PCFA A consists of all words $w \in \Sigma^*$, where $(s_{0,1}, w\vartriangleleft, s_{0,2}, w\vartriangleleft, ..., s_{0,k}, w\vartriangleleft) \vdash^* (p_1, a_1 y_1, p_2, a_2 y_2, ..., p_k, a_k y_k)$, so that $p_i \in F_i$ and $\delta(p_i, a_i)$ undefined for some $1 \leq i \leq k$.

The PCFA is referred to as a deterministic PCFA (abbreviated as DPCFA) if all components are deterministic automata. Formally, the criterion that $\delta_i(s, a)$ contains at most one state and is undefined for all $a \in \Sigma$ if $\delta_i(s, \lambda)$ is defined, needs to be fulfilled for all components, all states and all $a \in \Sigma$ $(1 \leq i \leq k)$.

The definition of a PCFA introduced above, corresponds to a fully connected network topology, where each component can directly communicate with every other component. Other network topologies can be emulated by limiting the amount of query states for the components. In this fashion, Martín-Vide et al. [12] introduced a CPCFA – a centralized PCFA, where only one designated component is allowed to query states. Bordihn et al. [1] formally define a CPCFA as a PCFA where $S_i \cap Q = \emptyset$ for $2 \leq i \leq k$.

In this paper, we use the approach of limiting the number of components each component is allowed to query to introduce new topologies. We define: A Token-Ring PCFA(k) is a PCFA(k) A where for each of its components A_i

$$S_i \cap Q = \begin{cases} \{q_{i-1}\} & \text{if } i > 1 \\ \{q_k\} & \text{if } i = 1 \end{cases}$$

for all $1 \leq i \leq k$. If $k = 1$, then $Q = \emptyset$.

Whereas a Token-Bus PCFA(k) is formally defined as a PCFA(k) A where for each of its components A_i

$$S_i \cap Q = \begin{cases} \{q_2\} & \text{if } i = 1 \\ \{q_{k-1}\} & \text{if } i = k \\ \{q_{i-1}, q_{i+1}\} & \text{otherwise} \end{cases}$$

When $k = 1$, then $Q = \emptyset$.

For the sake of completeness, we give the definition of a k-DFA [1]. We define a k-DFA or multi-head automaton as a deterministic finite automaton with k reading heads.

Formally a k-DFA is defined as a system $\langle S, \Sigma, k, \delta, \vartriangleleft, s_0, F \rangle$, where S is the finite state set. Σ is the set of input symbols and $\delta : S \times (\Sigma \cup \{\vartriangleleft\})^k \rightarrow S \times \{0, 1\}^k$ is the partial transition function. $\vartriangleleft \notin \Sigma$ is the end-of-input marker, $s_0 \in S$ is the initial state and $F \subseteq S$ is the set of accepting states.

The heads do not obstruct each other while moving, and can read the same input symbol at the same time. In each step, the automaton uses its current state and the k symbols read by the heads to determine its new state and to decide how to move the heads. That means that the result of the transition function is a $k + 1$-tuple. If the result of the transition function for a specific head is a one, it moves one step to the right. If it is a zero, the head waits in its current position for one time step. At the beginning of a computation, all heads start at the first symbol on the tape.

A k-NFA is the non-deterministic version of the k-DFA, where the partial transition function is generalized to $\delta : S \times (\Sigma \cup \{\triangleleft\})^k \to 2^{S \times \{0,1\}^k}$. The accepted language of a k-DFA or a k-NFA is the set of words w where the automaton starts with $w\triangleleft$ written on the input tape and ends the computation by halting while in an accepting state. A k-DFA and a k-NFA halt when no transition function is defined for the current state.

In the following, $\mathscr{L}(\mathrm{X}(k))$ and $\mathscr{L}(\mathrm{X})$ stand for the family of accepted languages by the type of automaton (or automata system) X.

3   Proof of Equivalence

3.1   Cycling-Token Method for Token-Ring Topology

This chapter introduces the *cycling-token method* for Token-Ring DRPCFA(k)s. The method transfers a finite data record – called the *token* – around a deterministic and returning Token-Ring PCFA(k) system. The token is a state that consists of a tuple of states. That means that the state space stays finite because it is a tuple with a finite domain.

When working in the returning mode, the state of a component is lost once it is reset to its initial state after communicating. The token is a way to preserve that information. For that reason, the cycling-token method helps us prove the equivalence of Token-Ring DRPCFA(k), Token-Ring DPCFA(k) and k-DFA. The idea for the proof, as well as for the cycling-token method, are taken from Bordihn et al. [1]. We have adapted the idea to the new topologies.

The components process the token in the order of their indices, starting from A_1. After being processed by A_k, the token has completed its first round and is handed back to A_1 to begin the next round. The predecessor of a component A_i is A_{i-1} ($2 \leq i \leq k$) and the predecessor of A_1 is A_k. The token that is passed around the Token-Ring DRPCFA(k) is called $t_{i,j}$ ($i \in \mathbb{N}$ and $1 \leq j \leq k$). In the denotation of the token $t_{i,j}$, the i indicates the round the token is in and j refers to the component that last processed the token. The initial token is called $t_{0,1}$. Every transition the components complete in the following is a λ-transition with the sole exception of the processing of the token. Processing the token refers to the component that holds the token changing its contents.

The cycling-token method consists of two phases: In the first phase all the components get "desynchronized". After that, the token can begin to circle the ring in the second phase.

The desynchronization phase can be compared to the setup of a relay race. The runners of one team need to be spread out over the track. If all of the runners start the race at the same time and position, nobody is waiting to take the baton when the first runner arrives at the handover point. Since all components would normally start at the same time, they have to be delayed as well. The higher their index, the more delay is necessary so that their predecessors have enough time to process the token. The desynchronization only needs to be completed once, as the components stay desynchronized throughout the whole process.

Table 1 shows an example of the cycling-token method for a deterministic, returning Token-Ring PCFA with four components. It shows the states of the components at each time step. The results of queries are entered in the rows without a time in the t-column.

A_1 produces the token, but to do that, it first enters the state T_1. T_1 is a placeholder for the token and its index refers to the component that is in possession of it. This placeholder is necessary, because A_1 can only produce and process the token in a non-λ-transition. The placeholder however is produced by A_1 and processed by the other components with a λ-transition. Everything else about passing the placeholder around the ring is identical to how the actual token behaves. Once the placeholder arrives back at A_1 after one round, A_1 replaces it with the token in a non-λ-transition.

Table 1. An example of the cycling-token method for the Token-Ring topology of deterministic returning PCFAs with four components.

t	A_1	A_2	A_3	A_4
1	$s_{0,1}$	$s_{0,2}$	$s_{0,3}$	$s_{0,4}$
2	1	1	1	1
3	2	2	2	2
4	3	3	3	3
5	4	4	4	4
6	T_1	q_1	q_2	q_3
	$s_{0,1}$	$s_{0,2}$	$s_{0,3}$	T_1
7	1	1	1	$1'$
8	2	2	2	$2'$
9	3	3	3	$3'$
10	4	4	4	$4'$
11	T_1	q_1	q_2	$5'$
	$s_{0,1}$	$s_{0,2}$	T_1	
12	1	1	$1'$	$6'$
13	2	2	$2'$	$7'$
14	3	3	$3'$	$8'$
15	4	4	$4'$	$9'$
16	T_1	q_1	$5'$	$10'$
	$s_{0,1}$	T_1		
17	1	T_2	$6'$	$11'$
18	2	T_2	q_2	$12'$
		$s_{0,2}$	T_2	
19	3	1	T_3	$13'$
20	4	2	T_3	q_3
			$s_{0,3}$	T_3
21	T_1	3	1	T_4
22	q_4	4	2	T_4
	T_4			$s_{0,4}$
23	$t_{0,1}$	q_1	3	1
	$s_{0,1}$	$t_{0,1}$		
24	1	$t_{0,2}$	4	2
25	2	$t_{0,2}$	q_2	3
		$s_{0,2}$	$t_{0,2}$	
27	4	2	$t_{0,3}$	q_3
			$s_{0,3}$	$t_{0,3}$
28	T_1	3	1	$t_{0,4}$
29	q_4	4	2	$t_{0,4}$
	$t_{0,4}$			$s_{0,4}$
30	$t_{1,1}$	q_1	3	1
	$s_{0,1}$	$t_{1,1}$		
31	1	$t_{1,2}$	4	2

In the desynchronization phase, all components start by counting to $z = 2(k-2)$. This is the time that all components have to wait in phase two so that the other components have enough time to process the token in the meantime. After counting, A_1 enters the state T_1. All other components now enter their query state to receive the state of their predecessor. At the end of the communication phase when the placeholder T_1 has arrived at the last component A_k, the other components are reset to their initial states. A_k now knows that it needs to wait for the other components to get desynchronized, before it receives the placeholder directly from its predecessor A_{k-1} and can process it. Therefore it starts counting to $(z + 1)(i - 2) + 2(i - 3) + 1$ $(2 \leq i \leq k)$. The i refers to the index of the component that received the placeholder from another component that is not their neighbour – in this case $i = k$. This formula describes how much time needs to pass until the second phase has begun and it is A_k's turn to request the placeholder. A_k can now be considered to be desynchronized.

Now the components A_1 to A_{k-1} start to count to z again. After that – in the example this would be time step 11 in Table 1 – A_1 produces the placeholder again and all other components enter their query state. As a result, A_{k-1} receives the placeholder T_1 and is now desynchronized and starts to count according to the formula given above. This process is repeated until all components up to A_2 are desynchronized.

Component A_2 does not need to be desynchronized. As soon as the placeholder T_1 arrives at A_2 – time step 16 in the example –, it changes to state T_2. Later this will correspond to the processing of the token (see time steps 23 and 24 in Table 1). A_2 changes to state T_2 because it receives the placeholder from its direct predecessor.

A_2 stays in this state until it is queried. In the example, this happens at time step 18. If a component A_3 exists (so $k > 2$), it requests the placeholder after A_2 waited for one time unit. Subsequently, A_3 adapts the index of the placeholder, waits until it is queried and then sends it to A_4. This process is repeated in the same way for the other components. Each component has to wait one time unit while holding the placeholder. This is due to the fact that A_1 can only produce the placeholder after it has counted to z. The other components each wait for one additional time unit with the placeholder in their possession to balance the time A_1 takes to produce it.

Once A_1 has counted to z and entered the T_1 state, it queries A_k for the placeholder (see $t = 22$ in the example). A_1 receives the placeholder T_k. It now knows to produce the first token $t_{0,1}$ in a non-λ-transition. After that the other components proceed as they did with the placeholder. The only difference is that they each process the token in a non-λ-transition. Once the first lap has been completed, the second one can begin.

It can be observed that this method works for an arbitrary number of components as long as there are at least two. If there is a different number of components in the ring, the method still works, as long as the components wait long enough for the desynchronization and count to the correct z. The formulas given above can be used to calculate those values for any k components ($k \geq 2$).

3.2 Deterministic Returning vs Non-returning Token-Ring Topology

Bordihn et al. [1] proved that the three language families $\mathscr{L}(\text{DPCFA})$, $\mathscr{L}(\text{DRPCFA})$ and $\mathscr{L}(k\text{-DFA})$ are equal. The goal of this section is to prove the same equivalences as mentioned above for Token-Ring PCFA(k) in an analogous way. To do that, we construct a cyclic proof.

Proposition 1. *For all $k \geq 1$,*
$\mathscr{L}(\text{Token-Ring DPCFA}(k)) \subseteq \mathscr{L}(\text{Token-Ring DRPCFA}(k))$.

Proof. To prove Proposition 1, a Token-Ring DRPCFA(k)
$A = \langle \Sigma, A_1, A_2, ..., A_k, Q, \triangleleft \rangle$ is constructed, that simulates a DPCFA(k) A'. In the case of $k = 1$, the two are trivially equal. With only one component there is no communication possible. If $k > 1$, the cycling-token method from Sect. 3.1 can be utilised. We use the token as a finite data buffer. It saves the current states of all components of A as well as the last component that processed it. Each component that receives the token processes it. To process the token means that A_i reads the state s of the ith component off the token ($1 \leq i \leq k$, $s \in S_i$). A_i simulates the transition that it would perform if it read its current input while in state s.

A_1 simulates the communication phase by resolving all query states within the token, in addition to processing it like the others.

In a non-returning Token-Ring PCFA(k), all components make their transition at the same time. Here, if two components have an undefined transition function in the same round, it can lead to problems. If only one of them is in an accepting state, the order in which these transitions are executed can change whether the word is accepted or not. To solve this, two new states s_r and s_a are introduced. A component immediately shifts to a new state whenever it ends up in a state where the transition function is undefined. The component enters s_a if it was in an accepting state. If not, it transitions to s_r.

Consequently, A_1 is able to anticipate that A will halt. If there are any components that are in state s_r or s_a at the beginning of a round, A terminates. In the case that there is at least one component in state s_a, A_1 shifts to an accepting state and no further transition is defined. Otherwise, if there is no component in state s_a but one or more components in the state s_r, A_1 changes to a non-accepting state with an undefined transition function. □

Proposition 2. *For all $k \geq 1$, $\mathscr{L}(k\text{-DFA}) \subseteq \mathscr{L}(\text{Token-Ring DPCFA}(k))$.*

Proof. As above, in the case of $k = 1$, the two are trivially equal. In order to get a proof for $k > 1$ for Proposition 2, a Token-Ring DPCFA(k) $A = \langle \Sigma, A_1, A_2, ..., A_k, Q, \triangleleft \rangle$ is constructed that simulates a k-DFA A'. To do that, we utilise the cycling-token method for non-returning Token-Ring DPCFA(k). Bordihn et al. [1] already introduced a cycling-token method for non-returning fully-connected DPCFA(k). Incidentally, the same method can be reused for Token-Ring DPCFA(k). The contents of Table 4 (see Appendix B) are taken from their paper (Figure 3 in their paper, some states are renamed) and illustrate the functionality of the non-returning version of the cycling-token method.

The token is created by A_1 after it leaves its initial state. To break the synchronisation, the other components count until the token has been processed by their predecessor. So every component A_i $(2 \leq i \leq k)$ has to count to $(i-2)'$. After that, the components enter their query state q_{i-1} and receive the processed token from their predecessor. Then they process it in one time step and send it to the next component. As before, the token being processed is the only step that a component performs as a non-λ-step. After a component forwards the token, it counts to $k-2$ and then requests the token again. With this, the behaviour of the Token-Ring DPCFA(k) becomes cyclic.

The cycling-token method is used to store three things in the token: The current state of A', the symbols read by each of the heads of A' and the moves of the heads that resulted from the last transition. So, at the beginning of the computation, the token saves: the initial state of A', arbitrary but fixed starting symbols for each head and the instruction for each head to move to the right. In addition to that, the token saves the component that last processed it.

When a component receives the token, it processes it. It moves its head according to what is saved in the token. If the token requests the component's head not to move (i.e. a zero is saved), it performs a λ-transition and transmits the token without any changes. If the token instead requests the component's head to move (i.e. a one is saved), it moves its head one step to the right. Afterwards, the component overwrites the input symbol that is stored in the token for the component, and sends the token on its way.

A_1 processes the token in the same way as the other components. Additionally, it simulates the state change of the multi-head automaton A'. Based on the current state and the current input symbols, A_1 computes the next state of A' and stores its next head movements and writes it to the token. This way, the first transition of A' is simulated at the beginning of the second round of the token.

Lastly, the constructed DPCFA(k) needs to halt exactly when A' halts. Therefore A_1 judges whether A' halts in a certain step. A_1 continues to send the token if A' does not halt. If A' halts, so does A_1. If A' is in an accepting state, so is A_1.   □

Theorem 1. *For all $k \geq 1$, the families $\mathscr{L}$(Token-Ring DRPCFA(k)), $\mathscr{L}$(Token-Ring DPCFA(k)) and $\mathscr{L}$(k-DFA) are equal.*

Proof. To prove Theorem 1, we give a construction that is specifically tailored to the topology. To do that, we construct a k-DFA A' that is equivalent to a Token-Ring DRPCFA(k) $A = \langle \Sigma, A_1, A_2, ..., A_k, Q, \triangleleft \rangle$. Doing that shows that $\mathscr{L}$(Token-Ring DRPCFA(k)) $\subseteq \mathscr{L}$(k-DFA). Together with Propositions 1 and 2, we will have completed a cyclic proof of the equivalence of all three automata systems.

A' simulates the states of the components of the Token-Ring DRPCFA(k) by encapsulating them in its own state. So the state of A' is a tuple $(s_1, s_2, ..., s_k)$, where all states s_i of A are condensed into a single state of A' ($1 \leq i \leq k$). A' transitions from the state $(s_1, s_2, ..., s_k)$ to $(p_1, p_2, ..., p_k)$ iff $p_i \in \delta_i(s_i, x_i)$ for every A_i with x_i being the input symbol read by the ith component of A. In other words, A' simulates the transitions of the components of A. Like this, all components of A are replicated at the same time by A'. If at least one component of A is in a query state, A' imitates the communication phase to resolve the query states. If A halts, so does A', because there is one component that has no defined transition function. If at least one "substate" of A' is an accepting state of A, A' is also in an accepting state.   □

3.3 Cycling-Token Method for Token-Bus Topology

The version of the cycling-token method for Token-Bus DRPCFA(k)s behaves according to the following rules:

After A_1 generates the token, the next component to process it is A_k and after that A_{k-1} and so on. The "predecessor" of a component A_i is A_{i+1} for $1 \leq i \leq k-1$. The predecessor of A_k is A_1. With the exception of the step in which components process the token, all transitions in the following are λ-transitions. After leaving their initial state, all A_i ($2 \leq i \leq k$) change into the query state q_{i-1}. The first component needs to produce the token, so it changes to the state $t_{0,1}$. The indices of the token $t_{i,j}$ mean the same as before. If a component receives a token that was last processed by its predecessor, it processes the token. A component A_i ($i > 2$) can also receive a token that was last processed by the neighbour that is *not* its predecessor. If that happens, A_i waits for $i-3$ time steps and then changes to the state q_{i-1}. A component A_i ($i \neq k$) in state $t_{0,1}$ assumes the state q_{i+1} afterwards.

This method works for any number of components, as long as there are at least two. Table 2 displays an example of this method with five components.

Table 2. An example of the cycling-token method for the Token-Bus topology of deterministic returning PCFAs with five components.

t	A_1	A_2	A_3	A_4	A_5
1	$s_{0,1}$	$s_{0,2}$	$s_{0,3}$	$s_{0,4}$	$s_{0,5}$
2	$t_{0,1}$	q_1	q_2	q_3	q_4
	$s_{0,1}$	$s_{0,2}$	$s_{0,3}$	$s_{0,4}$	$t_{0,1}$
3	$t_{0,1}$	q_1	q_2	q_3	$t_{0,5}$
	$s_{0,1}$	$s_{0,2}$	$s_{0,3}$	$t_{0,1}$	
4	$t_{0,1}$	q_1	q_2	q_5	$t_{0,5}$
	$s_{0,1}$	$s_{0,2}$	$t_{0,1}$	$t_{0,5}$	$s_{0,5}$
5	$t_{0,1}$	q_1	q_4	$t_{0,4}$	q_4
	$s_{0,1}$	$t_{0,1}$	$t_{0,4}$	$s_{0,4}$	$t_{0,4}$
6	$t_{0,1}$	q_3	$t_{0,3}$	q_3	1
		$t_{0,3}$	$s_{0,3}$	$t_{0,3}$	
7	q_2	$t_{0,2}$	q_2	1	2
	$t_{0,2}$	$s_{0,2}$	$t_{0,2}$		
8	$t_{1,1}$	q_1	q_2	q_3	q_4
	$s_{0,1}$	$s_{0,2}$	$s_{0,3}$	$s_{0,4}$	$t_{1,1}$
9	$t_{0,1}$	q_1	q_2	q_3	$t_{1,5}$
	$s_{0,1}$	$s_{0,2}$	$s_{0,3}$	$t_{0,1}$	

The token takes $k+1$ time steps to complete one round. The behaviour of the components is cyclic. Since the automata are deterministic and, after completing one round, arrive back at the beginning of the round, they continue passing the token around.

3.4 Deterministic Returning vs Non-returning Token-Bus Topology

Proposition 3. *For all $k \geq 1$, $\mathscr{L}$(Token-Bus DPCFA(k)) $\subseteq$ $\mathscr{L}$(Token-Bus DRPCFA(k)).*

Proof (Sketch). Analogously to the proof of Proposition 1, to prove Proposition 3, we need to construct a Token-Bus DRPCFA(k) that is equivalent to a Token-Bus DPCFA(k). To do that we use the cycling-token method for Token-Bus DRPCFA(k) as introduced in the last section. This method is used in the same way we use the cycling-token method for Token-Ring DRPCFA(k) in the proof of Proposition 1. A complete proof can be found in Appendix A.   □

Proposition 4. *For all $k \geq 1$, $\mathscr{L}$(k-DFA) $\subseteq$ $\mathscr{L}$(Token-Bus DPCFA(k)).*

Proof (Sketch). In the case of $k = 1$, they are trivially equal.

For the sake of proving Proposition 4 for $k > 1$, we introduce a cycling-token method for Token-Bus DPCFA(k). An example of this cycling-token method for six components can be found in Table 3.

Table 3. An example of the cycling-token method for deterministic non-returning Token-Bus PCFA(k) consisting of six components.

t	A_1	A_2	A_3	A_4	A_5	A_6
1	$s_{0,1}$	$s_{0,2}$	$s_{0,3}$	$s_{0,4}$	$s_{0,5}$	$s_{0,6}$
2	$t_{0,1}$	q_1 $t_{0,1}$	q_2 $t_{0,1}$	q_3 $t_{0,1}$	q_4 $t_{0,1}$	q_5 $t_{0,1}$
3	1	1	1	1	q_6 $t_{0,6}$	$t_{0,6}$
4	2	2	2	q_5 $t_{0,5}$	$t_{0,5}$	$1'$
5	3	3 q_3	q_4 $t_{0,4}$	$t_{0,4}$	$1'$	$2'$
6	4	q_3 $t_{0,3}$	$t_{0,3}$	$1'$	$2'$	$3'$
7	q_2 $t_{0,2}$	$t_{0,2}$	$1'$	$2'$	$3'$	$4'$
8	$t_{1,1}$	q_1 $t_{0,1}$	q_2 $t_{0,1}$	q_3 $t_{0,1}$	q_4 $t_{0,1}$	q_5 $t_{0,1}$

With the exception of processing the token, all transitions in the following are λ-transitions. At time step two, A_1 generates the initial token and all other components A_i enter the query state q_{i-1} ($2 \leq i \leq k$). As before, the components process the token in the opposite order of what their numbers suggest.

After the communication phase of time step two, all components are in state $t_{0,1}$. A_k needs to process the token since it came from its predecessor. In step three, A_{k-1} requests the token from A_k. All components A_i now count to $(k - 2) - (i - 1)$ ($1 \leq i \leq k - 1$). Afterwards, the components enter the query state q_{i+1} to receive the processed token from their right neighbour.

Directly after a component A_j ($2 \leq j \leq k$) sends a token to its left neighbour, it starts to count to $(j - 2)'$. Once a component has finished counting, it enters the query state q_{j-1}.

The full proof of Proposition 4 can be found in Appendix A. It is analogous to the proof of Proposition 2 in the way the specific cycling-token method is utilised.   □

Theorem 2. *For all* $k \geq 1$, *the families* $\mathscr{L}(\text{Token-Bus DRPCFA}(k))$, $\mathscr{L}(\text{Token-Bus DPCFA}(k))$ *and* $\mathscr{L}(k\text{-DFA})$ *are equal.*

Proof. To show Theorem 2, we give a construction that is specifically tailored to the topology. We do that by constructing a deterministic finite k-head automaton A', that emulates a Token-Bus DRPCFA(k) $A = \langle \Sigma, A_1, A_2, ..., A_k, Q, \triangleleft \rangle$. This concludes the cyclic proof of Proposition 3, 4 and Theorem 2.

The state of A' contains all states of the components of A. Due to this, the state of A' is a tuple that stores the state of component A'_i in the ith place. A' can only change from a state $(s_1, s_2, ..., s_k)$ to $(p_1, p_2, ..., p_k)$ if $\forall p_i.\ p_i \in \delta_i(s_i, x_i)$. The x_i refers to the input read by the ith reading head, so $x_i \in \Sigma$ and $s_i, p_i \in S_i$. Additionally $1 \leq i \leq k$. If at least one component of A is in a query state, A' imitates the communication phase to resolve the query states.

A state $(s_1, s_2, ..., s_k)$ where at least one s_i is an accepting state, is an accepting state. If at least one component A_i has no defined transition for the current input symbol of the ith head, the transition function of A' is undefined. The transition function δ_i for a component A_i in state s_i can be undefined for the input currently read by the ith head of A'. If that is the case, the transition function of A' is also undefined if the ith position of its state is s_i.   □

Now that the cyclic proofs are concluded, we deduce that the following corollary is also correct (A proof for Corollary 1 can be found in Appendix A):

Corollary 1. *For all* $k \geq 1$, *the families* $\mathscr{L}(\text{DPCFA}(k))$, $\mathscr{L}(\text{Token-Ring DPCFA}(k))$, $\mathscr{L}(\text{Token-Bus DPCFA}(k))$ *and* $\mathscr{L}(k\text{-DFA})$ *are equal. For all* $k \geq 1$, *the families* $\mathscr{L}(PCFA(k))$, $\mathscr{L}(\text{Token-Ring PCFA}(k))$, $\mathscr{L}(\text{Token-Bus PCFA}(k))$ *and* $\mathscr{L}(k\text{-NFA})$ *are also equal.*

4   Other Properties of Token-Ring and Token-Bus PCFA Systems

Bordihn et al. [1] prove some other characteristics of parallel communicating finite automata in their work. Due to the equivalences of Corollary 1, many of those characteristics are transferable to deterministic Token-Ring and Token-Bus PCFA systems with few adaptations. Additionally, if a proof is based on constructing a PCFA system with two components, the proof is transferable to Token-Ring or Token-Bus systems because they have the same topology. Proofs for the following Theorems can be found in the Appendix in Sect. A.

Theorem 3. *Non-deterministic* Token-Ring *and* Token-Bus PCFA *in the returning or non-returning mode are strictly more powerful than their deterministic versions.*

Theorem 4. *The strict inclusion* $\mathscr{L}$(Token-Ring/-Bus PCFA) $\subset$ **NL** *is valid.*

Theorem 5. *All language classes accepted by* Token-Ring/-Bus PCFA *are incomparable with the class of (deterministic) (linear) context-free languages.*

Theorem 6. *All language classes accepted by* Token-Ring *and* Token-Bus PCFA *are incomparable with the class of Church-Rosser languages.*

5   Outlook

There are still many questions open for investigation. Many other properties of PCFA have been investigated that could also be studied for Token-Ring and Token-Bus PCFA. One example of that is the question of how to measure the amount of communication that is necessary to solve problems. Additionally it was also researched how PCFA handle the decidability status of problems as well as the reversibility of PCFA – both of those can be explored as well for Token-Ring and Token-Bus PCFA.

Last but not least, looking into additional topologies and their computational power could turn out to be worthwhile. A topology that springs to mind is a star topology with a server in the middle. In that case, every component can only query that one component, which was chosen to be the server.

Acknowledgments. Thanks to Henning Bordihn for supporting me to publish my Bachelor thesis.

A   Appendix: Proofs

A.1   Proof of Proposition 3

Proof. Analogously to the proof of Proposition 1, to prove Proposition 3, we construct a Token-Bus DRPCFA(k) $A = \langle \Sigma, A_1, A_2, ..., A_k, Q, \lhd \rangle$ that is equivalent to a Token-Bus DPCFA(k). We use the cycling-token method for Token-Bus DRPCFA(k) as introduced in the last section.

However, the cycling-token method only works for $k > 1$. In the case of $k = 1$, the two are trivially equal.

The token is used as before: It buffers the states of all components and the component that last processed it. Every component that receives the token processes it. That means component A_i updates the ith state saved in the token based on its transition function and the input symbol it reads. A_1 is responsible for producing the token in the first step and also for resolving all query states.

Similarly to the proof of Proposition 1, the components now work sequentially instead of in parallel. To arrange the accepting state in a suitable manner, two new states s_r and s_a are defined. Each time the transition function for a component is undefined, we now define the following transition: If the component is in an accepting state, it changes its state representation in the token to s_a, otherwise to s_r. For those two states the transition function is undefined.

This way, A_1 knows that the system halts, if any of the components are in state s_a or s_r. If at least one component is in state s_a, A_1 shifts to an accepting state and there is no transition out of that state defined. Otherwise, if no component is in state s_a, but at least one component in the state s_r, A_1 changes to a non-accepting state with an undefined transition function.   □

A.2   Proof of Proposition 4

Proof. We use the introduced cycling-token method to simulate a k-DFA A' with a Token-Bus DPCFA(k) $A = \langle \Sigma, A_1, A_2, ..., A_k, Q, \lhd \rangle$. Analogously to the last proofs, the token saves the current input symbol for all k heads and the instruction for moving (a zero for not moving and a one for moving one step to the right). Additionally, the token stores the state of A'.

The token is processed by all components of A. That includes reading the move-instruction for the ith head off the token. If it is a zero, the component performs a λ-transition and sends the token unchanged to the next component while not moving. If, however, the arriving token saves a one, the component reads the next input symbol. It overwrites the symbol saved for the ith component with the one it just read.

The first component of A – called A_1 – simulates the state transitions of the k-DFA. A_1 overwrites the state of A' stored in the token with the new state based on the input symbols stored in the token. Additionally it stores the moves that the heads should perform in the next round.

The initial token is created with arbitrary but fixed input symbols, the instruction for every head to move to the right and the initial state of A'. Once every component performed the instructed move, the input symbols in the token are the first symbol of the input word.

A_1 also makes sure that A halts iff A' halts. A_1 halts when A_1 receives the token and the information on the token would make A' halt. The token does not get passed on. The states of A_1 are defined so that A_1 halts in an accepting state only if A' also halts in an accepting state. Whenever A_1 receives a token and A' does not halt with the corresponding state and input, A_1 processes the token and sends it to the next component.   □

A.3   Proof of Corollary 1

Proof. We know from Bordihn et al. [1], that the language families $\mathscr{L}(DPCFA(k))$ and $\mathscr{L}(k\text{-DFA})$ are equal.

Theorems 1 and 2 show that both $\mathscr{L}(\text{Token-Ring DPCFA}(k))$ as well as $\mathscr{L}(\text{Token-Bus DPCFA}(k))$ are also equal to $\mathscr{L}(k\text{-DFA})$. Therefore all four language families are equal.

Now it remains to be shown that the second sentence is also correct. It is shown by Martín-Vide et al. [12] that the language family $\mathscr{L}((R)PCFA(k))$ coincides with the family $\mathscr{L}(k\text{-NFA})$ in the non-centralized case based on the mode of acceptance introduced by Martín-Vide et al. Bordihn et al. [1] argue that the same equivalence is valid for the mode of acceptance introduced by them. In this paper, we also use the mode of acceptance introduced by Bordihn et al. We use their arguments to show that the family $\mathscr{L}(\text{Token-Ring/-Bus PCFA}(k))$ is also equivalent to the family $\mathscr{L}(k\text{-NFA})$: We showed in Theorem 1 and 2 that the equivalence described above is valid in the deterministic case. It can be observed that an analogous proof can be constructed in the non-deterministic case. Therefore, $\mathscr{L}(\text{Token-Ring/-Bus PCFA}(k)) = \mathscr{L}(k\text{-NFA})$ is also true.   □

A.4   Proof of Theorem 3

Proof. First of all, Bordihn et al. [1] prove that the non-deterministic PCFA are strictly more powerful than deterministic PCFA in all working modes. To do that, it is first shown that the language $L_{nopal} = \{\$\} \cdot \{w \mid w \in \{a, b, c\}^* \text{ and } w \neq w^R\}$ is an element of $\mathscr{L}(CPCFA)$ and of $\mathscr{L}(RCPCFA)$. This is done by constructing a CPCFA(2) and an RCPCFA(2) that accept L_{nopal} respectively. The already constructed automata systems in the paper of Bordihn et al. are already Token-Ring and Token-Bus PCFA. Therefore we can use the same transition function for the construction.

It remains to be shown that deterministic versions of Token-Ring and Token-Bus PCFA do not accept the language. Bordihn et al. use the equivalence of DPCFA(k) to k-DFA to prove that. Since Token-Ring and Token-Bus D(R)PCFA(k) are also equivalent to k-DFA (as shown in Sect. 3), this proof also extends to them.

In conclusion, the following strict inclusions are true:

1. $\mathscr{L}(\text{Token-Ring/-Bus DRPCFA}) \subset \mathscr{L}(\text{Token-Ring/-Bus RPCFA})$
2. $\mathscr{L}(\text{Token-Ring/-Bus DPCFA}) \subset \mathscr{L}(\text{Token-Ring/-Bus PCFA})$

□

A.5   Proof of Theorem 4

Proof. It is proven in [1] that $\mathscr{L}(\text{PCFA})$ is strictly included in the complexity class **NL**. **NL** is characterized by two-way k-NFA [15]. Bordihn et al. give an example language $L = \{ww^R \mid w \in \{a, b, c\}^+\}$ that is included in **NL**, because it is accepted by some 2-NFA. However, they argue that it is not accepted by any one-way k-NFA ($k \geq 1$). Because of that, L also does not belong to $\mathscr{L}(\text{Token-Ring/Token-Bus PCFA})$. □

A.6   Proof of Theorem 5

Proof. Bordihn et al. use the marked mirror language $\{w\$w^R \mid w \in \{a, b\}^+\}$ and the marked copy language $\{w\$w \mid w \in \{a, b\}^+\}$ to separate $\mathscr{L}(\text{PCFA})$ from the class of (deterministic) (linear) context-free languages.

The marked mirror language is deterministic linear context free. At the same time, it is not accepted by any k-NFAs. Since we argued that $\mathscr{L}(k\text{-NFA}) = \mathscr{L}(\text{Token-Ring/-Bus PCFA}(k))$ (see Corollary 1), the marked mirror language is also not accepted by any deterministic Token-Ring or Token-Bus PCFA.

The marked copy language, on the other hand, is accepted by DRPCFA (as shown in [1]). That means that it is also accepted by Token-Ring and Token-Bus DRPCFA, since the language families described by both are equal to $\mathscr{L}(k\text{-DFA})$. However it does not belong to the context free languages. □

A.7   Proof of Theorem 6

Proof. The language $\{a^{2^n} \mid n \geq 1\}$ is called the unary language and belongs to the Church-Rosser languages [13]. It is not an element of the language family $\mathscr{L}(\text{PCFA}(k))$ (see [1]) and thus also not an element of $\mathscr{L}(\text{Token-Ring/-Bus PCFA}(k))$ since they are all equal to $\mathscr{L}(k\text{-NFA})$.

On the other hand, the marked copy language $\{w\$w \mid w \in \{a, b\}^+\}$, belongs to $\mathscr{L}(\text{Token-Ring/-Bus DRPCFA})$ (see above). It is, however, not a Church-Rosser language [13]. □

B   Appendix: Tables

Table 4. The cycling-token method for deterministic non-returning Token-Ring PCFA(k).

t	A_1	A_2	A_3	A_4	$\cdots$	A_k
1	$s_{0,1}$	$s_{0,2}$	$s_{0,3}$	$s_{0,4}$	$\cdots$	$s_{0,k}$
2	$t_{0,1}$	q_1 $t_{0,1}$	$1'$	$1'$	$\cdots$	$1'$
3	1	$t_{0,2}$	q_2 $t_{0,2}$	$2'$	$\cdots$	$2'$
4	2	1	$t_{0,3}$	q_3 $t_{0,3}$	$\cdots$	$3'$
$\vdots$	$\vdots$	$\vdots$	$\vdots$	$\vdots$	$\vdots$	$\vdots$
$k-1$	$k-2$	$k-3$	$k-4$	$k-5$	$\cdots$	q_{k-1} $t_{0,k-1}$
k	q_k $t_{0,k}$	$k-2$	$k-3$	$k-4$	$\cdots$	$t_{0,k}$
$k+1$	$t_{1,1}$ $t_{1,1}$	q_1	$k-2$	$k-3$	$\cdots$	1

References

1. Bordihn, H., Kutrib, M., Malcher, A.: On the computational capacity of parallel communicating finite automata. In: Ito, M., Toyama, M. (eds.) DLT 2008. LNCS, vol. 5257, pp. 146–157. Springer, Heidelberg (2008). https://doi.org/10.1007/978-3-540-85780-8_11
2. Bordihn, H., Kutrib, M., Malcher, A.: Undecidability and hierarchy results for parallel communicating finite automata. In: Gao, Y., Lu, H., Seki, S., Yu, S. (eds.) DLT 2010. LNCS, vol. 6224, pp. 88–99. Springer, Heidelberg (2010). https://doi.org/10.1007/978-3-642-14455-4_10
3. Bordihn, H., Kutrib, M., Malcher, A.: Measuring communication in parallel communicating finite automata. Electron. Proc. Theor. Comput. Sci. **151** (2014). https://doi.org/10.4204/EPTCS.151.8
4. Bordihn, H., Kutrib, M., Malcher, A.: Returning parallel communicating finite automata with communication bounds: hierarchies, decidabilities, and undecidabilities. Int. J. Found. Comput. Sci. **26**, 1101–1126 (2015). https://doi.org/10.1142/S0129054115400146
5. Bordihn, H., Vaszil, G.: Reversible parallel communicating finite automata systems. Acta Inf. **58**(4), 263–279 (2021). https://doi.org/10.1007/s00236-021-00396-9

6. Brand, D., Zafiropulo, P.: On communicating finite-state machines. J. ACM **30**(2), 323–342 (1983). https://doi.org/10.1145/322374.322380

7. Buda, A.O.: Multiprocessor automata. Inf. Process. Lett. **25**(4), 257–261 (1987). https://doi.org/10.1016/0020-0190(87)90172-4

8. Harrison, M.A., Ibarra, O.H.: Multi-tape and multi-head pushdown automata. Inf. Control **13**(5), 433–470 (1968). https://doi.org/10.1016/S0019-9958(68)90901-7

9. Ibarra, O.H.: On two-way multihead automata. J. Comput. Syst. Sci. **7**(1), 28–36 (1973). https://doi.org/10.1016/S0022-0000(73)80048-0

10. IEEE standard for information processing systems – local area networks – part 4: standard for token-passing bus access method and physical layer specifications (1990). https://doi.org/10.1109/IEEESTD.1990.7229456

11. IEEE information technology - telecommunications and information exchange between systems - local and metropolitan area networks - part 5: token ring access method and physical layer specifications (1998). https://doi.org/10.1109/IEEESTD.1998.8941590

12. Martín-Vide, C., Mateescu, A., Mitrana, V.: Parallel finite automata systems communicating by states. Int. J. Found. Comput. Sci. **13**(05), 733–749 (2002). https://doi.org/10.1142/S0129054102001424

13. McNaughton, R., Narendran, P., Otto, F.: Church-Rosser Thue systems and formal languages. J. ACM **35**(2), 324–344 (1988). https://doi.org/10.1145/42282.42284

14. Rosenberg, A.L.: On multi-head finite automata. In: 6th Annual Symposium on Switching Circuit Theory and Logical Design (SWCT 1965), pp. 221–228 (1965). https://doi.org/10.1109/FOCS.1965.19

15. Wagner, K., Wechsung, G.: Computational Complexity. Mathematics and Its Applications. Springer (1986). https://books.google.de/books?id=KLF7LMzr3moC

Finite Automata with Sets of Translucent Words

Benedek Nagy^{1,2} and Friedrich Otto$^{3(\boxtimes)}$

^{1} Department of Mathematics, Faculty of Arts and Sciences, Eastern Mediterranean University, 99628 Famagusta, Mersin-10, North Cyprus, Turkey
^{2} Department of Computer Science, Institute of Mathematics and Informatics, Eszterházy Károly Catholic University, Eger, Hungary
^{3} Universität Kassel, Fachbereich Elektrotechnik/Informatik, 34109 Kassel, Germany
`f.otto@uni-kassel.de`

Abstract. Here we study some restrictions and extensions of deterministic and nondeterministic finite automata with translucent letters (DFAwtl and NFAwtl). On the one hand, we restrict the cardinality of the sets of translucent letters, while, on the other hand, we introduce finite automata for which each state has an associated set of words that are translucent for that state. Here we require that each such set is a finite prefix code. We expect that, based on the cardinality of the sets of translucent words and the length of the longest word admitted in any set of this form, a strictly increasing two-dimensional hierarchy of language classes is obtained. In addition, we study closure and non-closure properties for the resulting language classes.

Keywords: Finite automaton · Translucent letter · Language class · Hierarchy · Closure property

1 Introduction

While a finite automaton reads its input strictly from left to right, letter by letter, by now many types of automata have been considered in the literature that process their inputs in a different, more involved way. Under this aspect, the most extreme is the *jumping finite automaton* of Meduna and Zemek [4] (see also [1]), which, after reading a letter, jumps to an arbitrary position of the remaining input. Another example is the *restarting automaton* as introduced by Jančar, Mráz, Plátek, and Vogel in [3], which processes a given input in cycles, in each cycle scanning the remaining tape contents from left to right until it deletes one or more letters, returns its window to the left end of the tape, and reenters its initial state. Finally, there is the (deterministic and nondeterministic) finite automaton *with translucent letters* (or DFAwtl and NFAwtl) of Nagy and Otto [7]. For each state q of an NFAwtl, there is a set $\tau(q)$ of *translucent letters*, which is a subset of the input alphabet that contains those letters that the automaton cannot see when it is in state q. Accordingly, in each step, the NFAwtl

J. D. Day and F. Manea (Eds.): DLT 2024, LNCS 14791, pp. 236–251, 2024.
https://doi.org/10.1007/978-3-031-66159-4_17

just reads (and deletes) the first letter from the left that is not translucent for the current state. It has been shown that the NFAwtl accepts a class of semi-linear languages that properly contains all rational trace languages, while its deterministic variant, the DFAwtl, is properly less expressive [6, 8–10]. A recent survey on the various types of automata with translucent letters can be found in [11].

In the current paper, we propose and study some restrictions and extensions of the NFAwtl and the DFAwtl. On the one hand, we study NFAwtls and DFAwtls for which the sets of translucent letters are restricted to be of bounded cardinality. As we shall see in Sect. 2, this requirement restricts the expressive capacity of the resulting automata considerably. In fact, we obtain an infinite strictly ascending hierarchy of automata and language classes based on the cardinality of the sets of translucent letters. Then, in Sect. 3, we introduce non-deterministic and deterministic finite automata with sets of translucent words (NFAwtw and DFAwtw). These sets of translucent words are required to be finite prefix codes. As it turns out, these types of automata are strictly more expressive than the NFAwtl, but they still only accept languages that are semi-linear. Then we parameterize the NFAwtw using two parameters: the length of the longest word in any set of translucent words and the cardinality of these sets (Sect. 4). We claim that, based on these parameters, strictly increasing two-dimensional hierarchies of language classes are obtained for DFAwtws and for NFAwtws.

In Sect. 5, we study closure and non-closure properties of the language classes considered. We shall see in particular that none of these classes is closed under intersection (with regular sets). Furthermore, while the class $\mathcal{L}(\mathsf{NFAwtw})$ of languages accepted by NFAwtws is closed under union, it is not closed under complementation. Moreover, it turns out that, for each language accepted by a DFAwtw, the complement is accepted by an NFAwtw. However, it remains open whether the class $\mathcal{L}(\mathsf{DFAwtw})$ is closed under this operation. In addition, while $\mathcal{L}(\mathsf{NFAwtw})$ is closed under marked product, the class $\mathcal{L}(\mathsf{DFAwtw})$ is not closed under (marked) product.

The paper is completed by Sect. 6 in which the results obtained are summarized, the membership problem and other decision problems for DFAwtws are considered in short, and some open problems are listed.

2 Restricting the Cardinality of Translucency Sets

In order to use it as a reference, we restate the definition of the NFAwtl from [7].

Definition 1. *An NFAwtl is defined as a 7-tuple $A = (Q, \Sigma, \lhd, \tau, I, F, \delta)$, where Q is a finite set of states, Σ is a finite input alphabet, $\lhd \notin \Sigma$ is a special letter that is used as an end-of-tape marker, $\tau : Q \to \mathcal{P}(\Sigma)$ is a translucency mapping, $I \subseteq Q$ is a set of initial states, $F \subseteq Q$ is a set of final states, and $\delta : Q \times \Sigma \to \mathcal{P}(Q)$ is a transition function. Here we require that, for each state $q \in Q$ and each letter $a \in \Sigma$, if $a \in \tau(q)$, then $\delta(q, a) = \emptyset$. For each state $q \in Q$, the letters from the set $\tau(q)$ are translucent for q, that is, when A is in state q, then it cannot see these letters.*

An NFAwtl $A = (Q, \Sigma, \triangleleft, \tau, I, F, \delta)$ is a deterministic finite automaton with translucent letters *(or a* DFAwtl*), if $|I| = 1$ and $|\delta(q, a)| \leq 1$ for all $q \in Q$ and all $a \in \Sigma$.*

For an input word $w \in \Sigma^*$, an NFAwtl $A = (Q, \Sigma, \triangleleft, \tau, I, F, \delta)$ starts in a nondeterministically chosen initial state $q_0 \in I$ with the word $w \cdot \triangleleft$ on its tape. This configuration is denoted by $q_0 w \cdot \triangleleft$. On its set $Q \cdot \Sigma^* \cdot \triangleleft \cup \{\mathsf{Accept}, \mathsf{Reject}\}$ of configurations, A induces the following single-step computation relation:

$$qw \cdot \triangleleft \vdash_A \begin{cases} q'uv \cdot \triangleleft, & \text{if } w = uav, \ u \in (\tau(q))^*, \ a \in \Sigma \smallsetminus \tau(q), \text{ and } q' \in \delta(q, a), \\ \mathsf{Reject}, & \text{if } w = uav, \ u \in (\tau(q))^*, \ a \in \Sigma \smallsetminus \tau(q), \text{ and } \delta(q, a) = \emptyset, \\ \mathsf{Accept}, & \text{if } w \in (\tau(q))^* \text{ and } q \in F, \\ \mathsf{Reject}, & \text{if } w \in (\tau(q))^* \text{ and } q \notin F. \end{cases}$$

A word $w \in \Sigma^*$ is *accepted by A* if there exists an initial state $q_0 \in I$ and a computation $q_0 w \cdot \triangleleft \vdash_A^* \mathsf{Accept}$, where $\vdash_A^*$ denotes the reflexive transitive closure of the relation $\vdash_A$. Now $L(A) = \{ w \in \Sigma^* \mid w \text{ is accepted by } A \}$ is the *language accepted by A* and $\mathcal{L}(\mathsf{NFAwtl})$ denotes the class of all languages that are accepted by NFAwtls. Analogously, $\mathcal{L}(\mathsf{DFAwtl})$ denotes the class of all languages that are accepted by DFAwtls.

Now we consider a restricted variant of the NFAwtl and the DFAwtl.

Definition 2. *An NFAwtl $A = (Q, \Sigma, \triangleleft, \tau, I, F, \delta)$ is m-restricted or an m-rNFAwtl for some $m \geq 1$ if $|\tau(q)| \leq m$ for each state $q \in Q$, that is, for each state of A, there are at most m translucent letters. If A is deterministic, then it is called an m-rDFAwtl. To emphasize the cardinality of the underlying alphabet, we use m-rNFAwtl(n) and m-rDFAwtl(n) to denote those m-restricted NFAwtls and DFAwtls that have an input alphabet Σ of cardinality n.*

The following observations are straightforward.

Proposition 1. *For all $m \geq 1$,*

(a) $\mathcal{L}(m\text{-rDFAwtl}(1)) = \mathcal{L}(m - \text{rNFAwtl}(1)) = \mathsf{REG}(1) = \mathsf{CFL}(1)$.
(b) $\mathcal{L}(m\text{-rDFAwtl}(m + 1)) = \mathcal{L}(\mathsf{DFAwtl}(m + 1))$ *and*
$L(m\text{-rNFAwtl}(m + 1)) = \mathcal{L}(\mathsf{NFAwtl}(m + 1))$.

However, for alphabets of cardinality larger than two, we get proper inclusions. For each $m \geq 3$, we consider the example language

$$L_{=m} = \{ w \in \{a_1, a_2, \ldots, a_m\}^* \mid |w|_{a_1} = |w|_{a_2} = \cdots = |a|_{a_m} \}.$$

It is easily seen that $L_{=m} \in \mathcal{L}((m-1)\text{-rDFAwtl}(m))$. On the other hand, we have the following negative result.

Lemma 1. *For all $m \geq 3$, $L_{=m} \notin \mathcal{L}((m - 2)\text{-rNFAwtl})$.*

$$\mathcal{L}(\text{1-rNFAwtl}) \xrightarrow{} \mathcal{L}(\text{2-rNFAwtl}) \xrightarrow{} \cdots \xrightarrow{} \bigcup_{m \geq 1} \mathcal{L}(m\text{-rNFAwtl}) = \mathcal{L}(\text{NFAwtl})$$

$$\uparrow \qquad\qquad \uparrow \qquad\qquad\qquad\qquad \uparrow$$

$$\mathcal{L}(\text{1-rDFAwtl}) \xrightarrow{} \mathcal{L}(\text{2-rDFAwtl}) \xrightarrow{} \cdots \xrightarrow{} \bigcup_{m \geq 1} \mathcal{L}(m\text{-rDFAwtl}) = \mathcal{L}(\text{DFAwtl})$$

$$\uparrow$$

$$\text{REG}$$

Fig. 1. Hierarchy of language classes accepted by the various restricted types of finite automata with translucent letters.

Proof outline. To accept a word of the form $w = a_1^n a_2^n \cdots a_m^n \in L_{=m}$, an NFAwtl A needs to read and delete occurrences of the letter a_m already before all occurrences of any of the letters a_1 to a_{m-1} have been read and deleted completely. However, A can do this only if it has a state q with $\tau(q) = \{a_1, a_2, \ldots, a_{m-1}\}$, that is, A is not $(m-2)$-restricted. $\qquad\square$

This yields the following consequence.

Theorem 1. *With respect to the cardinality m of the sets of translucent letters, the m-restricted DFAwtls and the m-restricted NFAwtls yield infinite strictly ascending hierarchies of language classes.*

Actually, we can separate the above hierarchy of DFAwtls from the corresponding hierarchy of NFAwtls. Let $\Sigma = \{a, b\}$, and let

$$L_\vee = \{\, w \in \Sigma^* \mid |w|_b = |w|_a \text{ or } |w|_b = 2 \cdot |w|_a \,\}.$$

The language $L_\vee$ is a rational trace language, and hence, it is accepted by an NFAwtl, but it is not accepted by any DFAwtl [10]. In fact, $L_\vee$ is accepted by a 1-rNFAwtl by Proposition 1, as $|\Sigma| = 2$. Thus, all inclusions in the diagram in Fig. 1 are strict, and classes that are not connected by a sequence of arrows are incomparable under inclusion.

3   Admitting Sets of Translucent Words

Here we propose a type of finite automata for which each state has an associated set of words that are translucent for that state.

Definition 3. *A finite automaton with translucent words, or an NFAwtw, is defined by a 7-tuple $A = (Q, \Sigma, \triangleleft, \tau, I, F, \delta)$, where Q, Σ, $\triangleleft$, I, and F are as for an NFAwtl, $\tau : Q \to \mathcal{P}_{\text{fin}}(\Sigma^*)$ is a translucency mapping, and $\delta : Q \times \Sigma \to \mathcal{P}(Q)$ is a transition function. For $q \in Q$, let $\Sigma_q = \{\, a \in \Sigma \mid \delta(q, a) \neq \emptyset \,\}$. It is required that, for each $q \in Q$, if $\tau(q) \neq \emptyset$, then the set $\tau(q) \cup \Sigma_q$ is a finite prefix code. In particular, no word in $\tau(q)$ begins with a letter from the set Σ_q.*

The single-step computation relation $\vdash_A$ that A induces on its set of configurations $Q \cdot \Sigma^* \cdot \triangleleft \cup \{\text{Accept}, \text{Reject}\}$ is defined as follows, where $q \in Q$ and $w \in \Sigma^*$:

$$qw\cdot \triangleleft \vdash_A \begin{cases} q'uv \cdot \triangleleft, & \text{if } w = uav, \ u \in (\tau(q))^*, \ a \in \Sigma_q, \ v \in \Sigma^*, \ \text{and } q' \in \delta(q, a), \\ \text{Reject}, & \text{if } w = uav, \ u \in (\tau(q))^*, \ a \in \Sigma \setminus \Sigma_q, \ v \in \Sigma^*, av \notin \tau(q) \cdot \Sigma^*, \\ \text{Accept}, & \text{if } w \in (\tau(q))^* \ \text{and } q \in F, \\ \text{Reject}, & \text{if } w \in (\tau(q))^* \ \text{and } q \notin F. \end{cases}$$

A word $w \in \Sigma^*$ is accepted by A if there exist an initial state $q_0 \in I$ and a computation $q_0 w \cdot \triangleleft \vdash_A^* \text{Accept}$. Now $L(A) = \{ w \in \Sigma^* \mid w \text{ is accepted by } A \}$ is the language accepted by A and $\mathcal{L}(\text{NFAwtw})$ denotes the class of all languages that are accepted by NFAwtws.

An NFAwtw $A = (Q, \Sigma, \triangleleft, \tau, I, F, \delta)$ is a DFAwtw if $|I| = 1$ and $|\delta(q, a)| \leq 1$ for each $q \in Q$ and each letter $a \in \Sigma$. Then $\mathcal{L}(\text{DFAwtw})$ denotes the class of all languages that are accepted by DFAwtws.

As $\tau(q)$ is a prefix code for each state q, the factorization $w = uav$, where $u \in (\tau(q))^*$, $a \in \Sigma$, $v \in \Sigma^*$, and $av \notin \tau(q) \cdot \Sigma^*$ is uniquely determined. In particular, this means that, for a DFAwtw, the induced computation relation is deterministic. This is not the case without the requirement that $\tau(q)$ is a prefix code. We illustrate this situation by a simple example.

Example 1. Let A be a DFAwtw on $\Sigma = \{a, b, c, d\}$, and assume that there is a state $q \in Q$ such that $\tau(q) = \{ab, abcd\}$. Then $\tau(q)$ is a code that is not a prefix code. Furthermore, assume that $\delta(q, c) = q'$, and let $w = abcdcd$. Then $qw \cdot \triangleleft = qabcdcd \cdot \triangleleft = q(ab)c(dcd) \cdot \triangleleft$. where $ab \in (\tau(q))^*$, $c \in \Sigma_q$, and $cdcd \notin \tau(q) \cdot \Sigma^*$. Thus, $qw \cdot \triangleleft \vdash_A q'abdcd \cdot \triangleleft$.

On the other hand, $qw \cdot \triangleleft = qabcdcd \cdot \triangleleft = q(abcd)c(d) \cdot \triangleleft$, where $abcd \in (\tau(q))^*$, $c \in \Sigma_q$, and $cd \notin \tau(q) \cdot \Sigma^*$, which implies that $qw \cdot \triangleleft \vdash_A q'abcdd \cdot \triangleleft$, too. Thus, we see that, in this case, the single-step computation relation is not deterministic. This is similar to nondeterministic translucency studied in [5]. ∎

From the definition of the NFAwtw, we immediately obtain the following.

Lemma 2. Let $A = (Q, \Sigma, \triangleleft, \tau, I, F, \delta)$ be an NFAwtw and assume that $puav \cdot \triangleleft \vdash_A quv \cdot \triangleleft$, where $u \in (\tau(p))^*$, $a \in \Sigma_p$, and $v \in \Sigma^*$. Then $puavw \cdot \triangleleft \vdash_A quvw \cdot \triangleleft$ for each word $w \in \Sigma^*$.

The following example presents a language that is accepted by a DFAwtw, but not by any NFAwtl.

Example 2. Let $D_1 \subseteq \{a, b\}^*$ be the semi-Dyck language on $\Sigma = \{a, b\}$ (see, e.g., [2]) that is generated by the grammar

$$G = (\{S\}, \Sigma, S, \{S \to \lambda, S \to ab, S \to SS, S \to aSb\}).$$

Furthermore, let $\Gamma = \{a, b, c\}$, let $\varphi : \Sigma^* \to \Gamma^*$ be the morphism that is defined through $a \mapsto ab$ and $b \mapsto c$, and let $L_1 = \varphi(D_1)$.

The language L_1 is accepted by the DFAwtw $A_1 = (Q, \Gamma, \lhd, \tau_1, q_0, \{q_0\}, \delta_1)$ that is defined as follows, where $Q = \{q_0, q_1, q_2\}$:

$$\tau_1(q_0) \;=\; \emptyset, \;\; \tau_1(q_1) \;=\; \emptyset, \;\; \tau_1(q_2) \;=\; \{ab\},$$
$$\delta_1(q_0, a) = q_1, \; \delta_1(q_1, b) = q_2, \; \delta_1(q_2, c) = q_0, \; \delta_1(q_0, \lhd) = \mathsf{Accept}.$$

Here δ_1 is undefined for all other pairs from $Q \times \Gamma$.

On the other hand, it can be verified that $L_1 \notin \mathcal{L}(\mathsf{NFAwtl})$. Indeed, assume that $B = (Q, \Gamma, \lhd, \tau, I, F, \delta)$ is an NFAwtl such that $L(B) = L_1$. Then, for each $n \geq 1$, B has an accepting computation for the word $w_n = (ab)^n c^n \in L_1$. Thus, there exist an initial state $q_0 \in I$, a final state $q_f \in F$, and a word $z \in (\tau(q_f))^*$ such that $q_0 w_n \cdot \lhd \vdash_B^* q_f z \cdot \lhd$. As $\tau(q_f) \subseteq \Gamma$, we can conclude that $\tau(q_f) = \emptyset$ and $z = \lambda$, as, otherwise, B would accept the words $w_n d^k$ for all $k \geq 1$ and all letters $d \in \tau(q_f)$.

If n is sufficiently large, then the accepting computation of B on input w_n cannot possibly read (and delete) the prefix $(ab)^n$ completely before it reads the first occurrence of the letter c. It follows that this computation has the following form:

$$q_0 (ab)^n c^n \cdot \lhd \vdash_B^* puc^n \cdot \lhd \vdash_B quc^{n-1} \cdot \lhd \vdash_B^* q_f \lhd$$

for some word $u \in \{a, b\}^+$ and some states $p, q \in Q$.

Now we distinguish three cases.

(1) If $|u|_a > 0$ and $|u|_b > 0$, then $\tau(p) = \{a, b\}$, and it follows that $u = vab$ for some word $v \in \{a, b\}^*$. But then B can also execute the following accepting computation:

$$q_0 (ab)^{n-1} acbc^{n-1} \cdot \lhd \vdash_B^* pvacbc^{n-1} \cdot \lhd \vdash_B qvabc^{n-1} \cdot \lhd = quc^{n-1} \cdot \lhd \vdash_B^* q_f \lhd,$$

which contradicts our assumption that $L(B) = L_1$, as $(ab)^{n-1} acbc^{n-1} \notin L_1$.

(2) If $|u|_a > 0$, but $|u|_b = 0$, then $u = a^m$ for some $1 \leq m \leq n$, that is, the above computation of B has the following structure:

$$q_0 (ab)^n c^n \cdot \lhd \vdash_B^{2n-m} pa^m c^n \cdot \lhd \vdash_B qa^m c^{n-1} \cdot \lhd \vdash_B^* q_f \lhd.$$

If m is small, then, using pumping arguments, it can be shown that B also accepts words of the form $(ab)^{n+k} c^n$ for some $k \geq 1$, a contradiction. (Almost as many a's are deleted as b's, which means that a sufficiently large number of factors ab are deleted that can then be pumped.)

If m is sufficiently large, then, using pumping arguments, it can be shown that B also accepts words in which the number of occurrences of the letter b is larger that the number of occurrences of the letter a, again a contradiction. (Only a few occurrences of the letter a are deleted, but many occurrences of the letter b. Accordingly, there is a long subcomputation in which only occurrences of the letter b are deleted. This subcomputation can then be pumped.)

(3) Finally, if $|u|_a = 0$, but $|u|_b > 0$, then by arguing as in (2), it follows that B would accept words in which the number of occurrences of the letter a is larger that the number of occurrences of the letter b, again a contradiction.

It follows that L_1 is indeed not accepted by any NFAwtl. Thus, L_1 separates the DFAwtw from the NFAwtl. ∎

For NFAwtws, we have the following result that gives a kind of normal form.

Lemma 3. *From a given NFAwtw A, one can construct an NFAwtw B such that $L(B) = L(A)$ and B only accepts once it has read and deleted its input completely.*

Proof. Let $A = (Q, \Sigma, \lhd, \tau, I, F, \delta)$ be an NFAwtw. For each final state $q \in F$, we take a DFA $C_q = (Q_q, \Sigma, q_0^{(q)}, F_q, \delta_q)$ that accepts the language $L(C_q) = (\tau(q))^*$. As $\tau(q)$ is a finite prefix code, such a DFA is easily constructed from $\tau(q)$. Of course, we can assume without loss of generality that all the automata A and C_q ($q \in F$) have pairwise disjoint sets of states. Now we define the NFAwtw $B = (Q_B, \Sigma, \lhd, \tau_B, I_B, F_B, \delta_B)$ as follows:

$$Q_B \quad = Q \cup \bigcup_{q \in F} Q_q, I_B = I \cup \{\, q_0^{(q)} \mid q \in I \cap F \,\}, F_B = \bigcup_{q \in F} F_q,$$

$$\tau_B(p) \quad = \begin{cases} \tau(p), \text{ if } p \in Q, \\ \emptyset, \quad\ \text{ if } p \in Q_q \text{ for some } q \in F, \end{cases}$$

$$\delta_B(p, a) = \begin{cases} \delta(p, a) \cup \{\, q_0^{(q)} \mid q \in \delta(p, a) \cap F \,\}, \text{ if } p \in Q, \\ \delta_q(p, a), \qquad\qquad\qquad\qquad\quad\ \text{ if } p \in Q_q \text{ for some } q \in F. \end{cases}$$

Thus, B behaves just like A, but, whenever A can enter a final state $q \in F$, then B has the option of either entering the same state or of entering the initial state of the DFA C_q. It is now easily seen that $L(B) = L(A)$ and that B reads and deletes its input completely before it accepts. □

Based on this technical result, we can derive the following fact, which extends a result to NFAwtws that was established in [6] for NFAwtls.

Proposition 2. *If A is an NFAwtw, then there exists a regular sublanguage R of the language $L(A)$ such that R is letter-equivalent to $L(A)$.*

Proof. Let $A = (Q, \Sigma, \lhd, \tau, I, F, \delta)$ be an NFAwtw. By Lemma 3, we can assume that A reads and deletes its input completely before it accepts. From A, we obtain an NFA $B = (Q, \Sigma, I, F, \delta_B)$ by simply ignoring the sets of translucent letters, that is, we take $\delta_B(q, a) = \delta(q, a)$ for all $q \in Q$ and all $a \in \Sigma$. Then $L(B)$ is a regular language, and it is immediate that $L(B) \subseteq L(A)$ holds, as each accepting computation of B is also an accepting computation of A.

It remains to show that the language $L(B)$ is letter-equivalent to the language $L(A)$, that is, for each word $w \in L(A)$, there exists a word $z \in L(B)$ such that, for each letter $a \in \Sigma$, $|z|_a = |w|_a$.

Claim. If $q_{i_0} w \cdot \lhd \vdash_A q_{i_1} w_1 \cdot \lhd \vdash_A \cdots \vdash_A q_{i_m} w_m \cdot \lhd = q_{i_m} \cdot \lhd \vdash_A$ Accept is a final part of an accepting computation of A, then there exists a word $z \in \Sigma^*$ such that $q_{i_0} z \vdash_B^* q_{i_m} \in F$ holds and $\psi(z) = \psi(w)$, where $\psi : \Sigma^* \to \mathbb{N}^{|\Sigma|}$ is the Parikh mapping.

Proof. We proceed by induction on the number m of steps in the above computation. If $m = 0$, then $q_{i_0} = q_{i_m} \in F$ and $w = w_m = \lambda$, that is, we can take $z = w = \lambda$.

If $q_{i_0} w \cdot \lhd = q_{i_0} xay \cdot \lhd \vdash_A q_{i_1} xy \cdot \lhd = q_{i_1} w_1 \cdot \lhd$, then $x \in (\tau(q_{i_0}))^*$, $a \in \Sigma$, and $q_{i_1} \in \delta(q_{i_0}, a)$. Thus, $q_{i_1} \in \delta_B(q_{i_0}, a)$. By the induction hypothesis, there exists a word $z_1 \in \Sigma^*$ that is accepted by B starting from the configuation $q_{i_1} z_1$ and that is letter-equivalent to $w_1 = xy$. Hence, the word $z = az_1$ is accepted by B starting from the configuration $q_{i_0} az_1$, and az_1 is letter-equivalent to axy and therewith to $w = xay$.   □

This result has the following immediate consequence.

Corollary 1. *The language accepted by an NFAwtw is semi-linear, that is, its Parikh image is a semi-linear subset of $\mathbb{N}^m$, where m is the cardinality of the underlying alphabet.*

In addition, we can conclude that the language $L_{\text{lin}} = \{\, a^n b^n \mid n \geq 0 \,\}$ is not accepted by any NFAwtw. On the other hand, we shall see in Theorem 5 below that the language $L_{\text{lin}}^c = \{a, b\}^* \setminus L_{\text{lin}}$ is accepted by an NFAwtw, but not by any DFAwtw. This yields the following proper inclusion.

Corollary 2. $\mathcal{L}(\text{DFAwtw}) \subsetneq \mathcal{L}(\text{NFAwtw})$.

4   Parameterizing NFAs with Translucent Words

We now parameterize the NFAwtw using the following two parameters:

(I) the cardinality of the sets $\tau(q)$ $(q \in Q)$ in analogy to Def. 2 and
(II) the length of the longest word in the set $\tau(q)$ for each state q.

In this way, we expect to obtain two-dimensional hierarchies for the NFAwtw and for the DFAwtw.

Definition 4. *For $k \geq 1$, k-lr-DFAwtw and k-lr-NFAwtw denote the classes of those DFAwtws and NFAwtws that are k-length-restricted, that is, for which the longest word in each set of translucent words is of length at most k.*

Observe that DFAwtl = 1-lr-DFAwtw and NFAwtl = 1-lr-NFAwtw. Using the idea from the construction in Example 2, we can separate the k-lr-DFAwtw from the $(k-1)$-lr-NFAwtw.

Theorem 2. *For all $k \geq 2$, $\mathcal{L}(\text{k-lr-DFAwtw}) \setminus \mathcal{L}((\text{k-1})\text{-lr-NFAwtw}) \neq \emptyset$.*

Proof. Let $k \geq 2$, let $\Gamma_k = \{a, b, e_2, e_3, \ldots, e_{k-1}, c\}$, let $\varphi_k : \Sigma^* \to \Gamma_k^*$ be the morphism that is defined through $a \mapsto abe_2 e_3 \cdots e_{k-1}$ and $b \mapsto c$, and let $L_{k-1} = \varphi_k(D_1)$. The language L_1 is just the language from Example 2, which belongs to $\mathcal{L}(2\text{-lr-DFAwtw}) \setminus \mathcal{L}(1\text{-lr-NFAwtw})$. Thus, it remains to study the case $k \geq 3$.

Claim 1. $L_{k-1} \in \mathcal{L}(k\text{-lr-DFAwtw})$.

Proof. Let $A_k = (Q_k, \Gamma_k, \vartriangleleft, \tau_k, q_0, \{q_0\}, \delta_k)$ be the k-lr-DFAwtw that is defined by taking $Q_k = \{q_0, q_1, q_2, \ldots, q_k\}$ and by defining the functions τ_k and δ_k as follows:

$$
\begin{aligned}
\tau_k(q_i) \quad &= \emptyset \quad \text{for all } 0 \le i \le k - 1, \ \tau_k(q_k) \quad = \{abe_2 e_3 \cdots e_{k-1}\}, \\
\delta_k(q_0, a) &= q_1, \delta_k(q_1, b) = q_2, \qquad \delta_k(q_i, e_i) = q_{i+1} \text{ for all } 2 \le i \le k - 1, \\
\delta_k(q_k, c) &= q_0, \delta_k(q_0, \vartriangleleft) = \mathsf{Accept},
\end{aligned}
$$

where δ_k is undefined for all other pairs from $Q_k \times \Gamma_k$. Then it is easily checked that $L(A_k) = L_{k-1}$. The set $\tau_k(q_k) = \{abe_2 e_3 \cdots e_{k-1}\}$ guarantees that each occurrence of the letter c in a word accepted by A_k is preceded by a prefix from $(\varphi_k(a))^* = \{abe_2 e_3 \cdots e_{k-1}\}^*$. □

Claim 2. $L_{k-1} \notin \mathcal{L}((k-1)\text{-lr-NFAwtw})$.

Proof. Assume that $A = (Q, \Gamma_k, \vartriangleleft, \tau, I, F, \delta)$ is an NFAwtw such that $L(A) = L_{k-1}$, where $\tau(q) \subseteq \Gamma_k^{\le k-1}$ for each state $q \in Q$. For each integer $n \ge 1$, the word $w_n = \varphi_k(a^n b^n) = (abe_2 e_3 \cdots e_{k-1})^n c^n$ belongs to L_{k-1}, that is, A has an accepting computation for w_n. This computation looks as follows:

$$
p w_n \cdot \vartriangleleft \vdash_A^* q u_n \cdot \vartriangleleft \vdash_A \mathsf{Accept}.
$$

Here $p \in I$, $q \in F$, and u_n is a scattered subword of w_n such that $u_n \in (\tau(q))^*$.

We claim that $u_n = \lambda$. Indeed, if $u_n \ne \lambda$, then $u_n = yz$ for some word $y \in (\tau(q))^*$ and a non-empty word $z \in \tau(q)$. As $\tau(q) \subseteq \Gamma_k^{\le k-1}$, $1 \le |z| \le k - 1$. From Lemma 2, it now follows that A can also execute the following computation:

$$
p w_n z \cdot \vartriangleleft \vdash_A^* q u_n z \cdot \vartriangleleft = q y z z \cdot \vartriangleleft \vdash_A \mathsf{Accept},
$$

that is, A also accepts the word $w_n z$. However, as $|w_n z| = (k+1) \cdot n + |z|$ is not a multiple of $k + 1$, $w_n z$ is not an element of L_{k-1}, a contradiction. Hence, $u_n = \lambda$, that is, in the accepting computation of A on input w_n, the word w_n is completely read and deleted before A accepts.

The NFAwtw A cannot simply read (and delete) the word w_n from left to right, as, in this case, pumping arguments can be used to show that, together with $w_n = (abe_2 e_3 \cdots e_{k-1})^n c^n$, A would also accept a word of the form $(abe_2 e_3 \cdots e_{k-1})^{n+r} c^n$ for some integer $1 \le r \le |Q|$. Thus, after a finite number of steps in which a prefix of w_n is read and deleted, a transition of the following form is applied:

$$
q_1 u x v \cdot \vartriangleleft \vdash_A q_2 u v \cdot \vartriangleleft,
$$

where $q_1, q_2 \in Q$, $x \in \Gamma_k$, $u, v \in \Gamma_k^*$, $q_2 \in \delta(q_1, x)$, and $u \in (\tau(q_1))^+$. Thus, the accepting computation of A on input w_n has the following form:

$$
p w_n \cdot \vartriangleleft = p w' u x v \cdot \vartriangleleft \vdash_A^{|w'|} q_1 u x v \cdot \vartriangleleft \vdash_A q_2 u v \cdot \vartriangleleft \vdash_A^* q \cdot \vartriangleleft \vdash_A \mathsf{Accept}.
$$

$$\mathsf{LRAT} \twoheadrightarrow \mathcal{L}(\mathsf{NFAwtl}) \twoheadrightarrow \mathcal{L}(\text{2-lr-NFAwtw}) \twoheadrightarrow \cdots \twoheadrightarrow \bigcup_{k \geq 1} \mathcal{L}(k\text{-lr-NFAwtw}) = \mathcal{L}(\mathsf{NFAwtw})$$

$$\uparrow \qquad\qquad \uparrow \qquad\qquad\quad \uparrow \qquad\qquad\qquad\qquad\qquad\quad \uparrow$$

$$\mathsf{REG} \twoheadrightarrow \mathcal{L}(\mathsf{DFAwtl}) \twoheadrightarrow \mathcal{L}(\text{2-lr-DFAwtw}) \twoheadrightarrow \cdots \twoheadrightarrow \bigcup_{k \geq 1} \mathcal{L}(k\text{-lr-DFAwtw}) = \mathcal{L}(\mathsf{DFAwtw})$$

Fig. 2. Hierarchy of language classes accepted by the various types of finite automata with translucent words. Here LRAT denotes the class of all rational trace languages.

As $u \neq \lambda$, we have $u = yz$ for some word $y \in (\tau(q_1))^*$ and a non-empty word $z \in \tau(q_1)$. However, in this situation, A can also execute the following computation:

$$pw'yxzv \cdot \lhd \vdash_A^{|w'|} q_1 yxzv \cdot \lhd \vdash_A q_2 yzv \cdot \lhd = q_2 uv \cdot \lhd \vdash_A^* q \cdot \lhd \vdash_A \mathsf{Accept}.$$

Thus, A accepts the word $w'yxzv$ that is obtained from $w_n = w'uxv = w'yzxv$ by commuting the factor $z \in \tau(q_1)$ with the letter $x \in \Sigma_{q_1}$. Since $z \in \tau(q_1)$, $1 \leq |z| \leq k - 1$, and x does not coincide with the first letter of z. Furthermore, $yzxv$ is a suffix of w_n of length $n \cdot (k + 1) - |w'|$. However, by performing a commutation of the above form on a suffix of w_n that does not only consist of occurrences of the letter c, we necessarily obtain a word that does not belong to the language L_{k-1}. Thus, we see that L_{k-1} is not accepted by any $(k - 1)$-lr-NFAwtw. □

Together, Claims 1 and 2 prove the theorem. □

The language L_{lin}^c mentioned above is actually accepted by an NFAwtl. Together with Theorem 2, this yields the following infinite hierarchy.

Theorem 3. *All arrows in the diagram in Fig. 2 represent proper inclusions, and language classes that are not connected by a sequence of arrows are incomparable under inclusion.*

It remains to extend the above results to a two-dimensional hierarchy, where the second dimension is obtained by restricting the cardinality of the sets of translucent words.

Definition 5. *An NFAwtw $A = (Q, \Sigma, \lhd, \tau, I, F, \delta)$ is called m-restricted or an m-rNFAwtw for some $m \geq 1$, if $|\tau(q)| \leq m$ for each state $q \in Q$. If A is deterministic, then it is called an m-rDFAwtw.*

For example, the DFAwtw for the language L_{k-1} in the proof of Theorem 2 is a k-lr-1-rDFAwtw. Next, we present a language that is accepted by a 2-lr-2-rDFAwtw, but not by any 1-rNFAwtw.

Let $D_1 \subseteq \{a, b\}^*$ be the semi-Dyck language on $\Sigma = \{a, b\}$, let $\Gamma_2 = \{a, b, b_1, c\}$, let $\psi : \Sigma^* \to \mathcal{P}(\Gamma^*)$ be the substitution that is defined through $a \mapsto \{ab, ab_1\}$ and $b \mapsto \{c\}$, and let $L_{c1} = \psi(D_1)$.

Theorem 4. $L_{c1} \in \mathcal{L}(\text{2-lr-2-rDFAwtw}) \smallsetminus \mathcal{L}(\text{1-rNFAwtw})$.

Proof. Let $C_1 = (Q, \Gamma_2, \vartriangleleft, \tau_1, q_0, \{q_0\}, \delta_1)$ be the DFAwtw that is defined by taking $Q = \{q_0, q_1, q_2\}$ and by defining the functions τ_1 and δ_1 as follows:

$$\tau_1(q_0) \; = \emptyset, \; \tau_1(q_1) \; = \emptyset, \; \tau_1(q_2) \; = \{ab, ab_1\},$$
$$\delta_1(q_0, a) = q_1, \; \delta_1(q_1, b) = q_2, \; \delta_1(q_1, b_1) = q_2, \; \delta_1(q_2, c) = q_0, \; \delta_1(q_0, \vartriangleleft) = \mathsf{Accept},$$

where δ_1 is undefined for all other pairs from $Q \times \Gamma_2$. Then it is easily checked that $L(C_1) = L_{c1}$. The set $\tau_1(q_2) = \{ab, ab_1\}$ guarantees that each occurrence of the letter c in a word accepted by C_1 is preceded by a prefix from the set $(\psi(a))^* = \{ab, ab_1\}^*$. Thus, $L_{c1} \in \mathcal{L}(2\text{-lr-}2\text{-rDFAwtw})$. On the other hand, it can be shown that $L_{c1} \notin \mathcal{L}(1\text{-rNFAwtw})$, that is, L_{c1} separates the 2-rDFAwtw from the 1-rNFAwtw. $\qquad\square$

In fact, we expect that this result generalizes. For $k \geq 2$ and $l \geq 2$, let

$$\Gamma_{k,l} = \{a, b, b_1, b_2, \ldots, b_{l-1}, e_2, e_3, \ldots, e_{k-1}, c\},$$

let $\psi_{k,l} : \Sigma^* \to \Gamma_{k,l}^*$ be the substitution that is defined through

$$a \mapsto \{abe_2 e_3 \cdots e_{k-1}\} \cup \{ab_i e_2 e_3 \cdots e_{k-1} \mid 1 \leq i \leq l-1\} \text{ and } b \mapsto \{c\},$$

and let $L_{c,k-1,l} = \psi_{k,l}(D_1)$. Then it is easily verified that $L_{c,k-1,l}$ is accepted by a DFAwtw that is k-length-restricted and l-restricted. On the other hand, using the strategy from the proof of Claim 2 in the proof of Theorem 2, it can be checked that the language $L_{c,k-1,l}$ is not accepted by any $(k-1)$-lr-NFAwtw. In addition, we conjecture that this language is not accepted by any $(l-1)$-rNFAwtw, either. That is, we conjecture that we have strictly increasing two-dimensional hierarchies for NFAwtws and for DFAwtws that are based on the parameters k and l.

5   On Closure and Non-Closure Properties

As DFAwtl = 1-lr-DFAwtw and NFAwtl = 1-lr-NFAwtw, we know from [7,9] that, for all $l \geq 1$, the language classes $\mathcal{L}(1\text{-lr-}l\text{-rNFAwtw})$ are closed under union, product, and Kleene star, but that they are not closed under intersection with regular languages, nor under complementation, nor under non-erasing morphisms. From the observation that

$$L_{=2} = \{ w \in \{a, b\}^* \mid |w|_a = |w|_b \} \in \mathcal{L}(1\text{-lr-}1\text{-rDFAwtw}),$$

while $L_{=2} \cap \{a\}^* \cdot \{b\}^* = \{ a^n b^n \mid n \geq 0 \} \notin \mathcal{L}(\mathsf{NFAwtw})$, we immediately obtain the following results.

Proposition 3. *For all $k, l \geq 1$, the language class $\mathcal{L}(k\text{-lr-}l\text{-rNFAwtw})$ is closed under union, but it is neither closed under intersection (with regular sets) nor under complementation.*

Furthermore, as $L_{2eq2} = \{\, w \in \{a,b\}^* \mid |w|_b = 2 \cdot |w|_a \,\} \in \mathcal{L}(\text{1-lr-1-rDFAwtw})$, while it can be shown that $L_\vee = L_{=2} \cup L_{2eq2} \notin \mathcal{L}(\text{DFAwtw})$, we have the following non-closure properties.

Proposition 4. *For all $k, l \geq 1$, the language class $\mathcal{L}(k\text{-lr-}l\text{-rDFAwtw})$ is neither closed under union nor under intersection (with regular sets).*

For DFAwtls, closure under complementation is proved by turning the transition function of a DFAwtl into a complete function and by interchanging the final states with the non-final states [10]. However, for DFAwtws, the situation is more complicated.

Let $A = (Q, \Sigma, \lhd, \tau, q_0, F, \delta)$ be a k-length-restricted l-restricted DFAwtw for some $k \geq 2$ and $l \geq 1$, and let $w \in \Sigma^*$. If $w \in L(A)$, then the computation of A on input w reduces $q_0 w \cdot \lhd$ to $qv \cdot \lhd$, where $q \in F$ and $v \in (\tau(q))^*$. Hence, if we take $B = (Q, \Sigma, \lhd, \tau, q_0, Q \smallsetminus F, \delta)$, then w is not accepted by B. However, the case that $w \notin L(A)$ is more problematic, as there are several possible ways in which A may reject on input w:

1. A may reduce $q_0 w \cdot \lhd$ to $qv \cdot \lhd$ such that $q \notin F$ and $v \in (\tau(q))^*$.
2. A may reduce $q_0 w \cdot \lhd$ to $quav \cdot \lhd$ such that $u \in (\tau(q))^*$, the letter a is not the first letter of any word from $\tau(q)$, but $\delta(q, a)$ is undefined.
3. A may reduce $q_0 w \cdot \lhd$ to $quxay \cdot \lhd$ such that $u \in (\tau(q))^*$, x is a non-empty proper prefix of a word from $\tau(q)$, $a \in \Sigma$, but xa is not the prefix of any word from $\tau(q)$.

The third case causes problems because it is not clear of how to check the stated conditions without actually reading the factor uxa. Thus, it remains open whether or not the class $\mathcal{L}(k\text{-lr-}l\text{-rDFAwtw})$, $k \geq 2$, $l \geq 1$, or the class $\mathcal{L}(\text{DFAwtw})$ is closed under complementation. However, we do have the following weaker result, which can be proved by expanding the proof of Lemma 3.

Proposition 5. *For each $k \geq 2$ and $l \geq 1$, if $L \subseteq \Sigma^*$ is accepted by a k-lr-l-rDFAwtw, then its complement $L^c = \Sigma^* \smallsetminus L$ is accepted by a k-lr-l-rNFAwtw.*

This result allows to prove that a certain language is not accepted by any DFAwtw.

Theorem 5. $L_{\text{lin}}^c = \{a,b\}^* \smallsetminus L_{\text{lin}} \in \mathcal{L}(\text{1-lr-1-rNFAwtw}) \smallsetminus \mathcal{L}(\text{DFAwtw})$.

Proof. Since $L_{\text{lin}} = L_{=2} \cap (\{a\}^* \cdot \{b\}^*)$, we see that $L_{\text{lin}}^c = L_{=2}^c \cup (\{a\}^* \cdot \{b\}^*)^c$. As $L_{=2} \in \mathcal{L}(\text{DFAwtl})$, and as the class $\mathcal{L}(\text{DFAwtl})$ is closed under complementation [7], $L_{=2}^c$ is accepted by a DFAwtl. Moreover, $(\{a\}^* \cdot \{b\}^*)^c$ is a regular language, and hence, it is also accepted by a DFAwtl. It follows that $L_{\text{lin}}^c = L_{=2}^c \cup (\{a\}^* \cdot \{b\}^*)^c$ is accepted by an NFAwtl, and therewith, by a 1-rNFAwtl by Proposition 1.

Assume that L_{lin}^c is accepted by some DFAwtw. Then, by Proposition 5, its complement $(L_{\text{lin}}^c)^c = L_{\text{lin}} = \{\, a^n b^n \mid n \geq 0 \,\}$ is accepted by some NFAwtw. This, however, contradicts Proposition 2. Thus, L_{lin}^c is not accepted by any DFAwtw. $\square$

Let Σ_1 and Σ_2 be two disjoint alphabets, let $L_1 \subseteq \Sigma_1^*$, and let $L_2 \subseteq \Sigma_2^*$. Then the language $L_1 \cdot L_2 = \{\, uv \mid u \in L_1, v \in L_2 \,\}$ is the *disjoint product* of L_1 and L_2. Based on Lemma 3, the following closure property can be derived.

Theorem 6. *For each $k, l \geq 1$, the class $\mathcal{L}(k\text{-lr-}l\text{-rNFAwtw})$ is closed under disjoint product.*

It is currently still open whether the class $\mathcal{L}(\mathsf{NFAwtw})$ is closed under arbitrary (that is, non-disjoint) product. However, the class $\mathcal{L}(\mathsf{DFAwtw})$ is not closed under product.

Theorem 7. *For each $k, l \geq 1$, the class $\mathcal{L}(k\text{-lr-}l\text{-rDFAwtw})$ is not closed under (disjoint) product, and neither is the language class $\mathcal{L}(\mathsf{DFAwtw})$.*

Proof. The language $L_\geq = \{\, w \in \{a, b\}^* \mid |w|_a \geq |w|_b \,\}$ is accepted by a DFAwtl [10]. Also the regular language $L_c = \{c\}$ is accepted by a DFAwtl. We now consider the disjoint product

$$L_p = L_\geq \cdot L_c = \{\, wc \mid w \in \{a, b\}^*, |w|_a \geq |w|_b \,\}.$$

This language is not accepted by any DFAwtl [10]. In fact, we prove that this language is not even accepted by any DFAwtw.

Assume to the contrary that $A = (Q, \Sigma, \lhd, \tau, q_0, F, \delta)$ is a DFAwtw such that $L(A) = L_p$, where $\Sigma = \{a, b, c\}$. For each $n \geq 1$, the word $w_n = a^n c$ belongs to L_p, and hence, the computation of A on input w_n is accepting, that is, it has the following form:

$$q_0 w_n \cdot \lhd = q_0 a^n c \cdot \lhd \vdash_A^* qu \cdot \lhd \vdash_A \mathsf{Accept},$$

where $q \in F$ and $u \in (\tau(q))^*$.

We claim that $\tau(q) = \emptyset$ and, therewith, $u = \lambda$. If v is a non-empty word from $\tau(q)$, then A can execute the following accepting computation:

$$q_0 w_n v \cdot \lhd = q_0 a^n cv \cdot \lhd \vdash_A^* quv \cdot \lhd \vdash_A \mathsf{Accept},$$

as with $u \in (\tau(q))^*$, also $uv \in (\tau(q))^*$. Hence, we see that $w_n v = a^n cv \in L(A)$. This, however, contradicts our assumption above as $w_n v \notin L_p$. Thus, it follows that $\tau(p) = \emptyset$ and $u = \lambda$.

Next, we claim that, within the accepting computation above, the letter c is only read and deleted in the very last step. In fact, assume that the above computation has the following form:

$$q_0 w_n \cdot \lhd = q_0 a^n c \cdot \lhd \vdash_A^m q_1 a^{n-m} c \cdot \lhd \vdash_A q_2 a^{n-m} \cdot \lhd \vdash_A^{n-m} q \cdot \lhd \vdash_A \mathsf{Accept},$$

where $q_1, q_2 \in Q$ and $m < n$. Then A would also execute the following accepting computation:

$$q_0 a^m ca^{n-m} \cdot \lhd \vdash_A^m q_1 ca^{n-m} \cdot \lhd \vdash_A q_2 a^{n-m} \cdot \lhd \vdash_A^{n-m} q \cdot \lhd \vdash_A \mathsf{Accept},$$

as during the prefix $q_0 a^n c \cdot \lhd \vdash_A^m q_1 a^{n-m} c \cdot \lhd$ of the above computation, A simply reads and deletes the first m occurrences of the letter a strictly from left to right. This shows that $a^m c a^{n-m} \in L(A)$, which contradicts our assumption above as $a^m c a^{n-m} \notin L_p$. Thus, the accepting computation above has actually the following form:

$$q_0 w_n \cdot \lhd = q_0 a^n c \cdot \lhd \vdash_A^n q' c \cdot \lhd \vdash_A q \cdot \lhd \vdash_A \text{ Accept}$$

for some state $q' \in Q$.

If n is sufficiently large, then there exists a state $\hat{q} \in Q$ such that this state is visited twice while A processes the prefix a^n, that is, the accepting computation above can be written as follows:

$$q_0 w_n \cdot \lhd = q_0 a^n c \cdot \lhd \vdash_A^i \hat{q} a^{n-i} c \cdot \lhd \vdash_A^j \hat{q} a^{n-i-j} c \cdot \lhd \vdash_A^{n-i-j} q' c \cdot \lhd \vdash_A q \cdot \lhd \vdash_A \text{ Accept}$$

for some integers $i \geq 0$ and $j \geq 1$.

Together with the word $w_n = a^n c$, A also accepts the word $a^n b^n c \in L_p$. As A is deterministic, the corresponding accepting computation looks as follows:

$$q_0 a^n b^n c \cdot \lhd \vdash_A^i \hat{q} a^{n-i} b^n c \cdot \lhd \vdash_A^j \hat{q} a^{n-i-j} b^n c \cdot \lhd \vdash_A^{n-i-j} q' b^n c \cdot \lhd \vdash_A^* \text{ Accept}.$$

However, A would then also execute the following accepting computation:

$$q_0 a^{n-j} b^n c \cdot \lhd \vdash_A^i \hat{q} a^{n-i-j} b^n c \cdot \lhd \vdash_A^{n-i-j} q' b^n c \cdot \lhd \vdash_A^* \text{ Accept},$$

which shows that $a^{n-j} b^n c \in L(A)$. This contradicts our assumption above as $a^{n-j} b^n c \notin L_p$. Thus, we see that the language L_p is in fact not accepted by any DFAwtw. This proves that the classes $\mathcal{L}(k\text{-lr-}l\text{-rDFAwtw})$, $k, l \geq 1$, are not closed under disjoint product. Of course, this proof shows that the class $\mathcal{L}(\text{DFAwtw})$ is not closed under (disjoint) product, either. $\square$

Moreover, we have the following negative results.

Proposition 6. *For all $k, l \geq 1$, the language class $\mathcal{L}(k\text{-lr-}l\text{-rDFAwtw})$ is not closed under alphabetic morphisms, and neither is the language class $\mathcal{L}(\text{DFAwtw})$.*

Proof. It is not hard to see that the language

$$L'_\vee = \{\, cw \mid w \in \{a,b\}^*, |w|_a = |w|_b \,\} \cup \{\, dw \mid w \in \{a,b\}^*, 2 \cdot |w|_a = |w|_b \,\}$$

is accepted by some 1-rDFAwtl. By applying the alphabetic morphism $\varphi : \{a,b,c,d\}^* \to \{a,b\}^*$ that is defined through $a \mapsto a$, $b \mapsto b$, $c \mapsto b$, and $d \mapsto b$, we obtain the language

$$\varphi(L'_\vee) = \{\, bw \mid w \in \{a,b\}^*, |w|_b = |w|_a \text{ or } |w|_b = 2 \cdot |w|_a \,\},$$

which can be shown to be not accepted by any DFAwtw. $\square$

Finally, for length-restricted NFAwtws, we also have the following negative result.

Proposition 7. *For all* $k, l \geq 1$, *the language classes* $\mathcal{L}(k\text{-lr-}l\text{-rNFAwtw})$ *and* $\mathcal{L}(k\text{-lr-NFAwtw})$ *are not closed under non-erasing morphisms.*

Proof. Let $k \geq 2$, let $\Gamma_k = \{a, b, e_2, e_3, \ldots, e_{k-1}, c\}$, and let $L_{k-1} \subseteq \Gamma_k^*$ be the language from the proof of Theorem 2. As shown in the proof of that theorem,

$$L_{k-1} \in \mathcal{L}(k\text{-lr-1-rDFAwtw}) \smallsetminus \mathcal{L}((k-1)\text{-lr-NFAwtw}).$$

Now, let $\psi_k : \Gamma_k^* \to \Gamma_{k+1}^*$ be the morphism that is defined through $a \mapsto a$, $b \mapsto be_2$, $e_i \mapsto e_{i+1}$, $2 \leq i \leq k-1$, and $c \mapsto c$. Then $\psi_k(L_{k-1}) = L_k$, which shows that $\psi_k(L_{k-1}) \notin \mathcal{L}(k\text{-lr-NFAwtw})$. Hence, the language classes $\mathcal{L}(k\text{-lr-}l\text{-rNFAwtw})$ $(l \geq 1)$ and $\mathcal{L}(k\text{-lr-NFAwtw})$ are not closed under non-erasing morphisms. $\qquad \square$

However, it remains open whether the unrestricted class $\mathcal{L}(\text{NFAwtw})$ is closed under non-erasing morphisms.

6   Conclusion

We have generalized the finite automaton with translucent letters by replacing the sets of translucent letters by sets of translucent words, where we require that each set of translucent words is a finite prefix code. It turned out that, through this generalization, we extend the expressive capacity of these types of automata properly. In fact, we obtain two infinite strictly ascending hierarchies of language classes, where the one is based on the cardinality of the sets of translucent words (letters) and the other is based on the length of the longest word in any set of translucent words. Actually, we conjecture that, based on these two parameters, infinite strictly ascending two-dimensional hierarchies of language classes are obtained. To establish this claim, it would suffice to prove that, for all $k, l \geq 2$, the language $L_{c,k-1,l} \in \mathcal{L}(k\text{-lr-}l\text{-rDFAwtw})$ is not accepted by any $(l-1)$-restricted NFAwtw.

Moreover, we have provided some closure and non-closure properties for the language classes that are accepted by the various types of automata with translucent words, but it remains open whether $\mathcal{L}(\text{DFAwtw})$ is closed under complementation and whether $\mathcal{L}(\text{NFAwtw})$ is closed under product.

Concerning decision problems, it is easily seen that the membership problem for a DFAwtw is solvable in quadratic time. However, it remains open whether $O(n^2)$ is also a lower bound for the complexity of this problem, or whether there exists a more efficient algorithm. On the other hand, based on Proposition 2, it can be shown that the emptiness and the finiteness problems are decidable for NFAwtws. In addition, it follows from Proposition 5 that universality is decidable for DFAwtws. Finally, from the corresponding results for DFAwtls [10], we see that the inclusion problem is undecidable for DFAwtws. However, it is still open whether the regularity problem or the equivalence problem are decidable for these types of automata.

Instead of only admitting finite prefix codes as sets of translucent words, we could admit infinite (e.g., regular) prefix codes as sets of translucent words. Do the resulting types of DFAwtws and NFAwtws have a larger expressive capacity than the DFAwtws and NFAwtws considered in the current paper?

References

1. Fernau, H., Paramasivan, M., Schmid, M.L.: Jumping finite automata: character-izations and complexity. In: Drewes, F. (ed.) CIAA 2015. LNCS, vol. 9223, pp. 89–101. Springer, Cham (2015). https://doi.org/10.1007/978-3-319-22360-5_8
2. Harrison, M.A.: Introduction to Formal Language Theory. Addison-Wesley, Reading, M.A. (1978)978-0-201-02955-0
3. Jančar, P., Mráz, F., Plátek, M., Vogel, J.: Restarting automata. In: Reichel, H. (ed.) FCT 1995. LNCS, vol. 965, pp. 283–292. Springer, Heidelberg (1995). https://doi.org/10.1007/3-540-60249-6_60
4. Meduna, A., Zemek, P.: Jumping finite automata. Int. J. Found. Comput. Sci. **23**, 1555–1578 (2012). https://doi.org/10.1142/S0129054112500244
5. Nagy, B.: State-deterministic finite automata with translucent letters and finite automata with nondeterministically translucent letters. In: Gazdag, Z., Szabolcs, I., Kovásznai, G. (eds.) 16th International Conference on Automata and Formal Languages (AFL 2023), EPTCS, vol. 386, pp. 170–184 (2023). https://doi.org/10.4204/EPTCS.386.14
6. Nagy, B., Otto, F.: CD-systems of stateless deterministic R(1)-automata accept all rational trace languages. In: Dediu, A.-H., Fernau, H., Martín-Vide, C. (eds.) LATA 2010. LNCS, vol. 6031, pp. 463–474. Springer, Heidelberg (2010). https://doi.org/10.1007/978-3-642-13089-2_39
7. Nagy, B., Otto, F.: Finite-state acceptors with translucent letters. In: Bel-Enguix, G., Dahl, V., De La Puente, A.O. (eds.) BILC 2011: AI Methods for Interdisciplinary Research in Language and Biology, Proceedings, pp. 3–13. SciTePress, Portugal (2011). https://doi.org/10.5220/0003272500030013
8. Nagy, B., Otto, F.: Globally deterministic CD-systems of stateless R(1)-automata. In: Dediu, A.-H., Inenaga, S., Martín-Vide, C. (eds.) LATA 2011. LNCS, vol. 6638, pp. 390–401. Springer, Heidelberg (2011). https://doi.org/10.1007/978-3-642-21254-3_31
9. Nagy, B., Otto, F.: On CD-systems of stateless deterministic R-automata with window size one. J. Comput. Syst. Sci. **78**, 780–806 (2012). https://doi.org/10.1016/j.jcss2011.12.009
10. Nagy, B., Otto, F.: Globally deterministic CD-systems of stateless R-automata with window size 1. Int. J. Comput. Math. **90**, 1254–1277 (2013). https://doi.org/10.1080/00207160.2012.688820
11. Otto, F.: A survey on automata with translucent letters. In: Nagy, B. (ed.) Implementation and Application of Automata, CIAA 2023, LNCS, vol. 14151, pp. 21–50. Springer, Cham (2023). https://doi.org/10.1007/978-3-031-40247-0_2

Careful Synchronization of One-Cluster Automata

Jakub Ruszil[(⊠)]

Doctoral School of Exact and Natural Sciences, Faculty of Mathematics
and Computer Science, Jagiellonian University, Cracow, Poland
jakub.ruszil@uj.edu.pl

Abstract. In this paper, we investigate the careful synchronization of one-cluster partial automata. First, we prove that the shortest carefully synchronizing word for such automata can be of length $2^{\frac{n}{2}} + 1$, where n is the number of states of an automaton. Additionally, we prove that checking whether a given one-cluster partial automaton is carefully synchronizing is NP-hard, even for the binary alphabet.

1 Introduction

The concept of synchronization of an automaton involves driving it into the same state regardless of its initial state by applying a specific word, called a synchronizing word. The synchronization problem arises in coding theory [5,13], parts orienting in manufacturing [9,19], testing reactive systems [23], and Markov Decision Processes [7,8]. This idea has been extensively studied for complete deterministic finite automata (DFA) [2,4,9,14,15,20,22,25–29] and non-deterministic finite automata [11,12]. The Černý Conjecture, one of the most famous long-standing open problems in theoretical computer science, posits that for any given synchronizing DFA with n states, there is always a synchronizing word of length at most $(n-1)^2$. While this conjecture has been proven for numerous classes of automata, it remains unsolved in the general case.

Allowing no outgoing transitions from some states for certain letters helps us model a system for which specific actions cannot be accomplished while being in a specified state. This leads to the problem of finding a synchronizing word for a finite automaton, where the transition function is not defined for all states. Notice that this is the most frequent case if we use automata to model real-world systems. In practice, it rarely happens that a real system can be modeled with a DFA where the transition function is total. The transition function is usually a partial one. This fact motivated many researchers to investigate the properties of partial finite automata relevant to practical problems of synchronization.

We know that checking if a partial automaton can be carefully synchronized is PSPACE-complete [18], even for the binary alphabet [30]. In this paper, we investigate the case of deterministic finite automata such that the transition from state to state is not necessarily defined for all states. We refer to this model as a *partial finite automaton* (PFA). We will say that a word that synchronizes a

J. D. Day and F. Manea (Eds.): DLT 2024, LNCS 14791, pp. 252–265, 2024.
https://doi.org/10.1007/978-3-031-66159-4_18

PFA is a *carefully synchronizing word*. Different definitions of synchronization of PFAs also exist, for example, here [3]. The problem of estimating the length of the shortest carefully synchronizing word for PFA was considered, among others, by Rystsov [21], Ito and Shikishima-Tsuji [12], Martyugin [16,17], Gazdag et al. [10] and de Bondt et al. [6]. Martyugin established a lower bound on the length of such words of size $\Theta(3^{n/3})$, and Rystsov [21] established an upper bound of size $O((3 + \epsilon)^{n/3})$, where n is the number of states of the automaton.

One-cluster DFAs have a letter with only one simple cycle on the set of states. Synchronization of one-cluster DFAs was studied by Steinberg, Béal and Perrin and resulted in proving Černý Conjecture for one-cluster automata with prime length cycle in [24] and establishing a quadratic upper bound for the length of the synchronizing word in the general case in [1] using a linear algebra approach. However, we also know that deciding if an automaton is synchronizing is in P, which is not true for careful synchronization. Following that research, we extend the notion of one-cluster to partial automata. Our research is motivated by the fact that all former constructions providing a high careful synchronization threshold do not contain a letter, which creates a cluster on the set of states. Constructions from [16,17] have a linear alphabet size in terms of the cardinality of the set of states, but only one letter is defined for all states, and the number of clusters also grows linearly in the size of the set of states for that letter. Cases from [6,30] are more interesting as they deal with a constant alphabet size. Still, constructions from those papers also do not define any letter to be one-cluster (one of the families of automata from [6], used in the proof of Lemma 7, has one letter defined as a permutation on the set of states with three cycles).

On the other hand, binary partial one-cluster automata have a simple structural condition concerning a non-cluster letter (stated in Observation 1), which gives the impression that deciding the careful synchronizability of such automata is an easy problem in terms of the complexity theory. Surprisingly, we managed to show that this decision problem is hard even in such a restricted case. Our contribution is twofold. First, we prove that the shortest carefully synchronizing words for one-cluster partial automata can be of exponential length in terms of the number of states (but assuming the alphabet is also of exponential size). From the complexity perspective, we show that the problem of deciding if a given binary one-cluster PFA admits any carefully synchronizing word is NP-hard by a reduction from the 3-SAT problem.

2   Preliminaries

A *partial finite automaton* (PFA) is an ordered tuple $\mathcal{A} = (\Sigma, Q, \delta)$ where Σ is a finite set of letters, Q is a finite set of states and $\delta : Q \times \Sigma \to Q$ is the transition function, not everywhere defined. For a word $w \in \Sigma^*$ and $q \in Q$, we define $\delta(q, w)$ inductively: $\delta(q, \epsilon) = q$ and $\delta(q, aw) = \delta(\delta(q, a), w)$ for $a \in \Sigma$, where ϵ is the empty word and $\delta(q, a)$ is defined. Let $S \subseteq Q$. Define $\delta(S, w) = \bigcup_{q \in S} \delta(q, w)$. We write $q.w$ (respectively $S.w$ for an image of a set $S \subseteq Q$) instead of $\delta(q, w)$ (respectively $\delta(S, w)$) wherever it does not cause ambiguity. We also denote

$S.w^{-1} = \{q \in Q : q.w \in S\}$ as a preimage of S under the word w. A word $w \in \Sigma^*$ is called *carefully synchronizing* if there exists $\bar{q} \in Q$ such that for every $q \in Q$, $\delta(q, w) = \bar{q}$ and all transitions are defined. A PFA is called *carefully synchronizing* if it admits any carefully synchronizing word. For a given $\mathcal{A} = (Q, \Sigma, \delta)$, we define the *power-automaton* $\mathcal{P}(\mathcal{A}) = (2^Q, \Sigma, \tau)$, where 2^Q stands for the set of all subsets of Q. The transition function $\tau : 2^Q \times \Sigma \to 2^Q$ is defined as follows: $\tau(Q', a) = \bigcup_{q \in Q'} \delta(q, a)$, $a \in \Sigma$, $Q' \subseteq Q$ if $\delta(q, a)$ is defined for all states in $q \in Q'$. Otherwise, $\tau(Q', a)$ is not defined. We can now state the fact useful in deciding whether a given PFA is carefully synchronizing.

Fact 1. *Let $\mathcal{A}$ be a PFA and $\mathcal{P}(\mathcal{A})$ be its power automaton. Then $\mathcal{A}$ is synchronizing if, and only if, for some state $q \in Q$ there exists a labeled path in $\mathcal{P}(\mathcal{A})$ from Q to $\{q\}$. The shortest synchronizing word for $\mathcal{A}$ corresponds to the shortest such labeled path in $\mathcal{P}(\mathcal{A})$.*

An example of the carefully synchronizing automaton $\mathcal{A}_{car}$ is depicted in Fig. 1. Its shortest carefully synchronizing word w_{car} is $abc(ab)^2c^2a$, which can be easily checked via the power automaton construction.

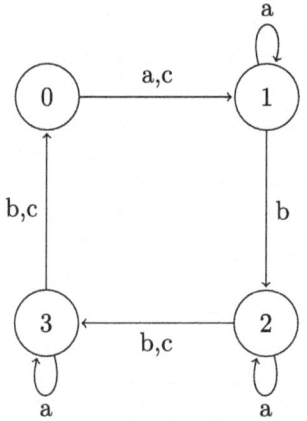

Fig. 1. A carefully synchronizing $\mathcal{A}_{car}$

Let $\mathcal{L}_n = \{\mathcal{A} = (\Sigma, Q, \delta) : \mathcal{A} \text{ is carefully synchronizing and } |Q| = n\}$. Notice that $\mathcal{L}_n$ does not depend on alphabet size. We define $d(\mathcal{A}) = \min\{|w| : w \text{ is a carefully synchronizing word for } \mathcal{A}\}$ and $d(n) = \max\{d(\mathcal{A}) : \mathcal{A} \in \mathcal{L}_n\}$.

For $\mathcal{A} = (Q, \Sigma, \delta)$ and $a \in \Sigma$ defined for all states, let $\mathcal{G}_a = (Q, E)$ be a directed graph such that $(p, q) \in E$ if, and only if, $p.a = q$. Put differently, $\mathcal{G}_a$ is a digraph made of only the edges labeled by a in $\mathcal{A}$. The graph $\mathcal{G}_a$ is a disjoint union of weakly connected components called a-*clusters*. Since each state has only one outgoing edge in $\mathcal{G}_a$, each a-cluster contains a unique cycle, called the a-*cycle* with trees attached to the cycle at their roots. For each $p \in Q$ we define a *level* of p as the distance between p and the root of the tree containing p in

$\mathcal{G}_a$. A *level* of $\mathcal{G}_a$ is the maximal level of its vertices. An example of the a-cluster is shown in Fig. 2.

One can easily deduce from Fact 1 that at least one letter $a \in \Sigma$ must be defined for all states of a PFA for it to be carefully synchronizing. We define a *one-cluster* PFA with respect to the letter a as a PFA which has only one a-cluster. We also refer to a *one-cluster* PFA as an automaton, which is one-cluster with respect to any of its letters.

Let $\mathcal{A} = (Q, \Sigma, \delta)$ be a PFA, $a \in \Sigma$, $S \subset Q$ and $G = (V, E)$ be a digraph. We say that S *induces G on the letter a in $\mathcal{A}$* if the set S induces in $\mathcal{G}_a$ a graph isomorphic to G. Whenever not stated differently in the paper, when we refer to one-cluster automata, we assume that it is one-cluster with respect to the letter a.

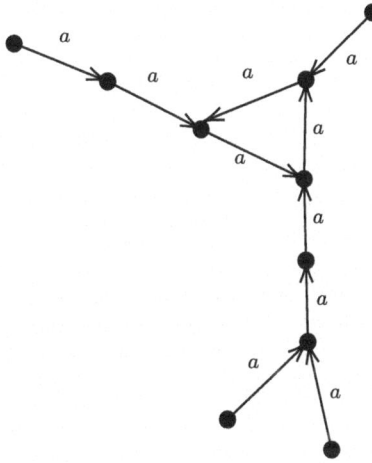

Fig. 2. An example of the a-cluster

Before proceeding, we state the preliminary observation and lemma that can shed more light on the synchronization and careful synchronization of one-cluster automata.

Observation 1. *If a one-cluster automaton $\mathcal{A} = (Q, \Sigma, \delta)$ is carefully synchronizing, C induces a directed cycle on the letter a, the letter a is the only one defined for all states and $|C| > 1$, then there exists a letter $b \in \Sigma$ such that $b \neq a$ and for any $q \in C$ it holds that $\delta(q, b)$ is defined.*

Proof. Obviously, since a is the only letter defined for all states, then every synchronizing word must start with it. On the other hand, for all $k \in \mathbb{N}$ holds $C \subseteq Q.a^k$ and if $k \geq l$, where l is the level of $\mathcal{G}_a$, then $Q.a^k = C$. If $|C| > 1$, then there must exist another letter $b \in \Sigma$ such that $C.b \neq C$ for $\mathcal{A}$ to be carefully synchronizing. □

Suppose that C induces a directed cycle on the set of states of a PFA. We state and prove the lemma that shows that if there exists a word w shrinking C, then there exists a word w' (whose length is polynomial in $|w|$) shrinking C by half.

Lemma 1. *Let $\mathcal{A} = (Q, \Sigma, \delta)$ be a one-cluster PFA, $|Q| = n$. Let C induce a directed cycle on the letter a and l be the level of $\mathcal{G}_a$. If there exists a word w such that $|C.w| < |C|$ then there exists a word w' such that $|Q.w'| \leq \lfloor \frac{1}{2}|C| \rfloor$ of length at most $n + \frac{1}{2}|C|(|w| + |C|)$.*

Proof. For any $p, q \in C$ define $dist_C(p, q) = \min\{k_1, k_2 : p.a^{k_1} = q \wedge q.a^{k_2} = p \wedge k_1, k_2 < |C|\}$. Since $p, q \in C$, then $dist_C(p, q)$ is always well defined and for all p, q holds $dist_C(p, q) \leq \lfloor \frac{1}{2}|C| \rfloor$. Since there exists w such that $|C| > |C.w|$, then there must exist $\bar{p}, \bar{q} \in C$ such that $\bar{p}.w = \bar{q}.w$. Denote $dist_C(\bar{p}, \bar{q}) = k$. We will show the claim:

Claim 1. *For any $C' \subset C$ such that $|C'| > \frac{1}{2}|C|$, there exist $p', q' \in C$, such that $dist(p', q') = k$.*

Proof. For the sake of contradiction, suppose that there exists $C' \subset C$ and $|C'| > \frac{1}{2}|C|$ such that for all $p, q \in C'$ holds $dist(p, q) \neq k$. Let $S = \{q_{i+k \bmod m} : q_i \in C'\}$. Since for any $q_i \in C'$ there exists exactly one state $q_{i+k \bmod m} \in C$, then $|C'| = |S|$. On the other hand, since for all $p, q \in C'$ holds $dist(p, q) \neq k$, then $C \cap S = \varnothing$. But since $|C'| = |S| > \frac{1}{2}|C|$ and $C' \subset C$ and $S \subset C$ then $|C'| + |S| > |C|$, so there must hold $C' \cap S \neq \varnothing$ and we have a contradiction. □

We can construct the word w' in the following way: Obviously $Q.a^l = C$ and $|Q.a^l w| < |C|$. If also $|Q.a^l w| \leq \lceil \frac{1}{2}|C| \rceil$ then the result holds. Otherwise, observe that $Q.a^l w a^l \subset C$ and $|Q.a^l w a^l| > \frac{1}{2}|C|$ so we can apply Claim 1 to find the states $p', q' \in Q.a^l w a^l$ such that $dist(p', q') = k = dist(\bar{p}, \bar{q})$. Denote $u = a^l w a^l$ and observe that, for some $m_1 \leq |C| < n$, it must hold that $\bar{p}, \bar{q} \in Q.u a^{m_1}$ so $|Q.u a^{m_1} w| < |Q.u|$. We can apply Claim 1 and the word $a^{m_1} w$ as long as the size of the resulting set is greater than $\frac{1}{2}|C|$. We obtain the word $w' = a^l w (a^{m_1} w)_2^m$ where $m_2 < \frac{1}{2}|C| - 1$ so the result holds. □

3   Long Shortest Carefully Synchronizing Word

In this section, we construct an infinite family of carefully synchronizing, one-cluster PFAs with exponential shortest carefully synchronizing words. The proof scheme is similar to the one given in [12] (Proposition 8) however, it differs in details. To simplify proofs, we assume that the size of the set of states is $n = 2k$, where $k \in \mathbb{N}$ but the arguments can be easily adapted to the case $n = 2k+1$. Let $C_k = \{c_1, \ldots, c_k\}$ and $T_k = \{t_1, \ldots, t_k\}$, $Q_k = C_k \cup T_k$ and $\Sigma_k = \{a\} \cup \Sigma_k'$ where Σ_k' is a set of letters specified later on. Finally let $\mathcal{B}_k = (Q_k, \Sigma_k, \delta_k)$. Define the action of the letter a on the set of states:

- $\delta_k(c_i, a) = c_{i+1}$ for $i < k$
- $\delta_k(c_n, a) = c_1$
- $\delta_k(t_i, a) = c_i$

It is easy to observe that $\mathcal{B}_k$ is one-cluster with respect to a for every k. An example of that cluster for $k = 3$ is depicted below (Fig. 3).

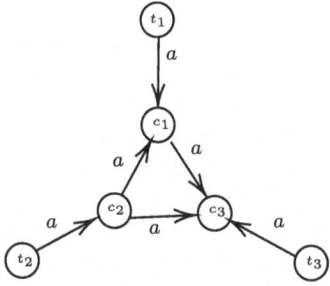

Fig. 3. The a-cluster of the automaton $\mathcal{B}_3$

Let $\mathcal{T}_k = \{T \cup (C_k \cap ((T_k \setminus T).a).a^{-1}) : T \in 2^{T_k} \setminus \{\varnothing\}\}$ Intuitively, it is the set of all sets T such that $|T'| = k$ and $T'.a = C_k$. First, we prove the following lemma.

Lemma 2. *Let $\mathcal{T}_k$ be the family of sets defined above. Then $|\mathcal{T}_k| = 2^{|T_k|} - 1$.*

Proof. Let $T_1', T_2' \in 2^{T_k}$ such that $T_1' \neq T_2'$. Denote $C_k \cap ((T_k \setminus T_1').a).a^{-1} = P_1$ and $C_k \cap ((T_k \setminus T_2').a).a^{-1} = P_2$ It suffices to show that, for any T_1', T_2', the sets $T_1' \cup P_1$ and $T_2' \cup P_2$ are different. But it is easy to notice that $P_1, P_2 \subset C_k$ and since $T_1' \neq T_2'$ and $T_1', T_2' \subset T_k$, then the lemma holds. $\qquad\square$

Using Lemma 2 we can enumerate sets of the family $\mathcal{T}_k$ arbitrarily from S_1 to $S_{2^{|T_k|}-1}$. Let us prove some properties of those sets in the next lemma.

Lemma 3. *For any $i \in \{1, \ldots, 2^{|T_k|} - 1\}$ holds $S_i.a = C_k$ and $|S_i| = \frac{n}{2}$.*

Proof. By the definition of S_i, we know that there exists $T \subseteq T_k$ such that $S_i = T \cup (C_k \cap ((T_k \setminus T).a).a^{-1})$. By the construction $T.a \subseteq C$ so it suffices to show that $(C \cap ((T_k \setminus T).a).a^{-1}).a = C \setminus T.a$. Indeed $(C_k \cap ((T_k \setminus T).a).a^{-1}).a = C_k.a \cap (T_k \setminus T).a = C_k \cap T_k.a \setminus T.a = C \cap C \setminus T.a = C \setminus T.a$. To show that $|S_i| = \frac{n}{2}$ first observe that $(T_k \setminus T).a = \{q \in C_k \setminus T.a\}$. Let $q \in C_k \setminus T.a$ and observe that the set $q.a^{-1}$ contains exactly two states $q_1 \in T_k$ and $q_2 \in C_k$ and by the construction we have that $|((T_k \setminus T).a).a^{-1}| = 2|C_k \setminus T.a|$ and the half of the states must be contained in C_k, which gives us $|S_i| = |T.a| + |C_k \setminus T.a|$, and that concludes the proof. $\qquad\square$

Let $\Sigma_k' = \{b_1, \ldots, b_{2^{|T_k|}-1}, c\}$ (this is the subalphabet used in the definition of Σ_k) and $S_i = \{s_1^i, \ldots, s_k^i\}$. Put the transition function δ_k (for Σ_k') as follows:

- $\delta_k(c_j, b_1) = s_j^1$, the letter b_1 not defined for other states
- for $0 < i < 2^{|T_k|} - 2$ let $\delta_k(s_j^i, b_{i+1}) = s_j^{i+1}$, the letter $b_i + 1$ not defined for other states
- $\delta_k(s_j^{2^{|T_k|}-1}, c) = c_1$, the letter c not defined for other states

Observe that, for any i, holds $\delta_k(S_i, b_{i+1}) = S_{i+1}$. Let us state and prove the main theorem of this section.

Theorem 1. *The automaton $\mathcal{B}_k$ is carefully synchronized by the word $w = ab_1 \ldots b_{2^{|T_k|-1}} c$, w is the shortest word that carefully synchronizes $\mathcal{B}_k$ and $|w| = 2^{\frac{n}{2}} + 1$.*

Proof. Since $\mathcal{B}_k$ is one-cluster with respect to a and the level of that cluster is 1, it is straightforward that $Q.a = C$. Immediately from the construction, we have that, for $w_i = ab_1 \ldots b_i$, the equality $Q.w_i = S_i$ holds. From that, we have that $Q.w = c_1$, so w carefully synchronizes $\mathcal{B}_k$. To prove the minimality of w, first notice that, for each i, holds that $S_i.b_j$ is undefined for $j \neq i - 1$ (because each $|S_i| = \frac{n}{2}$ from Lemma 3) and $S_i.a = C$ (by Lemma 3). Then $\mathcal{P}(\mathcal{B}_k)$ forms a path from Q to $\{c_1\}$ labeled with the consecutive letters of w, and for each state on the path, there is only one transition leading to C (visited in the first step of the path) and that ends the proof. $\qquad\square$

4   Complexity

This section is devoted to the proof that the problem of deciding whether a given one-cluster PFA is carefully synchronizing is NP-hard, even for the binary alphabet. Notice that Observation 1 implies that the letter b must be defined for at least all the states in the cycle induced by the letter a. Refer to that problem as 2-ONE-CLUSTER-CARSYNC. Let us state that theorem formally.

Theorem 2. *For a given PFA $\mathcal{A} = (Q, \{a, b\}, \delta)$ that is one-cluster with respect to a, the problem of deciding whether $\mathcal{A}$ has a carefully synchronizing word (2-ONE-CLUSTER-CARSYNC) is NP-hard.*

We construct a polynomial time reduction from 3-SAT to 2-ONE-CLUSTER-CARSYNC to prove the theorem. Let $\{x_1, \ldots, x_n\}$ be a set of variables and $C_1, \ldots, C_m$ be clauses. We also assume that for any pair $\{x_j, \bar{x}_j\}$ if $x_j \in C_i$ then $\bar{x}_j \notin C_i$. That is not really a problem since if both of them belong to C_i, then C_i is always true. For a given formula $\phi = C_1 \wedge \ldots \wedge C_m$, we define the PFA $\mathcal{A}_\phi = (Q, \{a, b\}, \delta_\phi)$. Let $Q = \bigcup_{i=1}^{m} (S_{C_i^t} \cup S_{C_i^f}) \cup P \cup R$ where:

- $P = \{p_1, \ldots, p_m\}$
- $R = \{r_1, \ldots, r_m\}$
- $S_{C_i^t} = \{c_i^{x_1}, \ldots, c_i^{x_n}, x_i^{end}\}$ for each i.
- $S_{C_i^f} = \{\bar{c}_i^{x_1}, \ldots, \bar{c}_i^{x_n}, \bar{x}_i^{end}\}$ for each i.

Define δ_ϕ in the following way:

1. $\delta_\phi(p_i, a) = p_{i+1 \bmod m}$
2. $\delta_\phi(p_i, b) = \bar{c}_i^{x_1}$
3. $\delta_\phi(r_i, a) = p_i$, $\delta_\phi(r_i, b)$ undefined
4. $\delta_\phi(c_i^{end}, a) = \delta_\phi(\bar{c}_i^{end}, a) = r_i$
5. $\delta_\phi(c_i^{end}, b) = p_1$ $\delta_\phi(\bar{c}_i^{end}, b)$ undefined

6. if $x_j \in C_i$ then $\delta_\phi(\bar{c}_i^{x_j}, a) = c_i^{x_{j+1}}$ otherwise $\delta_\phi(\bar{c}_i^{x_j}, a) = \bar{c}_i^{x_{j+1}}$ for $0 < j < n$
7. if $\bar{x}_j \in C_i$ then $\delta_\phi(\bar{c}_i^{x_j}, b) = c_i^{x_{j+1}}$ otherwise $\delta_\phi(\bar{c}_i^{x_j}, b) = \bar{c}_i^{x_{j+1}}$ for $0 < j < n$
8. $\delta_\phi(c_i^{x_j}, a) = \delta_\phi(c_i^{x_j}, b) = c_i^{x_{j+1}}$ for $0 < j < n$
9. if $x_n \in C_i$ then $\delta_\phi(\bar{c}_i^{x_n}, a) = c_i^{end}$ otherwise $\delta_\phi(\bar{c}_i^{x_n}, a) = \bar{c}_i^{end}$
10. if $\bar{x}_n \in C_i$ then $\delta_\phi(\bar{c}_i^{x_n}, b) = c_i^{end}$ otherwise $\delta_\phi(\bar{c}_i^{x_n}, b) = \bar{c}_i^{end}$
11. $\delta_\phi(c_i^{x_n}, a) = \delta_\phi(c_i^{x_n}, b) = c_i^{end}$

For the exemplary formula $\phi_{ex} = (x_1 \vee x_3 \vee x_4) \wedge (x_1 \vee x_2 \vee \bar{x}_3) \wedge (\bar{x}_1 \vee \bar{x}_2 \vee x_4)$, the corresponding $\mathcal{A}_{\phi_{ex}}$ is depicted in Fig. 4. Let us state and prove two lemmas before going further.

Lemma 4. *For any given ϕ, the automaton $\mathcal{A}_\phi$ is one-cluster with respect to a.*

Proof. First, observe that for any ϕ, the set P forms the a-cycle in the automaton $\mathcal{A}_\phi$ (Point 1 of the δ_ϕ definition). It is also easy to see from the definition of $\mathcal{A}_\phi$ that any set $S_{C_i^t}$ together with the states $r_i \in R$ and $p_i \in P$ create a directed path of the form $c_i^{x_1} \to c_i^{x_2} \to \dots \to c_i^{x_n} \to r_i \to p_i$ labeled with a (Points 3, 4 and 8 of the definition of δ_ϕ). Denote this path as p. Denote by v_j a variable of the formula ϕ and $v_j \in \{x_j, \bar{x}_j\}$. Consider now the clause $C_i \in \phi$ and the set $S_{C_i^f}$. Assume that $C_i = (v_{j_1} \vee v_{j_2} \vee v_{j_3})$ and $j_1 < j_2 < j_3$. Any of the variables $v_{j_3+1}, \dots, v_n$ do not belong to the clause C_i, so the set $\{\bar{c}_i^{x_{j_3+1}}, \bar{c}_i^{x_{j_3+2}}, \dots, \bar{c}_i^{x_n}, \bar{c}_i^{end}, r_i\}$ forms a directed path on the letter a (Points 4, 6, 7, 9, 10 of the definition of δ_ϕ). Denote it by p_1. Suppose that $v_{j_3} = \bar{x}_{j_3}$. Then, by Points 6 or 9, state $\bar{c}_i^{x_{j_3}}$ is attached to the first state of the path p_1 with the transition labeled by the letter a. Otherwise, by Points 6 or 9, $\bar{c}_i^{x_{j_3}}$ is attached to the path p by the letter a. Now, we can repeat our argument to the set $\{\bar{c}_i^{x_{j_2+1}}, \dots \bar{c}_i^{x_{j_3-1}}\}$ to show that it forms the path p_2 labeled with the letter a which begins in the state $\bar{c}_i^{x_{j_2+1}}$ and ends in the state $\bar{c}_i^{x_{j_3-1}}$. By Point 6, we obtain that this path is attached by its end to the state $\bar{c}_i^{x_{j_3}}$ by the letter a. Now, we can also repeat our arguments to the state $\bar{c}_i^{x_{j_2+1}}$ to show that it is either attached to p_2 or to p. The same reasoning for the sets $\{\bar{c}_i^{x_{j_1+1}}, \dots \bar{c}_i^{x_{j_2-1}}\}$ and $\{\bar{c}_i^{x_1}, \dots \bar{c}_i^{x_{j_2-1}}\}$ and the state $\bar{c}_i^{x_{j_1}}$ concludes the proof.   □

Lemma 5. *For any given ϕ, if $n_1 < n + 3$ then $\delta_\phi(Q, a^{n_1}b)$ is undefined.*

Proof. Observe that the letter a induces a path on $n + 3$ vertices on the set $S_{C_i^t} \cup \{r_i, p_i\} = S_i$ for each i. That path starts in $c_i^{x_1}$, ends in p_i and the one before last vertex on that path is r_i. Also $Q = \bigcup_{i=1}^{m} S_i$, so for each $n_1 < n + 3$ holds $r_i \in Q.a^{n_1}$ for any $0 < i < m+1$. On the other hand, $\delta_\phi(r_i, b)$ is undefined for each i (Point 3), so $\delta_\phi(Q, a^{n_1}b)$ is undefined and the lemma holds.   □

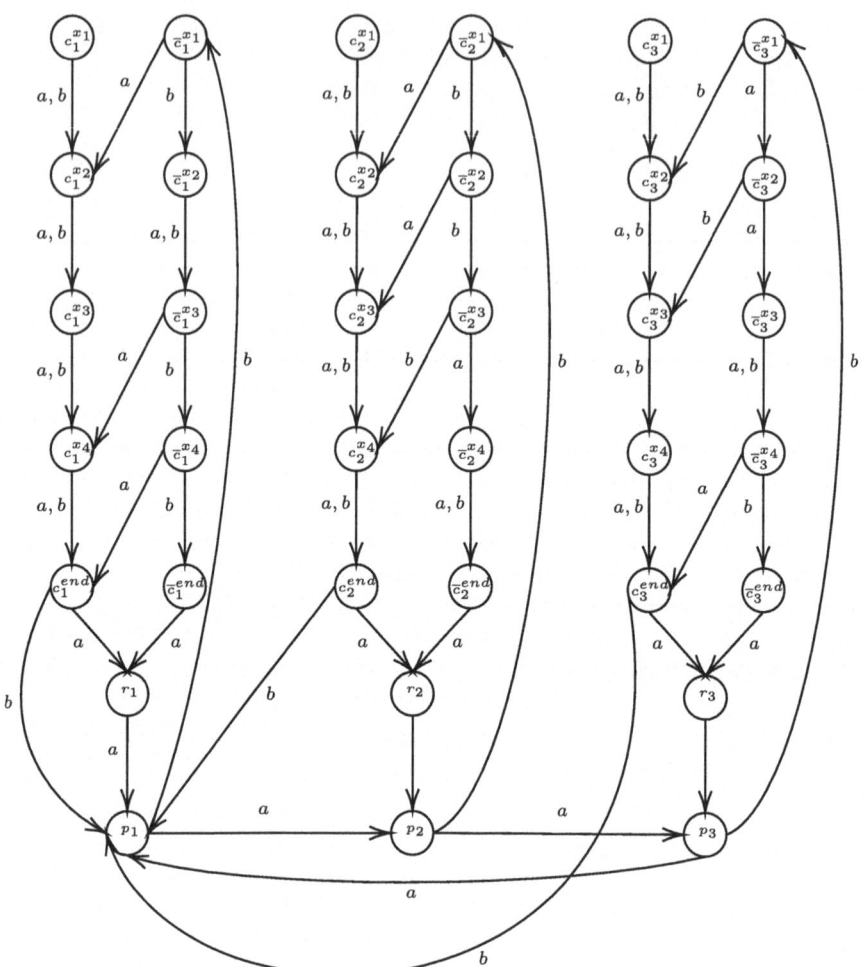

Fig. 4. The automaton $\mathcal{A}_{\phi_{ex}}$

Before moving further let us define, for a given $\mathcal{A}_\phi$, two subsets of its states $S_{init} = \{\bar{c}_1^{x_1}, \ldots, \bar{c}_m^{x_1}\}$ and $S_{end} = \{c_1^{end}, \ldots, c_m^{end}\}$ and state the following observation.

Observation 2. *For any $i \in \{1, \ldots, m\}$ and a formula ϕ, if $v \in \{a, b\}^*$ and $|v| = n$, then $\delta_\phi(\bar{c}_i^{x_1}, v) \in \{c_i^{end}, \bar{c}_i^{end}\}$.*

Proof. Immediate from Points 6 to 11 of the definition of δ_ϕ. □

Lemma 6. *For any given ϕ, the automaton $\mathcal{A}_\phi$ is carefully synchronizing if, and only if, there exists a word w of length n such that $\delta_\phi(S_{init}, w) = S_{end}$.*

Proof. First, assume that there exists a word w of length n such that $\delta_\phi(S_{init}, w) = S_{end}$. Observe that $\delta_\phi(Q, a^{n+3}) = P$ and $\delta_\phi(P, b) = S_{init}$. Also $\delta_\phi(S_{init}, w) = S_{end}$ and $\delta_\phi(S_{end}, b) = p_1$ so we conclude that the word $u = a^{n+3}bwb$ carefully synchronizes $\mathcal{A}_\phi$. Now, assume that $\mathcal{A}_\phi$ is carefully synchronizing. From Lemma 5 we obtain that any carefully synchronizing word for $\mathcal{A}_\phi$, say $u \in \{a, b\}^*$, must start with the word a^k where $k > n + 2$. Since $\delta_\phi(Q, a^{n+3}) = P$ and $\delta_\phi(P, a) = P$ we imply that u must be of the form $a^k bv$. We know that $\delta_\phi(Q, a^k b) = S_{init}$. From Observation 2, for any $v_1 \in \{a, b\}^*$ such that $|v_1| = n$ we have $\delta_\phi(\bar{c}_i^{x_1}, v_1) \in \{c_i^{end}, \bar{c}_i^{end}\}$. On the other hand, observe that for any set of the form $\{t_1, \ldots, t_n\}$, where $t_i \in \{c_i^{end}, \bar{c}_i^{end}\}$ but S_{end}, we have that $\delta_\phi(T, b)$ is undefined, $\delta_\phi(T, a) = R$ and $\delta_\phi(R, b)$ is undefined, and $\delta_\phi(R, a) = P$. So, since u carefully synchronizes $\mathcal{A}_\phi$, there must exist a desired word, and that concludes the proof. □

Lemma 7. *Let ϕ be a formula and $\mathcal{A}_\phi$ be the corresponding automaton. Then ϕ has a truth assignment if, and only if, there exists a word w of length n such that $\delta_\phi(S_{init}, w) = S_{end}$.*

Proof. First, assume that ϕ has a truth assignment $e : \{x_1, \ldots, x_n\} \to \{0, 1\}$. We define the word w of length n in the following way: if $e(x_i) = 1$ in that evaluation then the i-th letter of the word w is a, otherwise the i-th letter is b. Note w_i as the $i - 1$ letter prefix of w and $\bar{w}_i$ as the $i - 1$ letter suffix of w. Let $0 < k_1 < k_2 < k_3 < n + 1$ and consider the clause $C_j = y_{k_1} \vee y_{k_2} \vee y_{k_3}$ where $y_{k_i} \in \{x_{k_i}, \bar{x}_{k_i}\}$. Since e is a truth evaluation, at least one of the variables x_{k_1}, x_{k_2} or x_{k_3} must be evaluated as **true** in the sense that if $y_{k_i} = x_{k_i}$ then $e(x_{k_i}) = 1$, otherwise $e(x_{k_i}) = 0$. It is straightforward from the definition of δ_ϕ (Points 6 and 7) that $\delta_\phi(\bar{c}_j^{x_1}, w_{k_1}) = \bar{c}_j^{x_{k_1}}$. Now consider four pairwise exclusive cases:

1. $y_{k_1} = x_{k_1}$ and $e(x_{k_1}) = 1$
2. $y_{k_1} = x_{k_1}$ and $e(x_{k_1}) = 0$
3. $y_{k_1} = \bar{x}_{k_1}$ and $e(x_{k_1}) = 1$
4. $y_{k_1} = \bar{x}_{k_1}$ and $e(x_{k_1}) = 0$.

Now:

- if Case 1 holds, then the i-th letter of w is a and, by Point 6, we know that $\delta_\phi(\bar{c}_j^{x_{k_1}}, a) = c_j^{x_{k_1}+1}$
- if Case 2 holds, then the i-th letter of w is b and, by Point 7, we know that $\delta_\phi(\bar{c}_j^{x_{k_1}}, b) = \bar{c}_j^{x_{k_1}+1}$
- if Case 3 holds, then the i-th letter of w is a and, by Point 6, we know that $\delta_\phi(\bar{c}_j^{x_{k_1}}, a) = \bar{c}_j^{x_{k_1}+1}$
- if Case 4 holds, then the i-th letter of w is b and, by Point 7, we know that $\delta_\phi(\bar{c}_j^{x_{k_1}}, b) = c_j^{x_{k_1}+1}$.

If Case 1 or 4 hold, then $\delta_\phi(\bar{c}_j^{x_1}, w_{i+1}) = c_j^{x_{k_1}+1}$ and, from Points 8 and 11, we obtain that $\delta_\phi(c_j^{x_{k_1}+1}, \bar{w}_{n-k_1}) = c_j^{end}$. Otherwise, we obtain that $\delta(\bar{c}_j^{x_1}, w_{k_2}) =$

$\delta(\bar{c}_j^{x_{k_2}})$ and we can repeat our case analysis to obtain that either $\delta(\bar{c}_j^{x_1}, w_{k_2+1}) = c_j^{x_{k_2+1}}$ (if y_{k_2} is evaluated as **true**) or $\delta(\bar{c}_j^{x_1}, w_{k_2+1}) = \bar{c}_j^{x_{k_2+1}}$ (if y_{k_2} is evaluated as **false**). With the same argument applied to y_{k_3}, since at least one of the variables of C_j must be evaluated as **true**, we obtain that $\delta_\phi(\bar{c}_j^{x_1}, w) = c_j^{end}$. But since e is a truth evaluation, then any C_j must have at least one variable evaluated as **true**, so for any $j \in \{1, \ldots, m\}$ we have that $\delta_\phi(\bar{c}_j^{x_1}, w) = c_j^{end}$, so we obtain that $\delta(S_{init}, w) = S_{end}$.

Now, assume that there exists a word w of length n, such that $\delta(S_{init}, w) = S_{end}$. We define the evaluation of ϕ, $e_w : \{x_1, \ldots, x_n\} \to \{0, 1\}$ in the following way: if the i-th letter of w is a, then $e_w(x_i) = 1$ otherwise $e_w(x_i) = 0$. Since $\delta_\phi(S_{init}, w) = S_{end}$, then from Observation 2 for any $j \in \{1, \ldots, m\}$ holds $\delta_\phi(\bar{c}_j^{x_1}, w) = c_j^{end}$. Consider e_w, let $0 < k_1 < k_2 < k_3 < n + 1$ and the clause $C_j = y_{k_1} \vee y_{k_2} \vee y_{k_3}$ where $y_{k_i} \in \{x_{k_i}, \bar{x}_{k_i}\}$. It suffices to prove that for any j if $\delta_\phi(\bar{c}_j^{x_1}, w) = c_j^{end}$, then $e_w(y_{k_1}) = 1$ or $e_w(y_{k_1}) = 1$ or $e_w(y_{k_1}) = 1$ (we assume here that $e(\bar{x}) = 1 - e(x)$). For the sake of contradiction suppose, that there exists $j \in \{1, \ldots, m\}$ such that $\delta_\phi(\bar{c}_j^{x_1}, w) = c_j^{end}$ and $e_w(y_{k_1}) = 0$ and $e_w(y_{k_2}) = 0$ and $e_w(y_{k_3}) = 0$. First, observe that $\delta_\phi(\bar{c}_j^{x_1}, w_{k_1}) = \bar{c}_j^{x_{k_1}}$. Since $e(y_{k_1}) = 0$, then (from Points 6 and 7) we obtain that $\delta_\phi(\bar{c}_j^{x_1}, w_{k_1+1}) = \bar{c}_j^{x_{k_1+1}}$. The same argument for y_2 gives us $\delta_\phi(\bar{c}_j^{x_1}, w_{k_2+1}) = \bar{c}_j^{x_{k_2+1}}$ (from Points 6 and 7) and further for y_3 gives us $\delta_\phi(\bar{c}_j^{x_1}, w_{k_3+1}) = \bar{c}_j^{x_{k_3+1}}$ or $\delta_\phi(\bar{c}_j^{x_1}, w_{k_3+1}) = \bar{c}_j^{end}$ (respectively from Points 6 and 7 or 10 and 11). In the latter case, we obtain a contradiction straightforward. In the former case, we notice that, for any $v \in \{a, b\}^*$ such that $|v| = n - k_3$, we obtain that $\delta_\phi(\bar{c}_j^{x_{k_3}}, v) = \bar{c}_j^{end}$ (from Points 6, 7,10 and 11 and the observation that any variable with the index greater than k_3 does not belong to the clause C_j). That concludes the proof.   □

Combining Lemmas 6 and 7, we obtain that the automaton $\mathcal{A}_\phi$ is carefully synchronizing if, and only if, the formula ϕ has a truth evaluation. Taking Lemma 4 into account, the only thing we have to show to finish the proof of Theorem 2 is that the reduction can be performed in polynomial time. Observe that, for any ϕ with n variables and m clauses and corresponding $\mathcal{A}_\phi$, we have that $|P| = |R| = m$ and for any $i \in \{1, \ldots, m\}$ also holds $|S_{C_i^t}| = |S_{C_i^f}| = n + 1$, we obtain, that $|Q| = 2m(n + 2)$ and on the other hand $\mathcal{A}_\phi$ is a partial automaton, what means, that for any state there are at most two outgoing transitions (a and b), so computing the automaton $\mathcal{A}_\phi$ can be done in polynomial time of n and m and that concludes the proof of Theorem 2.

5   Conclusions

In Sect. 3, we have found an infinite family of PFAs with the shortest carefully synchronizing word of exponential length. However, the size of the alphabet used in the construction is also exponential in terms of the number of states, which makes the construction insufficient to analyze the decision problem stated in Sect. 4. An interesting problem to investigate is whether one can achieve the shortest carefully synchronizing word for one-cluster PFAs of exponential length using a smaller alphabet size.

In Sect. 4, we have proven NP-hardness of the problem of deciding whether a given binary one-cluster PFAs can be carefully synchronized. Let $\mathcal{A}$ be a one-cluster PFA with respect to the letter a and let C be the set of states that induces a cycle on the letter a. Further analysis of the proof leads to a conclusion that even the problem of deciding whether there exists a word w such that $|C.w| < |C|$ is NP-hard. Let us make a simple observation:

Observation 3. *Let $\mathcal{A} = (Q, \Sigma, \delta)$ be a PFA, and $w \in \Sigma^*$. There exists an algorithm that, in $O(|w||Q|)$ time complexity, decides whether w is a carefully synchronizing word for $\mathcal{A}$.*

Proof. Consider the algorithm defined below:

$P \leftarrow Q$
for $i = 0, i < |w|, i \leftarrow i + 1$ **do**
    $P \leftarrow P.w[i]$
end for
if $|P| = 1$ **then**
    return true
else
    return false
end if

Obviously the algorithm answers true if, and only if, w carefully synchronizes $\mathcal{A}$ and requires $O(|Q|)$ additional space complexity and $O(|Q||w|)$ time complexity, what concludes the proof.

From Observation 3, we can infer that one possibility to prove that the problem 2-ONE-CLUSTER-CARSYNC is in NP is to show that the shortest carefully synchronizing word for binary one-cluster automata is of polynomial length. It is a problem we want to investigate further. Lemma 1 can be utilized to obtain that result. One must show first that if there exists a word w such that $|C.w| < C$ where C is the set of states inducing a cycle on the letter a, then such a word is of polynomial length in terms of the set of states. The next question is how to synchronize the remaining half of the states of the cycle.

References

1. Béal, M.-P., Berlinkov, M.V., Perrin, D.: A quadratic upper bound on the size of a synchronizing word in one-cluster automata. Int. J. Found. Comput. Sci. **22**(2), 277–288 (2011). https://doi.org/10.1142/S0129054111008039
2. Berlinkov, M.V.: On two algorithmic problems about synchronizing automata. In: Shur, A.M., Volkov, M.V. (eds.) DLT 2014. LNCS, vol. 8633, pp. 61–67. Springer, Cham (2014). https://doi.org/10.1007/978-3-319-09698-8_6

3. Berlinkov, M.V., Ferens, R., Ryzhikov, A., Szyku, M.:? Synchronizing strongly connected partial DFAs. In: Bläser, M., Monmege, B. (eds.), 38th International Symposium on Theoretical Aspects of Computer Science, STACS 2021, 16–19 March 2021, Saarbrücken, Germany (Virtual Conference), volume 187 of *LIPIcs*, pp. 12:1–12:16. Schloss Dagstuhl - Leibniz-Zentrum für Informatik (2021). https://doi.org/10.4230/LIPIcs.STACS.2021.12, https://doi.org/10.4230/LIPICS.STACS.2021.12.

4. Berlinkov, M.V., Szyku, M.: Algebraic synchronization criterion and computing reset words. Inf. Sci. **3690**, 718–730 (2016). https://doi.org/10.1016/j.ins.2016.07.049

5. Biskup, M.T., Plandowski, W.: Shortest synchronizing strings for Huffman codes. Theor. Comput. Sci. **410**(38-40), 3925–3941 (2009). https://doi.org/10.1016/j.tcs.2009.06.005,

6. de Bondt, M., Don, H., Zantema, H.: Lower bounds for synchronizing word lengths in partial automata. Int. J. Found. Comput. Sci. **30**(1), 29–60 (2019)

7. Doyen, L., Massart, T., Shirmohammadi, M.: Robust synchronization in markov decision processes. In: Baldan, P., Gorla, D. (eds.) CONCUR 2014. LNCS, vol. 8704, pp. 234–248. Springer, Heidelberg (2014). https://doi.org/10.1007/978-3-662-44584-6_17

8. Doyen, L., Massart, T., Shirmohammadi, M.: The complexity of synchronizing Markov decision processes. J. Comput. Syst. Sci. **100**, 96–129 (2019). https://doi.org/10.1016/j.jcss.2018.09.004

9. Eppstein, D.: Reset sequences for monotonic automata. SIAM J. Comput. **19**(3), 500–510 (1990). https://doi.org/10.1137/0219033

10. Gazdag, Z., Iván, S., Nagy-György, J.: Improved upper bounds on synchronizing nondeterministic automata. Inf. Process. Lett. **109**(17), 986–990 (2009). https://doi.org/10.1016/j.ipl.2009.05.007

11. Imreh, B., Steinby, M.: Directable nondeterministic automata. Acta Cybern. **14**(1), 105–115 (1999). https://cyber.bibl.u-szeged.hu/index.php/actcybern/article/view/3514

12. Ito, M., Shikishima-Tsuji, K.: Some results on directable automata. In: Karhumäki, J., Maurer, H., Păun, G., Rozenberg, G. (eds.) Theory Is Forever. LNCS, vol. 3113, pp. 125–133. Springer, Heidelberg (2004). https://doi.org/10.1007/978-3-540-27812-2_12

13. Jürgensen, H.: Synchronization. Inf. Comput. **206**(9-10), 1033–1044 (2008). https://doi.org/10.1016/J.IC.2008.03.005

14. Kari, J.: A counter example to a conjecture concerning synchronizing words in finite automata. Bull. EATCS **73**, 146 (2001)

15. Kari, J.: Synchronizing finite automata on Eulerian digraphs. Theor. Comput. Sci. **295**, 223–232 (2003). https://doi.org/10.1016/S0304-3975(02)00405-X

16. Martyugin, P.V.: A lower bound for the length of the shortest carefully synchronizing words. Russian Math. **54**, 46–54 (2010)

17. Martyugin, P.V.: Careful synchronization of partial automata with restricted alphabets. In: Bulatov, A.A., Shur, A.M. (eds.) CSR 2013. LNCS, vol. 7913, pp. 76–87. Springer, Heidelberg (2013). https://doi.org/10.1007/978-3-642-38536-0_7

18. Martyugin, P.V.: Computational complexity of certain problems related to carefully synchronizing words for partial automata and directing words for nondeterministic automata. Theory Comput. Syst. **54**(2), 293–304 (2014). https://doi.org/10.1007/s00224-013-9516-6

19. Natarajan, B.K.: An algorithmic approach to the automated design of parts Orienters. In: 27th Annual Symposium on Foundations of Computer Science, Toronto,

Canada, 27-29 October 1986, pp. 132–142. IEEE Computer Society (1986). https://doi.org/10.1109/SFCS.1986.5

20. Pin, J.: On two combinatorial problems arising from automata theory. In: Berge, C., Bresson, D., Camion, P., Maurras, J.F., Sterboul, F. (eds.) Combinatorial Mathematics, volume 75 of North-Holland Mathematics Studies, pp. 535–548. North-Holland (1983). https://www.sciencedirect.com/science/article/pii/S0304020808734327, https://doi.org/10.1016/S0304-0208(08)73432-7

21. Rystsov, I.: Asymptotic estimate of the length of a diagnostic word for a finite automaton. Cybernetics **16**, 194–198 (1980)

22. Rystsov, I.: Reset words for commutative and solvable automata. Theor. Comput. Sci. **172**(1–2), 273–279 (1997). https://doi.org/10.1016/S0304-3975(96)00136-3

23. Sandberg, S.: 1 homing and synchronizing sequences. In: Broy, M., Jonsson, B., Katoen, J.-P., Leucker, M., Pretschner, A. (eds.) Model-Based Testing of Reactive Systems. LNCS, vol. 3472, pp. 5–33. Springer, Heidelberg (2005). https://doi.org/10.1007/11498490_2

24. Steinberg, B.: The Černý conjecture for one-cluster automata with prime length cycle. Theor. Comput. Sci. **412**(39), 5487–5491 (2011). https://doi.org/10.1016/j.tcs.2011.06.012

25. Szykuła, M.: Improving the upper bound on the length of the shortest reset word. In: Niedermeier, R., Vallée, B. (eds.) 35th Symposium on Theoretical Aspects of Computer Science, STACS 2018, February 28 to March 3, 2018, Caen, France, volume 96 of *LIPIcs*, pp. 56:1–56:13. Schloss Dagstuhl - Leibniz-Zentrum für Informatik (2018). https://doi.org/10.4230/LIPIcs.STACS.2018.56

26. Trahtman, A.: The Černý conjecture for aperiodic automata. Discret. Math. Theor. Comput. Sci. **9**(2) (2007). https://doi.org/10.46298/dmtcs.395

27. Černý, J.: Poznámka k homogénnym eksperimentom s konečnými automatami. Mat.-Fyz. Cas. Slovens. Akad. Vied. **14**, 208–216 (1964)

28. Volkov, M.V.: Synchronizing automata and the Černý conjecture. In: Martín-Vide, C., Otto, F., Fernau, H. (eds.) LATA 2008. LNCS, vol. 5196, pp. 11–27. Springer, Heidelberg (2008). https://doi.org/10.1007/978-3-540-88282-4_4

29. Volkov, M.V.: Slowly synchronizing automata with idempotent letters of low rank. J. Autom. Lang. Comb. **24**(2-4):375–386 (2019). https://doi.org/10.25596/jalc-2019-375

30. Vorel, V.: Subset synchronization and careful synchronization of binary finite automata. Int. J. Found. Comput. Sci. **27**(5), 557–578 (2016). https://doi.org/10.1142/S0129054116500167

Verifying and Interpreting Neural Networks Using Finite Automata

Marco Sälzer$^{(\boxtimes)}$ ⃝, Eric Alsmann ⃝, Florian Bruse ⃝, and Martin Lange ⃝

School of Electrical Engineering and Computer Science, University of Kassel,
Kassel, Germany
{marco.saelzer,eric.alsmann,florian.bruse,martin.lange}@uni-kassel.de

Abstract. Verifying properties and interpreting the behaviour of deep neural networks (DNN) is an important task given their ubiquitous use in applications, including safety-critical ones, and their black-box nature. We propose an automata-theoretic approach to tackling problems arising in DNN analysis. We show that the input-output behaviour of a DNN can be captured precisely by a (special) weak Büchi automaton and we show how these can be used to address common verification and interpretation tasks of DNN like adversarial robustness or minimum sufficient reasons.

Keywords: neural networks · finite state automata · verification · interpretation

1 Introduction

Deep Neural Networks (DNN), trained using task-oriented and precisely crafted techniques, are the driving force of all modern deep learning applications, which have produced astonishing results: highly-developed driving assistants [8], the overcoming of language barriers due to neural machine translation [18], far-reaching support in early disease detection [14], the creation of inspiring art from textual user inputs [19,20], etc.

Their striking performance comes with a downside: they are a black box. While it is easy to describe structure and parameters of a DNN, it is hard to obtain reliable predictions for or explanation of their behaviour. Deep learning techniques need to be reliable, though, especially in safety-critical applications. However, certifying that some DNN satisfies specific safety properties, formally called *verifying* these properties, is difficult. The verification of a common safety property for a DNN N is informally described by the question "is there an input $\bar{x}$ of interest such that the output $N(\bar{x})$ has some unwanted characteristics?" The corresponding decision problem, formally called *output reachability*, is NP-complete [11], even for completely shallow DNN and simple specifications of relevant inputs and outputs [21]. Furthermore, DNN-based applications require comprehensible explanations of the outputs generated by a DNN due to legal, safety and ethical concerns. There is a need for techniques giving understandable

J. D. Day and F. Manea (Eds.): DLT 2024, LNCS 14791, pp. 266–281, 2024.
https://doi.org/10.1007/978-3-031-66159-4_19

explanations for DNN behaviour; this is formally known as *interpreting* DNN. A typical interpretation task for some DNN N and an input-output pair $(\overline{x}, N(\overline{x}))$ is to answer the question "which features of $\overline{x}$ are the relevant ones leading to the output $N(\overline{x})$?" A corresponding decision problem, called the MINIMUMSUF-FICIENTREASON problem, is known to be Σ_2^P-complete [2].

We propose an approach based on finite-state-automata for tackling various formal reasoning problems, arising from the black-box nature of DNN. A DNN N computes a function of type $\mathbb{R}^m \to \mathbb{R}^n$ for some $m, n \in \mathbb{N}$, which induces a relation $R_N \subseteq \mathbb{R}^m \times \mathbb{R}^n$. Using an appropriate encoding, R_N can be represented by a set of infinite words over an alphabet of $(m+n)$-*track symbols* of the form $(a_1, \ldots, a_m, b_1, \ldots, b_n)$ where a_i, b_i are taken from an alphabet like $\{0, 1, ., +, -\}$. A finite-state automaton $\mathcal{A}$ over such $(m+n)$-*track words* can be seen as a (synchronised) transducer between input symbols $(a_1, \ldots, a_m)$ and output symbols $(b_1, \ldots, b_n)$. Synchronicity guarantees regularity of the automata's languages [3]. We present a complete construction of a weak Büchi automaton that recognises the input-output relation induced by a DNN. Weak Büchi automata are known to allow for more efficient algorithms than general Büchi automata, as they can also be seen as co-Büchi automata. In fact, we show that not even the full power of weak automata is needed but a special subclass suffices. We use this translation to show how relevant problems regarding the verification and interpretation of DNN like output reachability or finding a minimum sufficient reason can be addressed using this construction and standard automata-theoretic machinery.

In Sect. 2 we give preliminary definitions regarding DNN, encodings of reals and Büchi automata. In Sect. 3, we introduce common verification and interpretation problems regarding DNN and show that they can be tackled using the translation from DNN to automata described in Sect 4. In Sect. 5, we summarise and discuss possible future work. The details of some technical lemmas are omitted for reasons of space limitations, but a full version of this paper is available via ArXiv [23].

Related work. The work presented here falls into the intersection of neural-network-based machine learning on the one hand, and automata theory on the other. Extra focus is on the use of automata-theoretic tools for tackling challenges on the machine learning side. Most ongoing research in this area is focused on the combination of automata and so-called recurrent neural networks (RNN), a model for processing sequential data [1,13,16,24]. Additionally, there was extensive research on automata and RNN in earlier days of neural network analysis. A good overview of this is given in [10]. The common underlying theme there is to obtain finite-state automata, often DFA, which capture the dynamic behaviour of RNN. The goal of our approach here is similar, yet there are two fundamental differences: first, the techniques mentioned above only work for RNN, while our approach can be applied to more general neural network models, including linear layers with piece-wise linear activations. However, it is open how far our proposed approach generalises. Second, the automata derived from RNN work on sequences of data points, where each single data point is a symbol. Finite alphabets are obtained by finitely partitioning the real-valued input space of an

268 M. Sälzer et al.

RNN. Our approach yields automata working on single, encoded data points. By using nondeterministic Büchi automata (NBA), we retain full precision regarding the input space. Xu et al. [25] present an active-learning based algorithm for extracting DFA from neural network classifiers. Similar to our approach, these DFA work on encoded inputs of the neural network. Since they use abstraction techniques, the resulting on-tape automata only approximate the behaviour of the neural network.

Use cases of our translation from DNN to finite-state automata explored in this paper include verification and interpretation of DNN. A comprehensive survey on DNN verification is given by Huang et al. [9], one on the state-of-the-art regarding DNN interpretation is given by Zhang et al. [26].

It is also not hard to see that the problems in DNN verification and interpretation considered here can be expressed in the (decidable) theory of the reals with addition and multiplication by rational constants. Interestingly, weak Büchi automata – which avoid most intrinsically difficult constructions for general Büchi automata – can be used to decide this theory [4,5]. We remark, though, that DNN do not need the full power of this logic but only the existential-positive fragment. It is therefore reasonable to construct weak Büchi automata for DNN directly instead of going through the more powerful general theory of the reals.

2 Preliminaries

For a k-dimensional vector $v \in A^k$ with $k \geq 1$ and some set A, we denote its components by $v_1, \ldots, v_k$ respectively. Sometimes, we write vectors like $\overline{x}, \overline{v}, \ldots$ to stress their vector nature.

(Deep) Neural Networks. A *(DNN-)node* is a function $v \colon \mathbb{R}^k \to \mathbb{R}$ with $v(\overline{x}) = \sigma(\sum_{i=1}^{k} c_i x_i + b)$, where k is the *input dimension*, the $c_i \in \mathbb{Q}$ are called *weights*, $b \in \mathbb{Q}$ is the *bias* and $\sigma : \mathbb{R} \to \mathbb{R}$ is the *activation function* of v.[1] A common activation function is the piecewise linear *ReLU* function (for *Rectified Linear Unit*), defined as $relu(x) = \max(0, x)$. A *(DNN-)layer* l is a tuple of some n nodes $(v_1, \ldots, v_n)$ where each node has the same input dimension m and the same activation function. It computes the function $l \colon \mathbb{R}^m \to \mathbb{R}^n$ via $l(\overline{x}) = (v_1(\overline{x}), \ldots, v_n(\overline{x}))$. We call m the *input* and n the *output dimension* of l. A *Deep Neural Network (DNN)* N consists of k layers $l_1, \ldots, l_k$, where l_1 has input dimension m, the output dimension of l_i is equal to the input dimension of l_{i+1} for $i < k$ and the output dimension of l_k is n. The DNN N computes a function from $\mathbb{R}^m$ to $\mathbb{R}^n$ by $N(\overline{x}) = l_k(l_{k-1}(\ldots l_1(\overline{x}) \ldots))$.

In order to estimate the asymptotic complexity of the proposed translation from DNN to finite-state automata, we introduce the following (approximate) size measures. For $c \in \mathbb{Q}$ let $\|c\| := \log |n| + \log d$ where d is the smallest positive

---

[1] The literature allows weights and biases from $\mathbb{R}$. Since we study effective translations, DNN need to be finitely representable so we require the values to be rational.

natural number s.t. $\frac{n}{d} = c$ with $n \in \mathbb{Z}$. Accordingly, we define the size of a DNN-node v computing $\sum_{i=1}^{k} c_i x_i + b$ as $\|v\| = \sum_{i=1}^{k} \|c_i\| + \|b\|$ and the size of a DNN N with a total of k nodes $v_1, \ldots, v_k$ as $\|N\| = \sum_{i=1}^{k} \|v_i\|$.

Weak Büchi Automata. Let Σ be an alphabet. As usual, let Σ^* and Σ^ω denote the set of all finite, resp. infinite words over Σ. A *nondeterministic Büchi automaton* (NBA) is a tuple $\mathcal{A} = (Q, \Sigma, q_0, \delta, F)$ s.t. Q is a finite set of states, Σ is the underlying alphabet, $q_0 \in Q$ is a designated starting state, $\delta \subseteq Q \times \Sigma \times Q$ is the transition relation, and $F \subseteq Q$ is a designated set of accepting states. The *size* of $\mathcal{A}$ is measured as $|\mathcal{A}| := |Q|$. A run on an infinite word $w = a_0 a_1 \ldots$ is an infinite sequence $\rho = q_0, q_1, \ldots$ starting in the initial state and satisfying $(q_i, a_i, q_{i+1}) \in \delta$ for all $i \geq 0$. It is accepting if $q_i \in F$ for infinitely many i. The language of an NBA $\mathcal{A}$ is $L(\mathcal{A}) = \{w \in \Sigma^\omega \mid \text{there is an accepting run of } \mathcal{A} \text{ on } w\}$.

A *weak (nondeterministic) Büchi automaton* (WNBA) is an NBA whose state set Q can be partitioned into strongly connected components (SCC) such that for each SCC $S \subseteq Q$ we have $S \subseteq F$ or $S \cap F = \emptyset$, i.e. each SCC either consists of accepting states or non-accepting states only. It is known that WNBA are less expressive than NBA, for example there is no WNBA accepting $(a^* b)^\omega$. For the purposes developed here, namely the recognition of relations of real numbers defined by arithmetical operations, weak NBA suffice, which has been observed before [4]. The benefit of using WNBA comes from better algorithmic properties: whilst, for example, determinisation is notoriously difficult for general NBA, it is much simpler for WNBA as they can also be seen as co-Büchi automata that are easier to determinise [17]. Likewise, minimisation is quite important for practical applications, and just like determinisation, minimisation of general Büchi automata is more difficult than it is for automata on finite words, while algorithms for those can typically be lifted to weak Büchi automata, cf. [15].

In fact, it turns out that we do not even need the full power of WNBA either. An *eventually-always weak nondeterministic Büchi automaton* (WNBA$_{\mathsf{FG}}$) is a WNBA such that every path through its state set contains at most one transition from a non-accepting to an accepting state and no transitions from accepting to non-accepting ones. In other words, every accepting run is of the form $(Q \backslash F)^* F^\omega$. Furthermore, WNBA$_{\mathsf{FG}}$ are closed under unions and intersections, using the usual product construction and appropriate sets of accepting states. Later we reduce decision problems on DNN to automata-theoretic ones. We therefore need to argue that the corresponding problems on the automata side are (efficiently) decidable. For the DNN problems considered here, language emptiness suffices, and more complex problems like inclusion are not needed. The following is well-known for (weak) Büchi automata.

Proposition 1. *Emptiness of a WNBA$_{\mathsf{FG}}$ $\mathcal{A}$ is decidable in time linear in $|\mathcal{A}|$.*

Encodings of Reals. In the following, let $\Sigma = \{+, -, ., 0, 1\}$ unless stated explicitly otherwise. A word $w = s a_{n-1} \ldots a_0 . b_0 b_1 \ldots$ with $n \geq 0$, $s \in \{+, -\}$, $a_i, b_i \in \{0, 1\}$ uniquely encodes a real value $dec(w) := (-1)^{\text{sign}(s)} \cdot (\sum_{i=0}^{n-1} a_i \cdot$

$2^i + \sum_{i=0}^{\infty} b_i \cdot 2^{-(i+1)})$ where $\text{sign}(s) = 0$ if $s = +$ and $\text{sign}(s) = 1$ otherwise, and $\sum_{i=0}^{-1} f_i = 0$ for any subterm f_i. Note that the infinite sum on the right is always converging. Moreover, while the decoding $dec(w)$ of a word w is unique, the encoding $enc(r)$ of any $r \in \mathbb{R}$ as such a word in binary representation is not necessarily unique, for three reasons: leading zeros change the word representation but not the underlying value, both $+0.0^{\omega}$ and -0.0^{ω} represent the same value, namely 0, and any number whose representation has a suffix of the form 10^{ω} (possibly including a dot) also can be written with the suffix 01^{ω} instead. For instance, the number 12 has representations $+1100.0^{\omega}$ and $+1011.1^{\omega}$.

WNBA$_{\mathsf{FG}}$ for Relations of Reals. Let $k \geq 1$. We denote with Σ^k the alphabet consisting of all k-vectors of letters from Σ, using both vertical (as below) and horizontal vector notation (like $[s_1, \ldots, s_k]$) for convenience. A word over Σ^k is *well-formed* if it is of the form $\bar{s}\,\bar{a}_n \cdots \bar{a}_0\,\bar{d}\,\bar{b}_0\,\bar{b}_1 \cdots$ with $s_i \in \{+, -\}$, $a_{i,j}, b_{i,j} \in \{0, 1\}$ and $\bar{d}$ being the vector of k dot-symbols. I.e. it starts with signs on all tracks, and each track contains exactly one dot, and these are all aligned. Such a word induces a k-tuple $(w_1, \ldots, w_k)$ of words over Σ in the straightforward way: w_i is represented by the Σ-word $s_i a_{i,n-1} \ldots a_{i,0}.b_{i,0} b_{i,1} \ldots$ as above. For example, let $k = 2$ and

$$w = \begin{bmatrix} - \\ + \end{bmatrix} \begin{bmatrix} 0 \\ 1 \end{bmatrix} \begin{bmatrix} 1 \\ 0 \end{bmatrix} \begin{bmatrix} 1 \\ 1 \end{bmatrix} \begin{bmatrix} 0 \\ 0 \end{bmatrix} \begin{bmatrix} 1 \\ 0 \end{bmatrix} \begin{bmatrix} . \\ . \end{bmatrix} \begin{bmatrix} 0 \\ 1 \end{bmatrix} \left(\begin{bmatrix} 1 \\ 0 \end{bmatrix} \begin{bmatrix} 1 \\ 1 \end{bmatrix} \right)^{\omega}.$$

It induces words w_1, w_2 that represent the numbers $dec(w_1) = -13.5$ as well as $dec(w_2) = 20\frac{2}{3}$. In the following, we will restrict our attention to well-formed words and write WF_{Σ}^k for the set of all such well-formed k-track words. It is definable by a WNBA$_{\mathsf{FG}}$ for any $k \geq 1$, namely the following one.

Using that WNBA$_{\mathsf{FG}}$ are closed under intersection, w.l.o.g. we assume that all words are well-formed. By the correspondence of a (well-formed) word from $(\Sigma^k)^{\omega}$ to k words from Σ^{ω}, we can view the language of a WNBA$_{\mathsf{FG}}$ over the alphabet Σ^k as a k-ary *relation* of words $(w_1, \ldots, w_k)$ and, by the use of the decoding function dec, as a k-ary relation of real numbers $(dec(w_1), \ldots, dec(w_k))$. We will therefore write $R(\mathcal{A})$ instead of $L(\mathcal{A})$ to denote the *relation* of the automaton $\mathcal{A}$ which, technically, is just its language of the multi-track alphabet.

We will need closure of the class of WNBA$_{\mathsf{FG}}$-definable languages under several (arithmetical) operations which can be derived from two further basic ones: projections and products. Given a k-ary relation R and a tuple $\pi = (i_1, \ldots, i_n)$ with $i_j \in \{1, \ldots, k\}$ for all j, the π-*projection* of R is the n-ary relation $(R){\downarrow}_{\pi} := \{(w_{i_1}, \ldots, w_{i_n}) \mid (w_1, \ldots, w_k) \in R\}$. The following is proved by a standard construction that applies the projection pointwise to the tuples in each transition.

Lemma 1. *Let $\mathcal{A}$ be a $WNBA_{FG}$ s.t. $R(\mathcal{A})$ is k-ary for some $k \geq 1$. Let $\pi \in \{1, \ldots, k\}^n$. There is a $WNBA_{FG}$ $(\mathcal{A})\downarrow_\pi$ of size $\mathcal{O}(|\mathcal{A}|)$ s.t. $R((\mathcal{A})\downarrow_\pi) = (R(\mathcal{A}))\downarrow_\pi$.*

Whilst, technically, the projection operation can be used to duplicate and re-arrange tracks in a multi-tracked word, we mostly use it to delete tracks. For example, if R is a 3-ary relation, then $R\downarrow_{(1,3)}$ results from the collection of all tuples that are obtained by deleting the second component in a triple from R.

Next, let R_1 be a k_1-ary and R_2 be a k_2-ary relation. The *Cartesian product* is, as usual, the $(k_1 + k_2)$-ary relation $R_1 \times R_2 := \{(w_1, \ldots, w_{k_1}, v_1, \ldots, v_{k_2}) \mid (w_1, \ldots, w_{k_1}) \in R_1, (v_1, \ldots, v_{k_2}) \in R_2\}$. The following lemma can also be proved by a standard product construction.

Lemma 2. *Let $\mathcal{A}_1, \mathcal{A}_2$ be two $WNBA_{FG}$ recognising a k_1-, resp. k_2-ary relation. There is a $WNBA_{FG}$ $\mathcal{A}_1 \times \mathcal{A}_2$ of size $\mathcal{O}(|\mathcal{A}_1| \cdot |\mathcal{A}_2|)$ s.t. $R(\mathcal{A}_1 \times \mathcal{A}_2) = R(\mathcal{A}_1) \times R(\mathcal{A}_2)$.*

Let $1 \leq i, j \leq k$. It is easy to construct an automaton which accepts some $w \in WF_\Sigma^k$ iff $w_i = w_j$, i.e. that checks for equality in the word representation of two numbers in a tuple. We need a more relaxed operation, namely an automaton that accepts such a k-track word iff the i-th and j-th track represent the same number, possibly using different representations of it. Note for example that $+0.1^\omega$ and $+1.0^\omega$, or $+.0^\omega$ and $-.0^\omega$ represent the same number in each case. Luckily, these two examples already show all the possibilities to create different representations of the same number in well-formed multi-track words, and these situations can be recognised by a $WNBA_{FG}$. The following lemma is easily proved by constructing an $WNBA_{FG}$ that compares the i- and j-th tracks in an input word and accepts if they are either entirely equal, or equal up to a point after which one is 10^ω and the other is 01^ω.

Lemma 3. *Let $k \geq 2$, $1 \leq i < j \leq k$. There is a $WNBA_{FG}$ $\mathcal{A}_{i=j}^k$ of size $\mathcal{O}(1)$ such that $R(\mathcal{A}_{i=j}^k) = \{w \in WF_\Sigma^k \mid dec(w_i) = dec(w_j)\}$.*

The automata $\mathcal{A}_{i=j}^k$ are only used as auxiliary devices to form the closure of certain operations under different number representation. As such, they are distinguished from other automata that we construct in the sense that most of them operate on words of k tracks which can be divided into m *input* tracks and n *output* tracks, s.t. $k = m + n$. There is no technical difference between input and output tracks, though; the distinction is just useful in the specification of certain operations.

In such a setup it is natural to generalise the composition of two binary relations to ones of arbitrary, but matching arities. Suppose R_1 and R_2 are relations of arities k_1, resp. k_2, and $k \leq \min\{k_1, k_2\}$. We regard R_1's last k tracks as its output and R_2's k first tracks as its input. Then $R_1 \circ_k R_2$ is defined as $\{(u_1, \ldots, u_{k_1}, w_1, \ldots, w_{k_2}) \mid \exists v_1, \ldots, v_k \text{ s.t. } (u_1, \ldots, u_{k_1}, v_1, \ldots, v_k) \in R_1, (v_1, \ldots, v_k, w_1, \ldots, w_{k_2}) \in R_2\}$. We observe, for later constructions, that the class of WNBA_{FG}-definable languages is closed under such generalised composi-tions. The proof is, again, by a standard product construction, involving Lem-mas 3 and 1 to ensure equality and existence of the common factors $v_1, \ldots, v_k$.

Lemma 4. *For $i \in \{1, 2\}$ let $\mathcal{A}_i$ be a WNBA_{FG} recognising a k_i-ary relation, and let $k \leq \min\{k_1, k_2\}$. There is a WNBA_{FG} $\mathcal{A}_1 \circ_k \mathcal{A}_2$ of size $\mathcal{O}(|\mathcal{A}_1| \cdot |\mathcal{A}_2|)$ s.t. $R(\mathcal{A}_1 \circ_k \mathcal{A}_2) = R(\mathcal{A}_1) \circ_k R(\mathcal{A}_2)$.*

We remark that Propsition 1 – emptiness checks in time linear in the num-ber of states – is of course true for multi-track WNBA_{FG} as well. However, their alphabet Σ^k is of size exponential in k, and this can lead to a number of tran-sitions that is exponential in k. However, on n states there can be at most n^2 many different transitions which calls for symbolic representations (e.g. binary decision diagrams) of Σ^k in actual implementations. Also, the automata derived from DNN will be of exponential size in which case the possibly exponential size of the alphabet does not affect the statements made in the following on asymp-totic complexity. For the sake of simplicity, these are made with regards to the number of states of an automaton.

3   Reasoning About DNN Using WNBA_{FG}

We consider two topics – formal verification and interpretation of DNN. The former is concerned with different safety properties, among which adversarial robustness and output reachability guarantees belong to the most important ones. Interpretation of DNN is concerned with techniques generating human-understandable explanations for the behaviour of DNN, for example, an expla-nation why a DNN computes some specific output given some input. Let N be a DNN. Throughout this section, we assume that there is a WNBA_{FG} $\mathcal{A}_N$ that captures the relation of input-output pairs induced by N. We formally prove the existence of $\mathcal{A}_N$ in Sect. 4.

3.1   Verifying DNN Using WNBA_{FG}

Adversarial Robustness. This is exclusively concerned with *classifier DNN*. A classifier DNN is used to assign to a given input one of the *classes* $\{c_1, \ldots, c_k\}$. Typically, such a classifier N is built as follows: N consists of a DNN N' with output dimension k and an additional *softmax layer*. It consumes the output $(y_1, \ldots, y_k)$ of N' and computes a probability for each class c_i. The input $\overline{x}$ is then said to be classified into c_i if its probability is maximal. However, the actual assigned class can be directly inferred from the output of N' by assigning class c_j to $\overline{x}$ such that j is a maximal output dimension. A formal definition of adversarial robustness relies on a distance measure on real vectors. One that is commonly used is the one induced by the 1-norm of vectors [9]. Let $\overline{r} \in \mathbb{R}^m$. Its *1-norm* is $||\overline{r}||_1 = \sum_{i=1}^{m} |r_i|$. It induces the *Manhattan distance* of $\overline{r}$ and some $\overline{s} \in \mathbb{R}^m$, defined as $||\overline{r}, \overline{s}||_1 = \sum_{i=1}^{m} |r_i - s_i|$. The *d-neighbourhood* with $d \in \mathbb{R}^{\geq 0}$ of $\overline{r}$ is defined as the set $\{\overline{s} \in \mathbb{R}^m \mid ||\overline{r}, \overline{s}||_1 \leq d\}$. Let N be a classifier

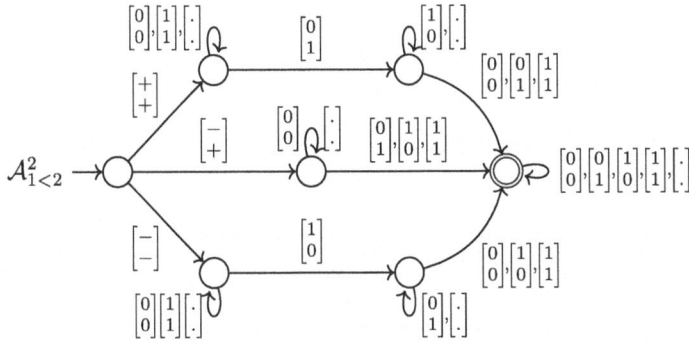

Fig. 1. Automaton recognizing that the number encoded on the first track is smaller than the number encoded on the second track.

with input dimension m and output dimension n and $1 \le h \le n$. We call a triple $P = (\bar{r}, d, h)$ with $\bar{r} \in \mathbb{Q}^m, d \in \mathbb{Q}$ an *adversarial robustness property* (ARP) and say that N satisfies P, written $N \models P$, if $N(\bar{r}')_h > N(\bar{r}')_{h'}$ for all $h' \ne h$ and all $\bar{r}'$ in the d-neighbourhood of $\bar{r}$. In other words, the entire d-neighbourhood is classified as belonging to class c_h. We measure the size of P by $\|P\| = \|\bar{r}\| + \|d\| + \|h\|$ where $\|\bar{r}\|$ is the sum of the measure of its elements $\|r_i\|$.

For a DNN N and and ARP P we combine three WNBA$_{\text{FG}}$ $\mathcal{A}^P_{\text{in}}, \mathcal{A}^P_{\text{out}}$ and $\mathcal{A}_N$ to a WNBA$_{\text{FG}}$ that can be used to check whether $N \models P$ holds. Based on the explanations above we disregard the softmax layer of N, giving a usual DNN. Then $\mathcal{A}_N$ is defined by Theorem 4. The automaton $\mathcal{A}^P_{\text{in}}$ should accept words corresponding to vectors $\bar{x}$ that are included in the d-neighbourhood of $\bar{r}$, i.e. for which $\sum_{j=1}^m |r_j + (-1 \cdot x_j)| \le d$ holds. To specify automata recognizing the validity ARPs, we need to formalise comparison of vector components using WNBA$_{\text{FG}}$.

Lemma 5. *Let $k \ge 2$, $1 \le i, j \le k$. There are WNBA$_{\text{FG}}$ $\mathcal{A}^k_{i<j}$ and $\mathcal{A}^k_{i \le j}$ of size $\mathcal{O}(1)$ s.t. for all $w \subset WF^k_\Sigma$ we have $w \in R(\mathcal{A}_<)$ iff $dec(w_i) < dec(w_j)$, resp. $dec(w_i) \le dec(w_j)$.*

Proof. We show how $\mathcal{A}^k_{i<j}$ can be built for $k = 2$, $i = 1$ and $j = 2$ in Fig. 1. It checks that the word on the second track encodes a greater number, depending on the sign, in the obvious way. If the preceding signs are $[-, +]$ then it only needs to check that not both tracks encode 0. If they are $[+, +]$ then the automaton needs to verify that the tracks differ at some point, and that, at the first point where they differ, the bit in the second track is set and the bit in the first track is not set. Moreover, the tracks can not continue with all following bits set in the first track, but none in the second, because then the numbers encoded in the tracks would be the same. Again, by padding the transition labels accordingly, one can create $\mathcal{A}^k_{i<j}$ for arbitrary k, i, j. $\mathcal{A}^k_{i \le j}$ is then simply obtained as $\mathcal{A}^k_{i<j} \cup \mathcal{A}^k_{i=j}$. All involved automata are of constant size. $\qquad \square$

Note the slight difference in the specification in Lemma 5 compared to the lemmas in Sect. 4. While the automata constructed there only accept well-formed words, the ones constructed in Lemma 5 also accept non-well-formed words. It would be easy to restrict the languages of $\mathcal{A}^k_{i<j}$ and $\mathcal{A}^k_{i\leq j}$ to well-formed words only by doubling the state space. This is, however, not necessary as they will only be used here in conjunction with other

Lemma 6. *Let $k \geq 2$, $1 \leq m < h \leq k$, $\bar{r} \in \mathbb{Q}^m$, $d \in \mathbb{Q}$ with $d > 0$ and $P = (\bar{r}, d, h)$. There is a WNBA$_{\text{FG}}$ $\mathcal{A}^{k,P}_{\text{in}}$ of size $2^{\mathcal{O}(\|P\|)}$ s.t. for all $w \in WF^k_\Sigma$ we have: $w \in R(\mathcal{A}^{k,P}_{\text{in}})$ iff $\sum_{i=1}^m |r_i + (-1 \cdot dec(w_i))| \leq d$.*

Proof. We start by arguing that one can construct a WNBA$_{\text{FG}}$ $\mathcal{A}^k_{j=\text{abs}(i)}$ (of constant size) that checks whether the j-th track in a k-track word contains the absolute value of the number encoded in the i-th track. It is easily obtained by swapping two transitions in $\mathcal{A}^k_{j=i}$, namely those out of the initial state with labels $[-,+]$ and $[-,-]$.

$\mathcal{A}^{k,P}_{\text{in}}$ can then be built by temporarily using $4m + 2$ tracks in addition to the k given ones which are checked to contain, respectively, for input values $x_1, \ldots, x_m$ encoded on the first m tracks, the values $-x_1, \ldots, -x_m$, then the values $r_1, \ldots, r_m$, then $r_1 - x_1, \ldots, r_m - x_m$, then their absolute values in the next m tracks, the sum of these in the next, and the constant d in the last. Using the WNBA$_{\text{FG}}$ from Lemmas 7.1., 7.3., 7.4., 5, the correctness of the tracks can be verified as follows. Let $\ell := k + 4m + 2$. Then $\mathcal{A}^{k,P}_{\text{in}}$ is defined via

$$\mathcal{A}^{k,P}_{\text{in}} := \left(\left(\bigcap_{i=1}^m \mathcal{A}^\ell_{k+i=\text{mult}(-1,i)} \cap \mathcal{A}^\ell_{k+m+i=\text{const}(r_i)} \cap \mathcal{A}^\ell_{k+2m+i=\text{add}(k+i,k+m+i)} \right. \right.$$
$$\left. \cap\, \mathcal{A}^\ell_{k+3m+i=\text{abs}(k+2m+i)} \right)$$
$$\left. \cap\, \mathcal{A}^\ell_{k+4m+1=\text{add}(k+3m+1,\ldots,k+4m)} \cap \mathcal{A}^\ell_{\ell=\text{const}(d)} \cap \mathcal{A}^\ell_{k+4m+1\leq\ell} \right) \!\downarrow_{1,\ldots,k}$$

This construction of $\mathcal{A}^{k,P}_{\text{in}}$ makes the size claim obvious: the important parts are the addition, multiplication and constant automata are exponential in their respective parameters, each a subparamter of P. The intersection of all these, leads to the size of $2^{\mathcal{O}(\|P\|)}$.

Given this construction and the fact that WNBA$_{\text{FG}}$ are closed under intersection, we get that for each ARP there is a WNBA$_{\text{FG}}$ which recognises its validity.

Theorem 1. *Let N be a DNN with m inputs and n outputs, and $P = (\bar{r}, d, i)$ be an ARP with $1 \leq i \leq n$. There is a WNBA$_{\text{FG}}$ $\mathcal{A}^{N,P}_{\text{arp}}$ of size $2^{\mathcal{O}(\|N\|+\|P\|)}$ s.t. $R(\mathcal{A}^{N,P}_{\text{arp}}) = \emptyset$ iff $N \models P$.*

Proof. Let $k := m + n$. Note that $\mathcal{A}_N$ is a k-track WNBA$_{\text{FG}}$ recognising the input-output behaviour of N. Let $\mathcal{A}^{N,P}_{\text{arp}} := \mathcal{A}_N \cap \mathcal{A}^{k,P}_{\text{in}} \cap \overline{\mathcal{A}}^{k,P}_{\text{out}}$ where $\overline{\mathcal{A}}^{k,P}_{\text{out}} := (\bigcup_{i'=1}^{i-1} \mathcal{A}^k_{m+i\leq m+i'}) \cup (\bigcup_{i'=i+1}^n \mathcal{A}^k_{m+i\leq m+i'})$ accepts a word with n output tracks if the number encoded in the i-th output is not greater than those in any other output track. The size claim is a straightforward result from the sizes of the subautomata $\mathcal{A}_N$, $\mathcal{A}_{\text{in}}^{k,P}$ and $\overline{\mathcal{A}}_{\text{out}}^{k,P}$. Consequently, $\mathcal{A}_{\text{arp}}^{N,P}$ accepts a k-track word if it encodes a vector $\overline{x} \in \mathbb{R}^m$ in its first m tracks that is within the d-neighborhood of $\overline{r}$, s.t. the following n tracks encode $N(\overline{x})$ and their i-th component is not strictly maximal. This is the case if and only if $N \not\models P$. □

Output Reachability. This is used to certify that specific "misbehaviour" of DNN does not occur. A formal definition hinges on a notion of valid inputs and outputs. Commonly, this is done using specifications defining (convex) sets of real vectors. A *vector specification* φ over variables $x_1, \ldots, x_k$ is a conjunction φ of statements of the form $(\sum_{i=1}^{k} c_i \cdot x_i) \leq b$ where $c_i, b \in \mathbb{Q}$. Let $\overline{r} = (r_1, \ldots, r_k) \in \mathbb{R}^k$. We say that $\overline{r}$ satisfies φ if each inequality $t \leq b$ in φ is satisfied in real arithmetic when each x_i is given the value r_i. Let N be a DNN with input dimension m and output dimension n, let φ_{in} be a vector specification over $x_1, \ldots, x_m$ and let φ_{out} be a vector specification over $y_1, \ldots, y_n$. We call the tuple $P = (\varphi_{\text{in}}, \varphi_{\text{out}})$ an *output reachability property (ORP)* and say that N satisfies $(\varphi_{\text{in}}, \varphi_{\text{out}})$, written $N \models P$, if there is $\overline{r} \in \mathbb{R}^m$ s.t. $\overline{r} \models \varphi_{\text{in}}$ and $N(\overline{r}) \models \varphi_{\text{out}}$. We define the size of P by $\|P\| = \|\varphi_{\text{in}}\| + \|\varphi_{\text{out}}\|$ where the measure of a specification φ is the sum of the measures of parameters c_i, b occurring in some inequality.

Theorem 2. *Let N be a DNN with m inputs and n outputs, and $P = (\varphi_{in}, \varphi_{out})$ be an ORP. There is a WNBA$_{FG}$ $\mathcal{A}_{orp}^{N,P}$ of size $2^{\mathcal{O}(\|N\| + \|P\|)}$ s.t. $R(\mathcal{A}_{orp}^{N,P}) = \emptyset$ iff $N \models P$.*

Proof. Similar to the constructions in Theorem 1, one can build, given k and a linear inequality $\psi = \sum_{i=1}^{k} c_i \cdot x_i \leq b$ with rational constants, a WNBA$_{FG}$ $\mathcal{A}_{\text{in}}^{\psi}$ that accepts a well-formed k-track word iff the first k tracks encode numbers $x_1, \ldots, x_k$ that satisfy ψ. Likewise, we can build such a WNBA$_{FG}$ $\mathcal{A}_{\text{out}}^{\psi}$ that does the same for the last n tracks. Note that the size of these automata is exponential in the measure of the parameters c_i, b. Then we get that $\mathcal{A}_{\text{orp}}^{N,P} := (\bigcap_{\psi \in \varphi_{\text{in}}} \mathcal{A}_{\text{in}}^{\psi}) \mid \mathcal{A}_N \mid (\bigcap_{\psi \in \varphi_{\text{out}}} \mathcal{A}_{\text{out}}^{\psi})$ accepts a word iff it encodes some $\overline{x}$ satisfying φ_{in}, s.t. $N(\overline{x})$ satisfies φ_{out}. The size claim about $\mathcal{A}_{\text{orp}}^{N,P}$ is a straightforward result from the intersection and the size of $\mathcal{A}_N$ and the specification automata. □

3.2 Interpreting DNN with WNBA$_{FG}$

Zhang et al. [26] present a three-dimensional taxonomy for interpretation techniques: *post-hoc* or *ad-hoc* interpretation either generates explanations for common neural network models or focuses on constructing neural network models that improve interpretability. *Examples, attribution, hidden semantics* or *rules* characterise the type of explanation. *Global, local* or *semi-local* explanations concern the model's overall behaviour, that of a single input value, resp. something in between. In the following, we introduce a widely considered post-hoc, attribution and local interpretation approach.

We start with an example. Assume some image-classification task, for instance the task to distinguish pictures of dogs and pigs. Commonly, such a task is addressed using a model like *Convolutional Neural Networks* [12] which processes a picture by computing layer-by-layer higher-order features of the picture and then classifies it based on these. A natural explanation for the CNN's decision is the image's regions that the CNN focuses on for making its decision, classifying it as either a picture of a dog or of a pig. For example, we would gain confidence in the CNN decision if we can prove that it focuses on the form of the snout of the animal (lengthy vs. flat) or the texture of its outer contours (fluffy vs. smooth). In technical terms, the task is to find the most important input dimensions, i.e. pixels of the image, that determine the output of the CNN. A widely used interpretation technique addressing this problem is called Integrated Gradient [22]. In the context of our general DNN model, we formulate the task of finding the most important features of an input as a decision problem: given a DNN N, some input $\bar{r} \in \mathbb{Q}^m$ and $I \subseteq \{1, \ldots, m\}$, decide whether for every $\bar{x} \in \mathbb{Q}^k$ which equals $\bar{r}$ on the dimensions in I we have $N(\bar{x}) = N(\bar{r})$. In correspondence to [2], we call this problem MSR (for *minimum sufficient reason*). An instance is of the form $P = (\bar{r}, I)$. As before, we write $N \models P$ to indicate that N satisfies the instance P. The measure $\|P\|$ is given by the sum of the measures of $\|r\|$ and $\|I\|$, which are defined in the obvious way.

Theorem 3. *Let N be a DNN with input dimension m and output dimension n, and some $P = (\bar{r}, I)$ with $\bar{r} \in \mathbb{Q}^m$ and $I \subseteq \{1, \ldots, m\}$. There is a WNBA$_{\mathsf{FG}}$ $\mathcal{A}_{\mathsf{msr}}^{N,P}$ of size $2^{\mathcal{O}(\|N\| + \|P\|)}$ s.t. $R(\mathcal{A}_{\mathsf{msr}}^{N,P}) = \emptyset$ iff $N \models P$.*

Proof. We need an auxiliary automaton $\mathcal{A}_{i \neq j}^k$ which accepts a k-track word iff its i-th and j-th track represent different numbers. Note that we cannot simply complement $\mathcal{A}_{i=j}^k$ from Lemma 3 since weak NBA, let alone WNBA$_{\mathsf{FG}}$, are not closed under complement. However, it is easy to construct $\mathcal{A}_{i \neq j}^k$ directly, or as $\mathcal{A}_{i<j}^k \cup \mathcal{A}_{j<i}^k$ using the WNBA$_{\mathsf{FG}}$ from Lemma 5. Let $\mathcal{A}_N$ be the l-track WNBA$_{\mathsf{FG}}$ with $k = m + n$ recognising N's input-output relation from Theorem 4. To construct $\mathcal{A}_{\mathsf{msr}}^{N,P}$ we use two copies that work in parallel, computing N's output on some $\bar{x}$ and on $\bar{r}$, checking whether the inputs agree on the dimensions in I and whether their outputs disagree. Define $\mathcal{A}_{\mathsf{msr}}^{N,P}$ as $(\mathcal{A}_N \bowtie \mathcal{A}_{\mathsf{wf}}^l) \cap (\mathcal{A}_{\mathsf{wf}}^l \bowtie \mathcal{A}_N) \cap (\bigcap_{i=1}^m \mathcal{A}_{l+i=\mathsf{const}(r_i)}^{2l}) \cap (\bigcap_{i \in I} \mathcal{A}_{l+i=i}^l) \cap (\bigcup_{i=1}^n \mathcal{A}_{m+i \neq l+m+i}^{2l})$. The size of this automaton is determined by the intersection of $\mathcal{A}_N$ and automata $\mathcal{A}_{l+i=\mathsf{const}(r_i)}^{2l}$ parametrized by parts of I. Take a $2l$-track word w. Define $\bar{x}$ as $(dec(w_1), \ldots, dec(w_m))$ and define $\bar{x'}$ as $(dec(w_{l+1}), \ldots, dec(w_{l+m}))$. Define $\bar{y}$ as $(dec(w_{m+1}), \ldots, dec(w_{m+n}))$ and define $\bar{y'}$ as $(dec(w_{l+m+1}), \ldots, dec(w_{2l}))$. Then $w \in R(\mathcal{A}_{\mathsf{msr}}^{N,P})$ iff $\bar{x'} = \bar{r}$, $x_i = r_i$ for all $i \in I$, and $N(\bar{x}) = \bar{y} \neq \bar{y'} = N(\bar{x'}) = N(\bar{r})$, i.e. if and only if $\bar{x}$ witnesses the fact that $N \not\models P$. $\qquad\square$

In practice, the dimensions in I are not explicitly given, but only a number $l \leq m$ is given with the proviso that a set I of input dimensions should be found s.t. $|I| = l$ and this set provides a minimum sufficient reason for the classification of $\bar{r}$. Clearly, by invoking Theorem 3, at most $\binom{m}{l}$ times a counterexample can be found using successive emptiness checks. It remains to be seen whether this can be improved, for instance by not enumerating all sets I in a brute-force way but to construct one from smaller ones for instance.

4 Translating DNN into WNBA$_{FG}$

The definition of DNN given in Sec. 2 implies an inductive view on DNN: each DNN node itself is a DNN with one layer consisting of one node, each DNN layer itself is a DNN with one layer and each subset of consecutive layers is a DNN with several layers. We use this inductive view to first argue that there are WNBA$_{FG}$ which capture the computation of each node and then that there are WNBA$_{FG}$ capturing whole layers and complete DNN.

Let v be a node computing $relu(\sum_{i=1}^{k} w_i x_i + b)$. From its functional form we can infer that the computation of v is built from multiple instances of three fundamental operations: 1. multiplication of some arbitrary value with a fixed constant, 2. summation of arbitrary values and 3. the application of $relu$ to some arbitrary value. For each operation, we define a corresponding WNBA$_{FG}$ and then combine these using the operations specified in Lemmas 1, 2, 4 and the closure under $\cap$ and $\cup$.

Lemma 7. *Let $k \geq 2$, $1 \leq i,j \leq k$ and $1 \leq i_1, \ldots, i_n \leq k$ where $i \neq i_h$ and $i_h \neq i_l$ for $h,l \in \{1, \ldots, n\}$. There is a WNBA$_{FG}$*

1. $\mathcal{A}_{i=\text{add}(i_1,\ldots,i_n)}^{k+1}$ *of size $2^{\mathcal{O}(k)}$ such that $R(\mathcal{A}_{i=\text{add}(i_1,\ldots,i_n)}^{k+1}) = \{w \in WF_{\Sigma}^{k+1} \mid dec(w_i) = \sum_{h=1}^{n} dec(w_{i_h})\}$,*
2. $\mathcal{A}_{j=\text{relu}(i)}^{k}$ *of size $\mathcal{O}(1)$ such that $R(\mathcal{A}_{j=\text{relu}(i)}^{k}) = \{w \in WF_{\Sigma}^{k} \mid dec(w_j) = relu(dec(w_i))\}$,*
3. $\mathcal{A}_{j=\text{mult}(c,i)}^{k}$ *of size $2^{\mathcal{O}(\|c\|)}$ such that $R(\mathcal{A}_{j=\text{mult}(c,i)}^{k}) = \{w \in WF_{\Sigma}^{k} \mid dec(w_j) = c \cdot dec(w_i)\}$ for every $c \in \mathbb{Q}$ and*
4. $\mathcal{A}_{i=\text{const}(c)}^{k-1}$ *of size $\mathcal{O}(2^{\|c\|})$ such that $R(\mathcal{A}_{i=\text{const}(c)}^{k}) = \{w \in WF_{\Sigma}^{k} \mid dec(w_i) = c\}$.*

The results of Lemma 7, although technically extensive, are straightforward implementations of the operations we presented in Sect. 2. The automaton $\mathcal{A}_{i=\text{add}(i_1,\ldots,i_n)}^{k+1}$ checks whether track i represents the usual binary sum of the words on tracks i_1 to i_n. As the carry bit is at most $k - 1$ in each step, this is obviously recognizable using a finite automaton. The automaton $\mathcal{A}_{j=\text{relu}(i)}^{k}$ simply checks if for tracks i and j holds $j = relu(i)$. As the value of $relu$ is the identity for positive values and zero for negative values, an automaton can distinguish these two cases based on the initial sign vector of the input word. The automaton $\mathcal{A}_{j=\text{mult}(c,i)}^{k}$ checks if for tracks i and j holds that $j = \text{mult}(c,i)$. This is recognizable using a finite automaton due to the fact that c is a predefined constant and, thus, $\mathcal{A}_{j=\text{mult}(c,i)}^{k}$ simply checks if the word on j is a shift of the word on i by a factor c. Otherwise, namely arbitrary multiplication, would not be recognizable using a finite WNBA$_{FG}$. Lastly, automaton $\mathcal{A}_{i=\text{const}(c)}^{k-1}$ simply checks if for track i holds that $i = \mathsf{const}(c)$. This means simply reading track i and verifying if it matches the word encoding c.

Next, we lift these constructions to build the WNBA$_\mathsf{FG}$ $\mathcal{A}_v$ representing the computation of a node v in a DNN.

Lemma 8. *Let $k \geq 2$, $h < j \leq k$, and v be a DNN-node computing $relu(b + \sum_{i=1}^{h} c_i x_i)$. There is a WNBA$_\mathsf{FG}$ $\mathcal{A}^k_{j=v(1,\ldots,h)}$ of size $2^{\mathcal{O}(\|v\|)}$ s.t. $R(\mathcal{A}^k_{j=v(1,\ldots,h)}) = \{w \in WF^k_\Sigma \mid dec(w_j) = relu(b + \sum_{i=1}^{h} c_i \cdot dec(w_i))\}$.*

Proof. Note that $\mathcal{A}^k_{j=v(1,\ldots,h)}$ is supposed to work over k-track words in which the first h tracks contain inputs $x_1, \ldots, x_h$, and the node's output is expected in the $(j-h)$-th output track which is the j-th overall. Let $\mathcal{A}^k_{j=v(1,\ldots,k)}$ be equal to

$$\left((\bigcap_{i=1}^{h} \mathcal{A}^{g+2}_{k+i=\mathsf{mult}(c_i,i)}) \cap \mathcal{A}^{g+2}_{g+1=\mathsf{const}(b)} \cap \mathcal{A}^{g+2}_{g+2=\mathsf{add}(k+1,\ldots,g+1)} \cap \mathcal{A}^{g+2}_{j=\mathsf{relu}(g+2)} \right) \downarrow_{1,\ldots,k}.$$

where $g = k + h$. It uses $h + 2$ additional and intermediate tracks that hold, respectively, for input values $x_1, \ldots, x_h$ in the first h tracks, the values $c_1 \cdot x_1, \ldots, c_h \cdot x_h$, the bias b, and their sum. By also insisting that the j-th track holds the ReLU-value of that sum, we model exactly the node's computation. Since it is constructed using $h + 2$ intersections of WNBA$_\mathsf{FG}$ of size that is either constant (for the addition) or of size bounded by the involved rational constants (for the multiplication and the bias), the overall size can be estimated as $2^{\mathcal{O}(\|v\|)}$. □

Using the inductive view on DNN described above, we are now set to provide the translation of DNN into input-output-equivalent WNBA$_\mathsf{FG}$.

Theorem 4. *Let N be a DNN with input dimension m and output dimension n. There is a WNBA$_\mathsf{FG}$ $\mathcal{A}_N$ of size $2^{\mathcal{O}(\|N\|)}$ s.t. $R(\mathcal{A}_N) = \{w \in WF^{m+n}_\Sigma \mid N(dec(w_1), \ldots, dec(w_m)) = (dec(w_{m+1}), \ldots, dec(w_{m+n}))\}$.*

Proof. Assume that N has k layers $l_1, \ldots, l_k$. For each layer l_i, we construct an WNBA$_\mathsf{FG}$ $\mathcal{A}_i$ recognising the relation between inputs to this layer and immediate outputs computed by it. Take a layer $l_i = (v^i_1, \ldots, v^i_{n_i})$, and assume that it takes m_i inputs (which must also be the number of outputs of the previous layer). Obviously, it produces n_i outputs as this is the number of nodes in this layer. Moreover, we have $m_1 = m$ and $n_k = n$, i.e. the inputs to the DNN are inputs to the first layer, and the outputs of the last layer are the outputs of the DNN. The desired WNBA$_\mathsf{FG}$ can be obtained from the WNBA$_\mathsf{FG}$ for the nodes v^i_j of this layer according to Lemma 8 as $\mathcal{A}_i := \bigcap_{j=1}^{n_i} \mathcal{A}^{m_i+n_i}_{m_i+j=v^i_j(1,\ldots,m_i)}$. This produces a WNBA$_\mathsf{FG}$ with m_i inputs and n_i outputs which contains, in the j-th output track, the result of the computation done by the j-th node in this layer on the inputs contained in the m_i input tracks. Finally, a WNBA$_\mathsf{FG}$ for the relation computed by the DNN N is then obtained simply as $\mathcal{A}_N := \mathcal{A}_1 \circ_{n_1} \cdots \circ_{n_{k-1}} \mathcal{A}_k$. Note that relation composition is in fact associative. The size of $\mathcal{A}_N$ can be bounded by $2^{\mathcal{O}(\|N\|)}$ because of the following observation: for a layer of n nodes v we need to form a product of n automata, each of size bounded by $2^{\mathcal{O}(\|v\|)}$, i.e. we get a size of $2^{\mathcal{O}(n \cdot \|v\|)}$ whose exponential corresponds to the size that a layer requires in a DNN representation. Likewise, forming the composition for k layers results in an WNBA_{FG} of size bounded by $2^{\mathcal{O}(k \cdot n \cdot \|v\|)} = 2^{\mathcal{O}(\|N\|)}$ where $k \cdot n$ is an upper bound for the number of nodes.   □

5 Discussion and Outlook

We presented an automata-theoretic framework that can be used to address a broad range of analysis tasks on DNN. Our key results (Thms. 1, 2 and 3) are that different particular verification and interpretation problems can be reduced to emptiness checks for eventually-always weak Büchi automaton, while using the transformation of a DNN N into such an automaton presented in Theorem 4.

The translation presented in Sect. 4 is conceptual rather than practical, mainly due to the exponential blowup. However, It turns out that the exponential blowup in the translation is unavoidable, (unless P = NP) as it can be used to decide output reachability, which is known to be NP-hard [11,21]. In order to obtain practically useful automata-based tools for DNN analysis, further work is needed. The exposition here is done w.r.t. to a particular neural network model. Hence, further work also consists of identifying other classes of NN which can be translated similarly into finite-state automata. The use of NFA and finite words, instead of WNBA_{FG} and infinite words, constitutes an abstraction of a DNN's behaviour in the form of a function of type $\mathbb{R}^m \rightarrow \mathbb{R}^n$ to functions on some subset. For instance, when cutting down all WNBA_{FG} to accept immediately rather than read dot symbols, we would obtain NFA over $\{+, -, 0, 1\}$ that approximate a DNN's behaviour as a function of type $\mathbb{Z}^m \rightarrow \mathbb{Z}^n$. We aim to investigate this idea more formally, making use of a well-developed theory of abstraction and refinement [6,7], with the aim of acquiring a better understanding of the possibilities to trade precision for efficiency in DNN analysis. Furthermore, implemented DNN usually work with fixed-length parameters, which makes NFA a relevant model from a practical point of view as well.

Besides, future research should focus on the identification of analysis problems for which the automata-theoretic framework is genuinely superior compared to other techniques, as one obtains, in the form of the automaton $\mathcal{A}_N$, a finite representation of the entire input-output behaviour of N. This may include transferring the comparison of two DNN N_1 and N_2 to their respective automata representations $\mathcal{A}_{N_1}$ and $\mathcal{A}_{N_2}$. We can compare the behaviour of N_1 and N_2 by investigating, for instance, the intersection of $R(\mathcal{A}_{N_1})$ and $R(\mathcal{A}_{N_2})$, or their symmetric difference (which will in general only be definable by an NBA rather than a WNBA_{FG}), to obtain notions of diverging behaviour or of equivalence between DNN.

Acknowledgements. We would like to thank Rüdiger Ehlers for fruitful discussions on this topic and helpful comments on an earlier draft of the paper.

280 M. Sälzer et al.

References

1. Ayache, S., Eyraud, R., Goudian, N.: Explaining black boxes on sequential data using weighted automata. In: Proceedings of the 14th International Conference on Grammatical Inference, ICGI'18. Proceedings of Machine Learning Research, vol. 93, pp. 81–103. PMLR (2018). http://proceedings.mlr.press/v93/ayache19a.html
2. Barceló, P., Monet, M., Pérez, J., Subercaseaux, B.: Model interpretability through the lens of computational complexity. In: Proceedings of the 33rd Annual Conference on Advances in Neural Information Processing Systems, NeurIPS 2020 (2020). https://proceedings.neurips.cc/paper/2020/hash/b1adda14824f50ef24ff1c05bb66faf3-Abstract.html
3. Berstel, J.: Transductions and Context-Free Languages. Teubner, Stuttgart (1979)
4. Boigelot, B., Jodogne, S., Wolper, P.: An effective decision procedure for linear arithmetic over the integers and reals. ACM Trans. Comput. Log. **6**(3), 614–633 (2005). https://doi.org/10.1145/1071596.1071601
5. Boigelot, B., Rassart, S., Wolper, P.: On the expressiveness of real and integer arithmetic automata. In: Larsen, K.G., Skyum, S., Winskel, G. (eds.) ICALP 1998. LNCS, vol. 1443, pp. 152–163. Springer, Heidelberg (1998). https://doi.org/10.1007/BFb0055049
6. Clarke, E.M., Grumberg, O., Jha, S., Lu, Y., Veith, H.: Counterexample-guided abstraction refinement for symbolic model checking. J. ACM **50**(5), 752–794 (2003)
7. Cousot, P., Cousot, R.: Abstract interpretation: a unified model for static analysis of programs by construction or approximation of fixpoints. In: Proceedings of the 4th ACM Symposium on Principles of Programming Languages, POPL 1977, pp. 238–252 (1977)
8. Grigorescu, S.M., Trasnea, B., Cocias, T.T., Macesanu, G.: A survey of deep learning techniques for autonomous driving. J. Field Robotics **37**(3), 362–386 (2020). https://doi.org/10.1002/rob.21918
9. Huang, X., et al.: A survey of safety and trustworthiness of deep neural networks: verification, testing, adversarial attack and defence, and interpretability. Comput. Sci. Rev. **37**, 100270 (2020). https://doi.org/10.1016/j.cosrev.2020.100270
10. Jacobsson, H.: Rule extraction from recurrent neural networks: a taxonomy and review. Neur. Comp. **17**(6), 1223–1263 (2005). https://doi.org/10.1162/0899766053630350
11. Katz, G., Barrett, C., Dill, D.L., Julian, K., Kochenderfer, M.J.: Reluplex: an efficient SMT solver for verifying deep neural networks. In: Majumdar, R., Kunčak, V. (eds.) CAV 2017. LNCS, vol. 10426, pp. 97–117. Springer, Cham (2017). https://doi.org/10.1007/978-3-319-63387-9_5
12. Khan, A., Sohail, A., Zahoora, U., Qureshi, A.S.: A survey of the recent architectures of deep convolutional neural networks. Artif. Intell. Rev. **53**(8), 5455–5516 (2020). https://doi.org/10.1007/s10462-020-09825-6
13. Khmelnitsky, I., et al.: Property-directed verification and robustness certification of recurrent neural networks. In: Hou, Z., Ganesh, V. (eds.) ATVA 2021. LNCS, vol. 12971, pp. 364–380. Springer, Cham (2021). https://doi.org/10.1007/978-3-030-88885-5_24
14. Litjens, G., et al.: A survey on deep learning in medical image analysis. Med. Image Anal. **42**, 60–88 (2017). https://doi.org/10.1016/j.media.2017.07.005
15. Löding, C.: Efficient minimization of deterministic weak ω-automata. Inform. Proc. Lett. **79**(3), 105–109 (2001)

16. Mayr, F., Yovine, S., Visca, R.: Property checking with interpretable error characterization for recurrent neural networks. Mach. Learn. Knowl. Extr. **3**(1), 205–227 (2021). https://doi.org/10.3390/make3010010
17. Miyano, S., Hayashi, T.: Alternating finite automata on omega-words. TCS **32**(3), 321–330 (1984)
18. Otter, D.W., Medina, J.R., Kalita, J.K.: A survey of the usages of deep learning for natural language processing. IEEE Trans. Neural Netw. Learn. Syst. **32**(2), 604–624 (2021). https://doi.org/10.1109/TNNLS.2020.2979670
19. Ramesh, A., Dhariwal, P., Nichol, A., Chu, C., Chen, M.: Hierarchical text-conditional image generation with CLIP latents. CoRR abs/2204.06125 (2022). https://doi.org/10.48550/arXiv.2204.06125
20. Ramesh, A., et al.: Zero-shot text-to-image generation. In: Proceedings of the 38th International Conference on Machine Learning, ICML 2021. Proceedings of the of Machine Learning Research, vol. 139, pp. 8821–8831. PMLR (2021)
21. Sälzer, M., Lange, M.: Reachability is NP-complete even for the simplest neural networks. In: Bell, P.C., Totzke, P., Potapov, I. (eds.) RP 2021. LNCS, vol. 13035, pp. 149–164. Springer, Cham (2021). https://doi.org/10.1007/978-3-030-89716-1_10
22. Sundararajan, M., Taly, A., Yan, Q.: Axiomatic attribution for deep networks. In: Proceedings of the 34th International Conference on Machine Learning, ICML 2017. Proceedings of Machine Learning Research, vol. 70, pp. 3319–3328. PMLR (2017). http://proceedings.mlr.press/v70/sundararajan17a.html
23. Sälzer, M., Alsmann, E., Bruse, F., Lange, M.: Verifying and interpreting neural networks using finite automata. CoRR abs/2211.01022 (2023). https://doi.org/10.48550/arXiv.2211.01022
24. Weiss, G., Goldberg, Y., Yahav, E.: Extracting automata from recurrent neural networks using queries and counterexamples. In: Proceedings of the 35th International Conference on Machine Learning, ICML 2018. Proceedings of Machine Learning Research, vol. 80, pp. 5244–5253. PMLR (2018). http://proceedings.mlr.press/v80/weiss18a.html
25. Xu, Z., Wen, C., Qin, S., He, M.: Extracting automata from neural networks using active learning. PeerJ Comput. Sci. **7**, e436 (2021). https://doi.org/10.7717/peerj-cs.436
26. Zhang, Y., Tiño, P., Leonardis, A., Tang, K.: A survey on neural network interpretability. IEEE Trans. Emerg. Top. Comput. Intell. **5**(5), 726–742 (2021). https://doi.org/10.1109/TETCI.2021.3100641

Around Don's Conjecture for Binary Completely Reachable Automata

Yinfeng Zhu$^{(\boxtimes)}$ (iD)

Institute of Natural Sciences and Mathematics, Ural Federal University,
620000 Ekaterinburg, Russia
`yinfeng.zhu7@gmail.com`

Abstract. A word w is called a *reaching* word of a subset S of states
in a deterministic finite automaton (DFA) if S is the image of the whole
state set under the action of w. A DFA is called *completely reachable*
if every non-empty subset of the state set has a reaching word. Don's
conjecture states that in every n-state completely reachable DFA, for
every s-element subset of states, there exists a reaching word of length
at most $n(n-s)$. We present infinitely many completely reachable DFAs
with two letters that violate this conjecture when $s = n - 2$. A subfamily
of completely reachable DFAs with two letters, called *standardized* DFAs,
was introduced by Casas and Volkov (2022). We prove that every s-
element subset of states in an n-state standardized DFA has a reaching
word of length $\leq n(n-s)+n-1$. Finally, we confirm the conjecture for
standardized DFAs with additional properties, thus generalizing a result
of Casas and Volkov (2023).

Keywords: Deterministic finite automaton · Complete reachability ·
Don's conjecture

1 Introduction

1.1 Completely Reachable DFAs and Don's Conjecture

A *deterministic finite automaton* (DFA) is a triple $\mathcal{A} = (Q, \Sigma, \delta)$ where Q and
Σ are finite non-empty sets and $\delta \colon Q \times \Sigma \to Q$ is a function. The elements of
Q are called *states*; the elements of Σ are called *letters*; and the function δ is
called the *transition function* of $\mathcal{A}$. Finite sequences over Σ (including the empty
sequence denoted by ϵ) are called *words*. Write Σ^* for the set of all words over
Σ. For a word $w \in \Sigma^*$, the *length* of w is denoted by $|w|$.

The transition function δ expands to a function $Q \times \Sigma^* \to Q$ (still denoted
by δ) via the recursion: for each $q \in Q$, $a \in \Sigma$, $w \in \Sigma^*$, set $\delta(q, \epsilon) := q$ and
$\delta(q, wa) := \delta(\delta(q, w), a)$. The power set of Q is denoted by $\mathcal{P}(Q)$. The transition
function δ can be further expanded to a function $\mathcal{P}(Q) \times \Sigma^* \to \mathcal{P}(Q)$ (denoted
by δ again) by setting $\delta(P, w) := \{\delta(p, w) : p \in P\}$, for all $P \subseteq Q$ and $w \in \Sigma^*$.
The *pre-image* of $P \subseteq Q$ via a word $w \in \Sigma$, denoted by $\delta^{-1}(P, w)$, is the set
$\{q : \delta(q, w) \in P\}$. It does not necessarily hold that $\delta(\delta^{-1}(P, w), w) = P$.

J. D. Day and F. Manea (Eds.): DLT 2024, LNCS 14791, pp. 282–295, 2024.
https://doi.org/10.1007/978-3-031-66159-4_20

A subset $P \subseteq Q$ is called *reachable* in $\mathcal{A}$ if there exists a word $w \in \Sigma^*$ such that $P = \delta(Q, w)$. For a reachable subset $P \subseteq Q$, a word w is called a *reaching* word of P if $P = \delta(Q, w)$. A DFA is called *completely reachable* if every non-empty subset of its state set is reachable.

The notion of complete reachability was first introduced by Bondar and Volkov [1]. This concept appears in the study of the descriptional complexity of formal languages [9] and is closely related to the Černy conjecture. The famous conjecture of Černy states that for an n-state DFA (Q, Σ, δ), if a singleton subset of Q is reachable, then there exists a word w of length at most $(n-1)^2$ such that $|\delta(Q, w)| = 1$. We refer to the chapter [8] of the "Handbook of Automata Theory" and the recent survey [10] for a summary of the state-of-the art around the Černý conjecture. Don [4, Conjecture 18] proposed a stronger conjecture: for an n-state DFA (Q, Σ, δ), if an s-element subset of Q is reachable, then it can be reached with a word of length at most $n(n-s)$. Don's conjecture was disproved in [7, Proposition 7]: for every $n \geq 6$, there exists a DFA (Q, Σ, δ) and a subset $S \subseteq Q$ such that the length of any word that reaches S is at least $\frac{2^n}{n}$. However, it is interesting to ask whether Don's conjecture holds for completely reachable DFAs. This problem was explicitly asked in [7, Problem 4]. Ferens and Szykuła [5, Theorem 30] proved that for a completely reachable DFA with n states, every s-element subset of states, $0 < s \leq n$, can be reached by a word of length at most $2n(n-s)$. Casas and Volkov [3, Theorem 1] verified Don's conjecture for a subfamily of completely reachable DFAs with two letters.

1.2   Binary Completely Reachable DFAs and Main Results

A DFA is called *circular* if the action of one letter is a cyclic permutation of all states. DFAs with two letters are called *binary*. For $n \geq 3$, a binary completely reachable n-state DFA is circular, see [2, Lemma 1]. In this article, for a binary circular DFA, we use b to denote the letter acting as a cyclic permutation and use a to denote the other letter. We will assume that all n-state circular DFAs have the state set $\mathbb{Z}_n$, which is the set of all residues modulo n, and the letter b acts on $\mathbb{Z}_n$ as $\delta(q, b) = q \oplus 1$ for each $q \in \mathbb{Z}_n$, where $\oplus$ stands for addition modulo n.

For a DFA $\mathcal{A} = (Q, \Sigma, \delta)$ and a word $w \in \Sigma^*$, we define the *excluded set* of w as

$$\mathrm{excl}(w) := \left\{ q : |\delta^{-1}(q, w)| = 0 \right\}$$

and the *duplicate set* of w as

$$\mathrm{dupl}(w) := \left\{ q : |\delta^{-1}(q, w)| > 1 \right\}.$$

For a binary completely reachable DFA $\mathcal{A} = (\mathbb{Z}_n, \{a, b\}, \delta)$, one can easily obtain that $|\mathrm{excl}(a)| = |\mathrm{dupl}(a)| = 1$. We may and will assume that $\mathrm{excl}(a) = \{0\}$. Further, $(\mathbb{Z}_n, \{a, b\}, \delta)$ is called *standardized* if $\mathrm{dupl}(a) = \{\delta(0, a)\}$. The notion of standardized DFA was introduced by Casas and Volkov [2].

For a standardized completely reachable DFA, we establish an upper bound for the length of shortest words that reach a non-empty subset of the state set.

Theorem 1. *Let $\mathcal{A} = (\mathbb{Z}_n, \{a, b\}, \delta)$ be a standardized completely reachable DFA. Every non-empty subset $S \subseteq \mathbb{Z}_n$ is reachable with a word of length at most $n(n - |S|) + n - 1$.*

Assume that $\mathcal{A} = (\mathbb{Z}_n, \{a, b\}, \delta)$ is a standardized completely reachable DFA. Denote the set $\{\delta(0, a^i) : i \geq 1\}$ by $O(\mathcal{A})$, which is called the *orbit* of $\mathcal{A}$. The subgroup of $(\mathbb{Z}_n, \oplus)$ generated by $O(\mathcal{A})$, denoted by $K(\mathcal{A})$ or simply K, is called the *orbit subgroup* of $\mathcal{A}$. We obtain a generalization of [3, Theorem 1].

Theorem 2. *Let $\mathcal{A} = (\mathbb{Z}_n, \{a, b\}, \delta)$ be a standardized completely reachable DFA. If $K(\mathcal{A}) \in \{2\,\mathbb{Z}_n, \mathbb{Z}_n\}$, then $\mathcal{A}$ fulfills Don's conjecture.*

To the best of our knowledge, it is still unknown whether Don's conjecture holds for standardized completely reachable DFAs or not. But we find infinitely many binary completely reachable DFAs do *not* fulfill Don's conjecture and thus we answer [7, Problem 4] negatively.

Theorem 3. *There are infinitely many binary completely reachable DFAs that do not fulfill Don's conjecture.*

The remaining of this article will proceed as follows. In Sect. 2, based on group theory, we prove Theorem 1. In Sect. 3, based on results of Casas and Volkov [3], we present a proof of Theorem 2. In Sect. 4, we describe experiments performed to gain some binary DFAs that violate Don's conjecture. Based on these experiments, we construct an infinite series of binary DFAs that do not fulfill Don's conjecture. At the end, we summarize our results and discuss the relation between them in Sect. 5.

2   An Upper Bound on the Length of Shortest Reaching Words in Standardized Completely Reachable DFAs

For a group G and a subset $S \subseteq G$, we write $\langle S \rangle$ for the *subgroup generated by* S, which is the smallest subgroup of G containing S. For a subset $X \subseteq \mathbb{Z}_n$ and $z \in \mathbb{Z}_n$, write $X \oplus z$ for $\{x \oplus z : x \in X\}$.

Let $\mathcal{A} = (\mathbb{Z}_n, \{a, b\}, \delta)$ be a standardized completely reachable DFA. Let H_0 be the trivial subgroup of $(\mathbb{Z}_n, \oplus)$. For every $i \geq 1$, define the subgroup $H_i(\mathcal{A})$ (or for short H_i if the corresponding automaton is clear from the context) of $(\mathbb{Z}_n, \oplus)$ as $\langle \delta(H_{i-1}, a) \cup H_{i-1} \rangle$ and define $U_i := \delta(H_{i-1}, a) \setminus H_{i-1}$.

Since $\mathcal{A}$ is standardized completely reachable, by [2, Proposition 1], there exists a positive integer $\ell(\mathcal{A})$ (or for short ℓ if the corresponding automaton is clear from the context), such that

$$H_0 \subsetneq H_1 \subsetneq \cdots \subsetneq H_\ell = (\mathbb{Z}_n, \oplus).$$

For a non-empty proper subset S of $\mathbb{Z}_n$, we define

- $m(S) := \min \{i : H_i \cap S \notin \{\emptyset, H_i\}\}$;
- $t(S) := \max \{i : H_{m(S)} \cap S \text{ is a union of } H_i\text{-cosets}\}$.

Since $H_{m(S)} \cap S \neq H_{m(S)}$, it holds that

$$0 \leq t(S) < m(S) \leq \ell(\mathcal{A}). \tag{1}$$

Let H be a subgroup of a group G. A subset of G of the form $gH = \{gh : h \in H\}$ for some $g \in G$ is called a *(left) H-coset* of G.

Lemma 1. *Let $\mathcal{A} = (\mathbb{Z}_n, \{a, b\}, \delta)$ be a standardized completely reachable DFA. Let S be a non-empty proper subset of $\mathbb{Z}_n$. Let $t = t(S)$ and $m = m(S)$. There exists an H_t-coset C and a state $u \in U_t$ such that $C \subseteq H_m \backslash S$ and $C \oplus u \subseteq H_m \cap S$.*

Proof. By the definition of m and t, there exist some H_t-cosets in $H_m \backslash S$.

Assume, for a contradiction, that for every $u \in U_t$ and H_t-coset $C \subseteq H_m \backslash S$, we have $C \oplus u$ is not a subset of $H_m \cap S$. By Eq. (1), H_t is a subgroup of H_m. Since both H_m and $H_m \cap S$ are unions of H_t-cosets, it holds that $H_m \backslash S$ is also a union of H_t-cosets. For every $u \in U_t$, the set $C \oplus u$ is an H_t-coset. And then $(C \oplus u) \cap (H_m \backslash S) \in \{\emptyset, C \oplus u\}$. Due to the assumption, $(C \oplus u)$ is a subset of $H_m \backslash S$. Since H_{t+1} is generated by $H_t \cup U_t$, the set $H_m \backslash S$ is a union of H_{t+1}-cosets. This implies $H_m \cap S$ is a union of H_{t+1}-cosets which contradicts with the maximality of t. □

Lemma 2. *Let H and H' be two subgroups of $(\mathbb{Z}_n, \oplus)$ such that $H \subseteq H' \subseteq \mathbb{Z}_n$. Let C be an H-coset such that $C \subseteq H'$. Then there exists an integer i such that $H \oplus i = C$ and $0 \leq i \leq \frac{n}{|H|} - \frac{n}{|H'|}$.*

Proof. Since C is an H-coset, there exists a non-negative integer $i \leq \frac{n}{|H|} - 1$ such that $H \oplus i = C$. Meanwhile, $C \subseteq H'$ and $C \oplus (n - i) = H \subseteq H'$ implies that i is a multiple of $\frac{n}{|H'|}$. Then $0 \leq i \leq \frac{n}{|H|} - \frac{n}{|H'|}$. □

For two subsets $P, P' \subseteq \mathbb{Z}_n$, if there exists a word $w \in \Sigma^*$ such that $\delta(P, w) = P'$, then we say that P is a *w-predecessor* of P', or simply a *predecessor* of P'.

Proposition 1. *Let $\mathcal{A} = (\mathbb{Z}_n, \{a, b\}, \delta)$ be a standardized completely reachable DFA. Let S be a non-empty proper subset of $\mathbb{Z}_n$. Then there exists a word w of length at most $\frac{n}{|H_{t(S)}|} - \frac{n}{|H_{m(S)}|} + 1$ such that there exists a w-predecessor R of S and one of the following conditions holds:*

(1) $|R| = |S| + 1$;
(2) $m(R) \leq t(S) < m(S)$.

Proof. Using Lemma 1, we can take an $H_{t(S)}$-coset C and $u \in U_t$ such that $C \subseteq H_{m(S)} \backslash S$ and $C \oplus u \subseteq H_{m(S)} \cap S$. By Lemma 2, let $C = H_{t(S)} \oplus i$ such that $0 \leq i \leq \frac{n}{|H_{t(S)}|} - \frac{n}{|H_{m(S)}|}$. Let $w = ab^i$ and $R = \delta^{-1}(S, w)$. Note that the length of w is at most $\frac{n}{|H_{t(S)}|} - \frac{n}{|H_{m(S)}|} + 1$. Since $C \cap S = \emptyset$, we have $0 \notin \delta^{-1}(S, b^i)$ and then the set R is a predecessor of S.

Now we need to check that R satisfies the required condition. If $|R| > |S|$ (this can only happen when $t(S) = 1$), we are done. Otherwise, we have $\delta(0, a) \notin \delta(R, a) = \delta^{-1}(S, b^i)$ which implies $0 \notin R$ and then

$$R \cap H_{t(S)} \neq H_{t(S)}. \tag{2}$$

Since $C \oplus u \subseteq S$, we have $H_{t(S)} \oplus u \subseteq \delta^{-1}(S, b^i) = \delta(R, a)$. Then $\delta^{-1}(u, a) \in R \cap H_{t(S)}$ which implies

$$R \cap H_{t(S)} \neq \emptyset. \tag{3}$$

Combining Eq. (1) to (3), $m(R) \leq t(S) < m(S)$. $\qquad\square$

Now we are ready to prove Theorem 1.

Proof of Theorem 1. Since $\mathbb{Z}_n$ is reachable by the empty word, we can assume that $S \neq \mathbb{Z}_n$. Write s for the size of S.

We will construct a sequence of subsets $S_0, S_1, \ldots, S_{n-s}$ of $\mathbb{Z}_n$ and a sequence of words $w_1, \ldots w_{n-s}$ such that

- $S_0 = S$, $S_{n-s} = \mathbb{Z}_n$;
- and $|S_i| = s + i$, for $0 \leq i \leq n - s$;
- and $\delta(S_i, w_i) = S_{i-1}$, for $1 \leq i \leq n - s$.

Let i be an integer such that $1 \leq i < n - s$. Assume that S_j and w_j are already defined for $0 \leq j \leq i$. Now we will define S_{i+1} and w_{i+1} as follows.

Let $T_{i,0} := S_i$. For $j > 1$, if $T_{i,j-1}$ has been defined and $|T_{i,j-1}| = |T_{i,0}| = s+i$, then, applying Proposition 1 to $T_{i,j-1}$, there exist a subset $R \subseteq \mathbb{Z}_n$ and a word w' of length at most $\frac{n}{|H_{t(T_{i,j-1})}|} - \frac{n}{|H_{m(T_{i,j-1})}|} + 1$ such that $\delta(R, w') = T_{i,j-1}$ and one of the following conditions holds

- $|R| = |T_{i,j-1}| + 1$,
- $m(R) \leq t(T_{i,j-1}) < m(T_{i,j-1})$.

Define $T_{i,j} := R$ and $w_{i,j} := w'$.

Since $m(X)$ is an integer with $0 < m(X) \leq \ell(A)$ for all sets

$$X \subseteq \mathbb{Z}_n$$

we know that there is only a finite number of sets $T_{i,j}$ such that the second condition is fulfilled. Therefore we use k_i to be the smallest integer j such that $T_{i,j}$ fulfills the first condition.

Observe that $|T_{i,k_i}| = |S_i| + 1$. Define $S_{i+1} := T_{i,k_i}$ and $w_i := w_{i,k_i} \cdots w_{i,1}$.

Set $w := w_{n-s} \cdots w_1$. It is clear that $\delta(\mathbb{Z}_n, w) = S$. To complete the proof, we estimate the length of w as follows:

$$
\begin{aligned}
|w| &= \sum_{i=1}^{n-s} \sum_{j=1}^{k_i} |w_{i,j}| \\
&\leq \sum_{i=1}^{n-s} \sum_{j=1}^{k_i} \frac{n}{|H_{t(T_{i,j-1})}|} - \frac{n}{|H_{m(T_{i,j-1})}|} + 1 \\
&\leq \sum_{i=1}^{n-s} \left(\frac{n}{|H_{t(T_{i,k_i-1})}|} - \frac{n}{|H_{m(T_{i,0})}|} + \sum_{j=1}^{k_i} 1 \right) \quad\quad (4)\\
&\leq \sum_{i=1}^{n-s} \left(n - 1 + \sum_{j=1}^{k_i} 1 \right) \\
&= (n-1)(n-s) + \sum_{i=1}^{n-s} \sum_{j=1}^{k_i} 1 \\
&= (n-1)(n-s) + \sum_{i=1}^{n-s} |\{H_t : t = t(T_{i,j-1}), 1 \leq j \leq k_i\}| \quad\quad (5)\\
&\leq (n-1)(n-s) + \sum_{t=0}^{\ell(\mathcal{A})} \frac{n}{|H_t|} \\
&\leq (n-1)(n-s) + 2n - 1 = n(n-s) + n - 1.
\end{aligned}
$$

The key step is Eq. (4) and (5). Using Proposition 1, we have $m(T_{i,j}) \leq t(T_{i,j-1})$ for all $1 \leq i \leq n - s$ and $1 \leq j \leq k_i - 1$ which implies Eq. (4). Observe that for a fixed i and $j \neq j'$, we have $t(T_{i,j}) \neq t(T_{i,j'})$. This implies Eq. (5).   □

3   DFAs that Fulfill Don's Conjecture

Let $\mathcal{A} = (\mathbb{Z}_n, \{a, b\}, \delta)$ be a circular DFA. We say that a word $w \in \{a, b\}^*$ *expands* a proper non-empty subset $S \subseteq \mathbb{Z}_n$ if there exists a subset $R \subseteq \mathbb{Z}_n$ such that $|R| > |S|$ and $\delta(R, w) = S$. We say S is *k-expandable* if there exists a word w of length at most k expands S. Recall that the orbit $O(\mathcal{A})$ of $\mathcal{A}$ is the set $\{\delta(0, a^i) : i \geq 1\}$. The orbit subgroup $K = K(\mathcal{A})$ of $\mathcal{A}$ is the subgroup of $(\mathbb{Z}_n, \oplus)$ generated by $O(\mathcal{A})$.

The following result is established in [3, Proposition 7].

Proposition 2. *Let $\mathcal{A} = (\mathbb{Z}_n, \{a, b\}, \delta)$ be a standardized completely reachable DFA. Every non-empty subset S of $\mathbb{Z}_n$ that is not a union of K-cosets is n-expandable.*

Moreover, the proof of the above proposition in [3] shows a stronger result as follows.

Proposition 3. *Let $\mathcal{A} = (\mathbb{Z}_n, \{a, b\}, \delta)$ be a standardized completely reachable DFA and K its orbit subgroup. Let S be a non-empty subset of $\mathbb{Z}_n$. If there exists a K-coset C such that $S \cap C \notin \{\emptyset, C\}$, then there exists a word $w = a^s b^t$ of length $\leq n$ that expands $S \cap C$.*

If one drops the condition that S is not a union of K-cosets in Proposition 2, then S is not necessarily n-expandable, see [3, Example 2]. In the cases when n is even and $K = (2\,\mathbb{Z}_n, \oplus)$, where $2\,\mathbb{Z}_n := \{z \oplus z : z \in \mathbb{Z}_n\}$, with a little more effort, we can obtain the following lemma.

Proposition 4. *Let $\mathcal{A} = (\mathbb{Z}_n, \{a, b\}, \delta)$ be a standardized completely reachable DFA and K its orbit subgroup. If n is even and $K = (2\,\mathbb{Z}_n, \oplus)$, then*

(1) for a subset $S \subseteq \mathbb{Z}_n$ which is not a union of K-cosets, S is n-expandable;
(2) $K \oplus 1$ is n-expandable. Moreover there exists a word $a^s b^{t-1} a$ with $s + t \leq n$ that expands $K \oplus 1$;
(3) K is $(n + 1)$-expandable.

Proof. (1) This follows from Proposition 2 directly.

(2) Let $T = \delta^{-1}(K \oplus 1, a)$. Since $0 \notin K \oplus 1$, the subset T is a predecessor of $K \oplus 1$. It is clear that $\delta(T, b)$ is not a union of K-cosets. Note that $(K \oplus 1) \cap \delta(T, b) \notin \{\emptyset, K \oplus 1\}$. Applying Proposition 3 for $S = \delta(T, b)$ and $C = K \oplus 1$, there exists a word $w = a^s b^t$ that expands $(K \oplus 1) \cap \delta(T, b)$. Note that $\delta(0, a) \in K \oplus 1$ due to the definition of K. With w expands $(K \oplus 1) \cap \delta(T, b)$, we obtain $t \geq 1$. And then $a^s b^{t-1}$ expands $K \cap T$. Hence, the word $a^s b^{t-1} a$ expands $K \oplus 1$.

(3) Since $\delta(K \oplus 1, b) = K$, K is $(n + 1)$-expandable.   □

Let $\mathcal{A} = (\mathbb{Z}_n, \{a, b\}, \delta)$ be a standardized completely reachable DFA such that its orbit group $K = (2\,\mathbb{Z}_n, \oplus)$. Let s be an integer such that $1 \leq s \leq n$. Using Proposition 4 repeatedly, one can deduce that every s-element subset of $\mathbb{Z}_n$ is reachable with a word of length at most $n(n - s) + 1$. In the rest of this section, we will improve this upper bound to $n(n - s)$.

Lemma 3. *Let p and q be two distinct states in $\mathbb{Z}_n$. If n is even, $K = (2\,\mathbb{Z}_n, \oplus)$ and $|O(\mathcal{A})| > 1$, then $\mathbb{Z}_n \setminus \{p, q\}$ is $(n - 1)$-expandable.*

Proof. Without loss of generality, we assume that $p < q$.
Case 1. $p \leq n - 3$

Let $T = \delta^{-1}(\mathbb{Z}_n \setminus \{p, q\}, b^p)$. Note that $0 \notin T$ and $\{\delta(0, a), \delta(0, a^2)\} \cap T \neq \emptyset$ because excl(a). Then either ab^p or $a^2 b^p$ expands $\mathbb{Z}_n \setminus \{p, q\}$. Hence $\mathbb{Z}_n \setminus \{p, q\}$ is $(n - 1)$-expandable, since $2 + p \leq n - 1$.
Case 2. $p = n - 2$

Note that $q = n - 1$ and $\delta(0, a) \neq 1$ because $K = 2\,\mathbb{Z}_n$. We have

$$\mathbb{Z}_n \setminus \{p, q\} = \delta\left(\{2, 3, \ldots, n - 1\}, b^{n-2}\right)$$

and then ab^{n-2} expands $\mathbb{Z}_n \setminus \{p, q\}$. Hence $\mathbb{Z}_n \setminus \{p, q\}$ is $(n - 1)$-expandable.   □

For a positive integer n and an integer a, we denote the residue of a modulo n by $\bar{a}$.

Lemma 4. *Assume that $O(\mathcal{A}) = \{d\}$ and $K = (2\,\mathbb{Z}_n, \oplus)$. Let $S \subseteq \mathbb{Z}_n$ such that $|S| > \frac{n}{2}$. If S is not $(n-1)$-expandable, then $S = \mathbb{Z}_n \setminus \{\overline{n-1-jd} : 0 \le j \le n - |S| - 1\}$.*

Proof. Since $|S| = |\mathbb{Z}_n \setminus \{\overline{n-1-jd} : 0 \le j \le n - |S| - 1\}|$, it is sufficient to prove $\mathbb{Z}_n \setminus S \subseteq \{\overline{n-1-jd} : 0 \le j \le n - |S| - 1\}$.

Take an arbitrary state $x \notin S$. Assume $x \in K \oplus i$, where $i \in \{0, 1\}$. Since $O(\mathcal{A}) = \{d\}$, we have $K \oplus i = \{\overline{x + td} : 0 \le t \le \frac{n}{2}\}$. Since $|S| > \frac{n}{2}$, $S \cap (K \oplus i) \ne \emptyset$. Let j be the least non-negative integer such that $\overline{x + (j+1)d} \in S$. It is clear that $j \le n - |S| - 1$. Let $T = \delta^{-1}(S, b^{\overline{x+jd}})$. Note that $0 \notin T$ and $d \in T$. And then $ab^{\overline{x+jd}}$ expands S. Since S is not $(n-1)$-expandable, this implies $\overline{x + jd} = n - 1$ which is equivalent to $x \in \{\overline{n-1-jd} : 0 \le j \le n - |S| - 1\}$. $\square$

Lemma 5. *Assume that $O(\mathcal{A}) = \{d\}$, $K = (2\,\mathbb{Z}_n, \oplus)$ and $n \ge 10$. There exists a word w that reaches K such that $|w| \le \frac{1}{2}n^2$.*

Proof. Let w be a shortest word that reaches K. Since $K \not\subseteq \delta(\mathbb{Z}_n, a)$, we have $w = w'b$ and $\delta(\mathbb{Z}_n, w') = K \oplus 1$.

Assume, for a contradiction, that $|w| \ge \frac{1}{2}n^2 + 1$. By applying the first and second statements of Proposition 4, we have

$$K \oplus 1 \xleftarrow{a} T \xleftarrow{a^m b^{t-1}} S_{\frac{n}{2}-1} \xleftarrow{a^{m\frac{n}{2}-1}b^{t\frac{n}{2}-1}} \cdots \xleftarrow{a^{m_2}b^{t_2}} S_1 \xleftarrow{a^{m_1}b^{t_1}} \mathbb{Z}_n$$

such that $|S_i| = n - i$ for all $i \in \{1, 2, \ldots, \frac{n}{2} - 1\}$ and

$$m + t \le n, \tag{6}$$
$$m_i + t_i \le n \tag{7}$$

for all $i \in \{1, 2, \ldots, \frac{n}{2} - 1\}$. Recall that w is a shortest word such that $\delta(\mathbb{Z}_n, w) = K$. Then

$$\left| a^{m_1}b^{t_1} \cdots a^{\frac{n}{2}-1}b^{t\frac{n}{2}-1}a^m b^{t-1}ab \right| = (1+m+t) + \sum_{i=1}^{\frac{n}{2}-1}(m_i + t_i) \ge |w| \ge n\frac{n}{2} + 1. \tag{8}$$

Due to $O(\mathcal{A}) = \{d\}$, we have $m = 1$ and $m_i = 1$ for all $i \in \{1, 2, \ldots, \frac{n}{2} - 1\}$. Then inequalities 6, 7 and 8 only hold when $t = t_i = n - 1$ for all $i \in \{1, 2, \ldots, \frac{n}{2} - 1\}$.

Observe that K is not n-expandable and S_i is not $(n-1)$-expandable for $i \in \{1, \ldots, \frac{n}{2} - 1\}$ (Otherwise, by Proposition 3, we can construct a word u of length at most $\frac{1}{2}n^2 - 1$ such that $\delta(\mathbb{Z}_n, u) = K \oplus 1$). By Lemma 4, $S_i = \mathbb{Z}_n \setminus \{\overline{n-1-jd} : 0 \le j \le i - 1\}$ for all $i \in \{1, \ldots, \frac{n}{2} - 1\}$.

Inducting on i, we will establish the following claim:

Claim. For each $i \in \{1, 2, \ldots, \frac{n}{2} - 2\}$, $\delta\left(\overline{n-1-(i-1)d}, a\right) = \overline{-id}$.

Proof of Claim. Define T_i to be the set $\delta(S_i, a)$, for all $i \in \{1, 2, \ldots, \frac{n}{2} - 2\}$. Since $\delta(T_i, b^{n-1}) = S_{i+1}$, we have $T_i = \mathbb{Z}_n \setminus \{0, \overline{-d}, \ldots, \overline{-id}\}$. Write $Q' = \mathbb{Z}_n \setminus \{0\}$ and $S_i' = S_i \setminus \{0\}$ for all $i \in \{1, 2, \ldots, \frac{n}{2} - 1\}$. Since a acts on Q' as a permutation, the letter a sends S_i' to T_i as a bijection for all $i \in \{1, 2, \ldots, \frac{n}{2} - 1\}$. And then a sends $Q' \setminus S_i$ to $Q' \setminus T_i$ as a bijection.

For $i = 1$, consider that a sends $Q' \setminus S_1 = \{n - 1\}$ to $Q' \setminus T_1 = \{n - d\}$ as a bijection. That is $\delta(\overline{n - 1}, a) = \overline{-d}$.

For $i \geq 2$, consider that a sends $Q' \setminus S_i = \{\overline{n - 1 - (j - 1)d} : 1 \leq j \leq i\}$ to $Q' \setminus T_i = \{\overline{-jd} : 1 \leq j \leq i\}$ as a bijection. By induction hypothesis, we have $\delta(\overline{n - 1 - (j - 1)d}, a) = \overline{-jd}$ for each $j \in \{1, \ldots, i - 1\}$. Then $\delta(\overline{n - 1 - (i - 1)d}, a) = \overline{-id}$.   □

By the above Claim and $\delta(T, a) = K \oplus 1$, $\overline{n - 1 - (j - 1)d} \notin T$ for all $j \in \{1, \ldots, \frac{n}{2} - 2\}$. Observe that $0, d \notin T$ because $\delta(0, a) \notin K \oplus 1$. Then $T = \{\overline{d - 1}, \overline{2d - 1}\} \cup \{\overline{jd} : 2 \leq j \leq \frac{n}{2} - 1\}$. Observe that ab^d expands T and then ab^dab expands K. Since K is not n-expandable, it holds that $d \geq n - 2$. Recall that $K = (2\mathbb{Z}_n, \oplus)$ and $\mathcal{O}(\mathcal{A}) = \{d\}$. Hence d is an even number and $d = n - 2 = t - 1$. Let T' be the set such that $\delta(T', b^{t-1}) = T$. One can calculate that $T' = \{n - 1, d - 1\} \cup \{\overline{jd} : 1 \leq j \leq \frac{n}{2} - 2\}$. Since $S_{\frac{n}{2} - 1}$ has only one odd number $d - 1$ and T' has at least two (because $n \geq 10$) even numbers besides d, by the Claim, this is a contradiction for $\delta(S_{\frac{n}{2} - 1}, a) = T'$.   □

Proof of Theorem 2. In the case that $|O(\mathcal{A})| > 1$, the claim of the theorem follows from Proposition 4 and Lemma 3. In the case that $|O(\mathcal{A})| = 1$ and $n \geq 10$, it follows from Proposition 4 and Lemma 5. In the case that $|O(\mathcal{A})| = 1$ and $n \leq 8$, it is proved by enumeration, see Table 1 below.   □

4   DFAs that Do Not Fulfill Don's Conjecture

Using a brute force search, we have checked whether binary completely reachable DFAs and standardized completely reachable DFAs with at most 10 states fulfill Don's conjecture. The number of non-isomorphic DFAs that violate Don's conjecture is presented in Table 1. The DFA $\mathcal{B}_8$, shown in Fig. 1, is one of the 8-state binary completely reachable DFAs that do not fulfill Don's conjecture.

Table 1. The number of n-state non-isomorphic standardized completely reachable DFAs and the number of n-state non-isomorphic binary completely reachable DFAs that violate Don's conjecture, for $n \leq 10$.

the number of states	≤ 7	8	9	10
standardized completely reachable	0	0	0	0
binary completely reachable	0	68	0	9210

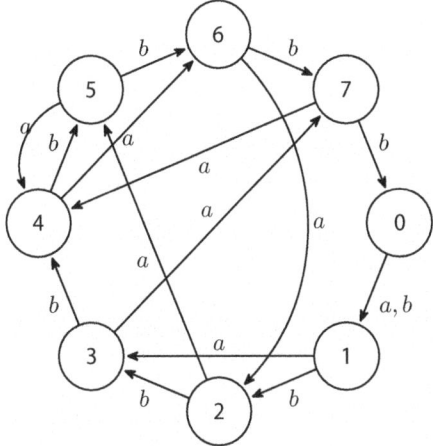

Fig. 1. The 8-state DFA $\mathcal{B}_8$. The shortest word that reaches $\{1, 2, 3, 5, 6, 7\}$ is the word $a^2 b^5 a b^5 a b^3$.

Example 1. Let $n \geq 10$ be an even positive integer. Let $\mathcal{A}_n = (\mathbb{Z}_n, \{a, b\}, \delta)$ be the circular DFA such that

- $\delta(q, b) = q \oplus 1$, for every $q \in \mathbb{Z}_n$;
- $\delta(0, a) = n - 3, \delta(n - 3, a) = 1, \delta(1, a) = n - 2, \delta(n - 2, a) = n/2, \delta(n/2, a) = n - 4, \delta(n - 4, a) = n - 1, \delta(n - 1, a) = n/2$;
- $\delta(q, a) = q$, for every $q \in \mathbb{Z}_n \setminus \{0, 1, n/2, n - 1, n - 2, n - 3, n - 4\}$.

The action of a is shown in Fig. 2. In the rest of this section, we will show that the DFA $\mathcal{A}_n$ is completely reachable (Lemma 6) and does not fulfill Don's conjecture (Lemma 8).

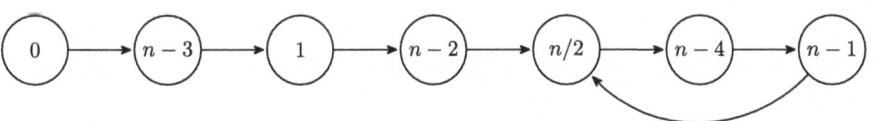

Fig. 2. The action of a in the DFA $\mathcal{A}_n$. The states that are fixed by a are omitted.

For a DFA, a non-empty subset of states is called a *witness* if it is not reachable and has the maximal size among all unreachable subsets of states. This concept was introduced by Ferens and Szykuła [5, Definition 4]. The following lemma is based on an important observation on witnesses [5, Corollary 6].

Lemma 6. *For each positive even integer n, the DFA $\mathcal{A}_n$ is completely reachable.*

Proof. Assume that $\mathcal{A}_n$ is not completely reachable. Take S to be a witness of $\mathcal{A}$. Observe that $s \in S$ if and only if $s \oplus \frac{n}{2} \in S$, otherwise, $b^s a$ expands S. Since $S \oplus (n-s)$ is also a witness of $\mathcal{A}$, without loss of generality, we can assume that $0, n/2 \notin S$. Let $T = \delta^{-1}(S, b)$ and $T' = \delta^{-1}(S, a)$. It is clear that both T and T' are predecessors of S. And then both T and T' are witnesses and $n - 1 \notin T \cup T'$ which contradicts [5, Corollary 6]: the union of any two distinct witnesses is the whole state set. $\qquad\square$

Let Π be the set of words $\{ab^i : i \geq 0\}$. For $S, S' \subseteq \mathbb{Z}_n$, the subset S is called a Π-*predecessor* of S' if S is a w-predecessor of S' for some word $w \in \Pi$. The following lemma contains some elementary facts of Π-predecessors in $\mathcal{A}_n$.

Lemma 7. *In $\mathcal{A}_n$, the following statements hold.*

(1) $\mathbb{Z}_n \setminus \{n-1, n-2\}$ is the unique Π-predecessor of $\mathbb{Z}_n \setminus \{\frac{n}{2} - 1, n - 1\}$.
(2) If $\mathbb{Z}_n \setminus X$ is a Π-predecessor of $\mathbb{Z}_n \setminus \{n-1, n-2\}$, then the minimum element of X is either $n - 3$ or $n - 4$.
(3) Let x be the minimum element in a non-empty proper subset $X \subseteq \mathbb{Z}_n$. The length of a word that reaches $\mathbb{Z}_n \setminus X$ is at least $x + 1$.

Proof. Let $S = \mathbb{Z}_n \setminus \{\frac{n}{2} - 1, n - 1\}$ and $T = \mathbb{Z}_n \setminus \{n-1, n-2\}$.

(1) Let R be a Π-predecessor of S. Let ab^i be a word such that $\delta(R, ab^i) = S$. Then $\delta(R, a) = \delta^{-1}(S, b^i)$ contains neither 0 nor $n/2$. Hence, $R = \mathbb{Z}_n \setminus \{n-1, n-2\}$ as wanted.

(2) Let R be a Π-predecessor of T. Let ab^i be a word such that $\delta(R, ab^i) = S$. Let $P = \delta(R, a) = \delta^{-1}(S, b^i)$. The set P is either equal to $\mathbb{Z}_n \setminus \{0, 1\}$ or $\mathbb{Z}_n \setminus \{0, n-1\}$. Since $\delta(\{n-1, n-2\}, a) = n/2$, the set R equals one of the six subsets: $\mathbb{Z}_n \setminus \{n-3\}$, $\mathbb{Z}_n \setminus \{n-3, n-1\}$, $\mathbb{Z}_n \setminus \{n-3, n-2\}$, $\mathbb{Z}_n \setminus \{n-4\}$, $\mathbb{Z}_n \setminus \{n-4, n-1\}$ and $\mathbb{Z}_n \setminus \{n-4, n-2\}$.

(3) Let R be an ab^i-predecessor of $\mathbb{Z}_n \setminus X$ for some integer i. It is clear that $0 \notin \delta(R, a) = \delta^{-1}(\mathbb{Z}_n \setminus X, b^i)$. Then $i \geq x$. Since $X \neq \emptyset$, every word that reaches $\mathbb{Z}_n \setminus X$ contains the letter a. Hence, the length of a word that reaches $\mathbb{Z}_n \setminus X$ is at least $i + 1 \geq x + 1$. $\qquad\square$

Lemma 8. *The length of a word that reaches $\mathbb{Z}_n \setminus \{n-1, \frac{n}{2} - 1\}$ in $\mathcal{A}$ is at least $\frac{5}{2}n - 3$.*

Proof. Let w be a shortest word that reaches T. Then w can be decomposed into several words in Π, that is, $w = w_1 w_2 \cdots w_t$ for some integer t and $w_1, \ldots, w_t \in \Pi$. Write S for $\mathbb{Z}_n \setminus \{n-1, \frac{n}{2} - 1\}$. Let $T = \delta^{-1}(S, w_t)$ and $R = \delta^{-1}(T, w_{t-1})$. By Lemma 7 (1), we have $T = \mathbb{Z}_n \setminus \{n-1, n-2\}$ and then $w_t = ab^i$ where $i \geq \frac{n}{2} - 1$ because $0 \notin \delta(T, a)$. By Lemma 7 (2), we have $\min(\mathbb{Z}_n \setminus R) \in \{n-3, n-4\}$. The rest of the proof is divided into two cases.

Case 1. $\min(\mathbb{Z}_n \setminus R) = n - 3$

Since $n - 3 \notin R$, $0, 1 \notin \delta(R, a)$. Then $w_{t-1} = ab^j$ such that $j \geq n - 2$. Using Lemma 7 (3), $|w_1 \cdots w_{t-2}| \geq n - 3 + 1 = n - 2$. Hence $w \geq (n-2) + (1 + n - 2) + (1 + \frac{n}{2} - 1) = \frac{5}{2}n - 3$.

Case 2. $\min(\mathbb{Z}_n \setminus R) = n - 4$

Since $n - 4 \notin R$, $0, n - 1 \notin \delta(R, a)$. Then $w_{t-1} = ab^j$ such that $j \geq n - 1$. Using Lemma 7 (3), $|w_1 \cdots w_{t-2}| \geq n - 4 + 1 = n - 3$. Hence $w \geq (n - 3) + (1 + n - 1) + (1 + \frac{n}{2} - 1) = \frac{5}{2}n - 3$. □

Proof of Theorem 3. Let n be any integer at least 10. Since $\frac{5n}{2} - 3 > n(n - s) = n(n - (n - 2)) = 2n$, by Lemmas 6 and 8, $\mathcal{A}_n$ is a binary completely reachable DFA and $\mathcal{A}_n$ does not fulfill Don's conjecture. □

Remark 1. Recently, Ferens and Szykuła [6, Theorem 5.6] presented a new upper bound for the length of the shortest words that reach a subset S of states in a binary completely reachable DFA. In the case that $|S| = n - 2$, the new upper bound is $5n/2$. Due to the above discussion on the properties of $\mathcal{A}_n$, this bound is almost tight.

Let $\mathcal{A} = (\mathbb{Z}_n, \{a, b\}, \delta)$ be a binary completely reachable DFA. For a positive integer k, set $\mathcal{A}^k$ to be the binary DFA $(\mathbb{Z}_n, \{a_k, b\}, \delta_k)$ such that $\delta_k(q, a_k) = \delta(q, b^k a)$ and $\delta_k(q, b) = \delta(q, b)$ for every $q \in \mathbb{Z}_n$. One can easily show that $\mathcal{A}^k$ is completely reachable. Let q_1 and q_2 be the two distinct elements in $\mathrm{dupl}(a)$. Both $\mathcal{A}^{q_1}$ and $\mathcal{A}^{q_2}$ are standardized completely reachable DFAs and each of these two DFAs is called the *standardization* of $\mathcal{A}$.

Remark 2. The standardizations of $\mathcal{A}_n$ fulfill the Don's conjecture, for every even integer $n \geq 10$.

Proof. The orbits of the standardizations of $\mathcal{A}_n$ are $\{n/2, n/2 - 2\}$ and $\{n/2, n/2 - 1\}$. Their orbit subgroups are $(2\mathbb{Z}_n, \oplus)$ and $(\mathbb{Z}_n, \oplus)$. By Theorem 2, the standardizations of $\mathcal{A}_n$ fulfill the Don's conjecture. □

5   Conclusions and Discussions

We have confirmed Don's conjecture for a subfamily of standardized completely reachable DFAs. Moreover, we have proved that in an n-state standardized completely reachable DFA, every s-element subset of states has a reaching word of length at most $n(n-s)+n-1$. On the other hand, we have constructed infinitely many binary completely reachable DFAs which do not fulfill Don's conjecture.

For a given DFA $\mathcal{A}$ and a subset S of states, to determine whether they fulfill Don's conjecture can be seen as a "qualitative analysis" of the length of a shortest word that reaches S in $\mathcal{A}$. More important and more difficult part is its "quantitative analysis". We will do some brief discussions.

For an n-state completely reachable DFA $\mathcal{A} = (Q, \Sigma, \delta)$ and a subset S of its state set, let $f_{\mathcal{A}}(S)$ be the length of a shortest word that reaches S. Let C be a subclass of completely reachable DFAs. For positive integers n and s, define

$$f_{\mathsf{C}}(n, s) := \max_{\substack{\mathcal{A}=(Q,\Sigma,\delta)\in\mathsf{C} \\ |Q|=n}} \max_{\substack{S \subseteq Q \\ |S|=s}} \{f_{\mathcal{A}}(S)\}$$

where $\mathcal{A}$ runs over all n-state DFA in C and S runs over all s-element subset of states in $\mathcal{A}$. The restriction of Don's conjecture to C can be stated as $f_{\mathsf{C}}(n, s) \leq n(n - s)$ for all $0 < s \leq n$.

Write BCR for the class of binary completely reachable DFAs and Std for the class of standardized completely reachable DFAs. Theorem 1 shows that

$$f_{\mathsf{Std}}(n, s) \leq n(n - s) + n - 1$$

for all $0 < s \leq n$. In Sect. 4, by constructing $\mathcal{A}_n$ in Example 1, we obtain that

$$f_{\mathsf{BCR}}(n, n - 2) \geq \frac{5}{2}n - 3$$

for $n \geq 10$. Unfortunately, our construction cannot give any non-trivial lower bound for $f_{\mathsf{BCR}}(n, s)$ when $s \neq n - 2$.

Finally, we present a relation between the reaching words in a binary completely reachable DFA and its standardization. Let $\mathcal{A} = (\mathbb{Z}_n, \{a, b\}, \delta)$ be a binary completely reachable DFA and let $\mathcal{A}^k$ be a standardization of $\mathcal{A}$. Let S be an s-element non-empty proper subset of $\mathbb{Z}_n$. In the proof of Theorem 1, we construct a word $w \in \{a_k, b\}^*$ of length $\leq (n - s)n + n - 1$ such that w reaches S in $\mathcal{A}^k$. Further, w contains at most $2n - s - 1$ occurrences of a_k. Let w' be the word over $\{a, b\}$ which is obtained by replacing a_k in w with $b^k a$. Since $\delta_k(\cdot, a_k) = \delta(\cdot, b^k a)$, it holds that $\delta_k(Q, w) = \delta(Q, w') = S$. And then we have

$$f_{\mathcal{A}}(S) \leq (n - s)n + n - 1 + k(2n - s - 1). \tag{9}$$

Recall that Ferens-Szykuła bound [5, Theorem 30] shows that

$$f_{\mathcal{A}}(S) \leq 2(n - s)n - n \ln(n - s) - n/(n - s). \tag{10}$$

Observe that for any positive real ϵ, in the case that $k < (\frac{1}{2} - \epsilon)n$, we have

$$(n - s)n + n - 1 + k(2n - s - 1) < (2 - 2\epsilon)(n - s)n + o(n^2)$$

and

$$2(n - s)n - n \ln(n - s) - n/(n - s) = 2(n - s)n + o(n^2).$$

Hence, in the case that $k < (\frac{1}{2} - \epsilon)n$, Eq. (9) provides an asymptotically better bound of $f_{\mathcal{A}}(S)$ than Eq. (10).

Acknowledgments. I thank Prof. Mikhail V. Volkov for valuable discussions, feedback and research suggestions.

References

1. Bondar, E.A., Volkov, M.V.: Completely reachable automata. In: Câmpeanu, C., Manea, F., Shallit, J. (eds.) DCFS 2016. LNCS, vol. 9777, pp. 1–17. Springer, Cham (2016). https://doi.org/10.1007/978-3-319-41114-9_1
2. Casas, D., Volkov, M.V.: Binary completely reachable automata. In: Castañeda, A., Rodríguez-Henríquez, F. (eds.) LATIN 2022. LNCS, vol. 13568, pp. 345–358. Springer, Cham (2022). https://doi.org/10.1007/978-3-031-20624-5_21
3. Casas, D., Volkov, M.V.: Don's conjecture for binary completely reachable automata: an approach and its limitations (2023). https://arxiv.org/abs/2311.00077
4. Don, H.: The Černý conjecture and 1-contracting automata. Electron. J. Combin. **23**(3), Paper 3.12, 10 (2016). https://doi.org/10.37236/5616
5. Ferens, R., Szykuła, M.: Completely reachable automata: a polynomial algorithm and quadratic upper bounds. In: Etessami, K., Feige, U., Puppis, G. (eds.) 50th International Colloquium on Automata, Languages, and Programming (ICALP 2023). Leibniz International Proceedings in Informatics (LIPIcs), vol. 261, pp. 59:1–59:17. Schloss Dagstuhl – Leibniz-Zentrum für Informatik, Dagstuhl, Germany (2023). https://doi.org/10.4230/LIPIcs.ICALP.2023.59
6. Ferens, R., Szykuła, M.: A polynomial algorithm deciding the complete reachability and quadratic reaching thresholds. https://arxiv.org/abs/2208.05956
7. Gonze, F., Jungers, R.M.: Hardly reachable subsets and completely reachable automata with 1-deficient words. J. Autom. Lang. Comb. **24**(2-4), 321–342 (2019). https://doi.org/10.25596/jalc-2019-321
8. Kari, J., Volkov, M.V.: Černý's conjecture and the road colouring problem. In: Pin, J.-É (ed.) Handbook of Automata Theory. Vol. I. Theoretical Foundations, pp. 525–565. EMS Press, Berlin (2021).https://doi.org/10.4171/AUTOMATA-1/15
9. Maslennikova, M.I.: Reset complexity of ideal languages. In: Bieliková, M., Friedrich, G., Gottlob, G., Katzenbeisser, S., Špánek, R., Turán II, G. (eds.) SOFSEM 2012, pp. 33–34. Institute of Computer Science Academy of Sciences of the Czech Republic (2012). http://arxiv.org/abs/1404.2816
10. Volkov, M.V.: Synchronization of finite automata. Russ. Math. Surv. **77**(5), 819–891 (2022). https://doi.org/10.4213/rm10005e

Correction to: Polyregular Functions: Characterisations and Refutations

Sandra Kiefer [iD]

Correction to:
Chapter 2 in: J. D. Day and F. Manea (Eds.): *Developments*
in Language Theory, **LNCS 14791,**
https://doi.org/10.1007/978-3-031-66159-4_2

In the originally published version of chapter 2 the chapter title and subtitle were erroneously merged. The chapter title and subtitle have been corrected by adding seperator.

The updated version of this chapter can be found at
https://doi.org/10.1007/978-3-031-66159-4_2

Author Index

J. D. Day and F. Manea (Eds.): DLT 2024, LNCS 14791, pp. 297–298, 2024.
https://doi.org/10.1007/978-3-031-66159-4

GPSR Compliance

The European Union's (EU) General Product Safety Regulation (GPSR) is a set of rules that requires consumer products to be safe and our obligations to ensure this.

If you have any concerns about our products, you can contact us on ProductSafety@springernature.com

In case Publisher is established outside the EU, the EU authorized representative is:

Springer Nature Customer Service Center GmbH
Europaplatz 3
69115 Heidelberg, Germany

The manufacturer's authorised representative in the EU is Springer
Nature Customer Service Centre GmbH, Europaplatz 3, 69115 Heidelberg,
Germany. If you have any concerns regarding our products, please
contact ProductSafety@springernature.com

Printed and bound by CPI Group (UK) Ltd, Croydon, CR0 4YY
29/04/2026
02099532-0003